Communications in Computer and Information Science 1443

More information about this series at http://www.springer.com/series/7899

Bernabé Dorronsoro · Lionel Amodeo ·
Mario Pavone · Patricia Ruiz (Eds.)

Optimization and Learning

4th International Conference, OLA 2021
Catania, Italy, June 21–23, 2021
Proceedings

Editors
Bernabé Dorronsoro
University of Cádiz
Cádiz, Spain

Lionel Amodeo
University of Technology of Troyes
Troyes, France

Mario Pavone
University of Catania
Catania, Italy

Patricia Ruiz
University of Cádiz
Cádiz, Spain

ISSN 1865-0929 ISSN 1865-0937 (electronic)
Communications in Computer and Information Science
ISBN 978-3-030-85671-7 ISBN 978-3-030-85672-4 (eBook)
https://doi.org/10.1007/978-3-030-85672-4

This Springer imprint is published by the registered company Springer Nature Switzerland AG
The registered company address is: Gewerbestrasse 11, 6330 Cham, Switzerland

Preface

This book compiles the best papers submitted to the Fourth International Conference on Optimization and Learning (OLA 2021). The papers were selected according to the scores awarded by the Program Committee during the blind review process. OLA 2021 was going to take place in Catania, Italy, during June 21–23, but it finally had to be organized in online mode. The main objective of the OLA conference series is to attract influential researchers from all over the world working in the fields of complex problems optimization, machine learning, and deep learning to identify synergies and discuss applications for real-world problems. The conference builds a welcoming atmosphere where researchers can present innovative work in the relevant fields.

Three categories of papers were considered for OLA 2021, namely ongoing research works and high impact journal publications (both in the shape of an extended abstract) along with regular papers with novel contents and important contributions. A selection of the best papers in this latter category is published in this book.

A total of 49 papers were presented at OLA 2021, which were arranged into nine sessions, covering topics such as the use of optimization methods to enhance learning techniques, the use of learning techniques to improve the performance of optimization methods, advanced optimization methods and their applications, machine and deep learning techniques, and applications of learning and optimization tools to transportation and routing, scheduling, or other real-world problems.

The conference received a total of 62 papers, from which only 27 (43.5% of all submissions) were selected for publication in this book.

June 2021

Bernabé Dorronsoro
Lionel Amodeo
Mario Pavone
Patricia Ruiz

Organization

Conference Chairs

Mario Pavone	University of Catania, Italy
Lionel Amodeo	Université de Technologie de Troyes, France

Conference Program Chairs

Bernabe Dorronsoro	University of Cadiz, Spain
Vincenzo Cutello	University of Catania, Italy

Conference Steering Committee Chair

El-Ghazali Talbi	University of Lille and Inria, France

Organization Committee

Jeremy Sadet	Université Polytechnique des Hauts-de-France, France
Rachid Ellaia	EMI and Mohamed V University of Rabat, Morocco
Rocco A. Scollo	University of Catania, Italy
Antonio M. Spampinato	University of Catania, Italy
Georgia Fargetta	University of Catania, Italy
Carolina Crespi	University of Catania, Italy
Francesco Zito	University of Catania, Italy

Publicity Chairs

Juan J. Durillo	Leibniz Supercomputing Center, Germany
Grégoire Danoy	University of Luxembourg, Luxembourg

Program Committee

Lionel Amodeo	Université de Technologie de Troyes, France
Mehmet-Emin Aydin	University of Bedfordshire, UK
Mathieu Balesdent	ONERA, France
Pascal Bouvry	University of Luxembourg, Luxembourg
Loïc Brevault	ONERA, France
Matthias R. Brust	University of Luxembourg, Luxembourg
Grégoire Danoy	University of Luxembourg, Luxembourg
Bernabe Dorronsoro	University of Cadiz, Spain
Patrick de Causmaecker	KU Leuven, Belgium
Krisana Chinnasarn	Burapha University, Thailand

Abdellatif El Afia	Mohamed V University of Rabat, Morocco
Rachid Ellaia	EMI, Morocco
Bogdan Filipic	Jožef Stefan Institute, Slovenia
José Manuel García Nieto	University of Málaga, Spain
Domingo Jimenez	University of Murcia, Spain
Hamamache Kheddouci	University of Lyon, France
Peter Korosec	Jožef Stefan Institute, Slovenia
Dmitri E. Kvasov	University of Calabria, Italy
Andrew Lewis	Griffith University, Australia
Francisco Luna	University of Malaga, Spain
Teodoro Macias Escobar	Technological Institute of Ciudad Madero, Mexico
Roberto Magán-Carrión	University of Granada, Spain
Renzo Massobrio	Universidad de la República, Uruguay
Gonzalo Mejia Delgadillo	Pontificia Universidad Católica de Valparaíso, Chile
Nouredine Melab	University of Lille, France
Edmondo Minisci	University of Strathclyde, UK
Antonio J. Nebro	University of Málaga, Spain
Sergio Nesmachnow	Universidad de la República, Uruguay
Eneko Osaba Icedo	TECNALIA, Spain
Gregor Papa	Jožef Stefan Institute, Slovenia
Apivadee Piyatumrong	NECTEC, Thailand
Helena Ramalhinho	Universitat Pompeu Fabra, Spain
Ángel Ruiz Zafra	University of Cadiz, Spain
Roberto Santana	University of the Basque Country, Spain
Alejandro Santiago	Polytechnic University of Altamira, Mexico
Javier del Ser	TECNALIA, Spain
Yaroslav D. Sergeyev	University of Calabria, Italy
Andrei Tchernykh	CICESE, Mexico
Sartra Wongthanavasu	Khon Kaen University, Thailand
Alice Yalaoui	Université de Technologie de Troyes, France
Farouk Yalaoui	Université de Technologie de Troyes, France
Xin-She Yang	Middlesex University London, UK
Nicolas Zufferey	Université de Genève, Switzerand

Contents

Synergies Between Optimization and Learning

Embedding Simulated Annealing within Stochastic Gradient Descent

Matteo Fischetti(✉) and Matteo Stringher

Department of Information Engineering, University of Padova, via Gradenigo 6/A, 35100 Padova, Italy
matteo.fischetti@unipd.it

Abstract. We propose a new metaheuristic training scheme for Machine Learning that combines Stochastic Gradient Descent (SGD) and Discrete Optimization in an unconventional way. Our idea is to define a discrete neighborhood of the current SGD point containing a number of "potentially good moves" that exploit gradient information, and to search this neighborhood by using a classical metaheuristic scheme borrowed from Discrete Optimization. In the present paper we investigate the use of a simple Simulated Annealing (SA) metaheuristic that accepts/rejects a candidate new solution in the neighborhood with a probability that depends both on the new solution quality and on a parameter (the *temperature*) which is modified over time to lower the probability of accepting worsening moves.

Computational results on image classification (CIFAR-10) are reported, showing that the proposed approach leads to an improvement of the final validation accuracy for modern Deep Neural Networks such as ResNet34 and VGG16.

Keywords: Simulated annealing · Stochastic gradient descent · Deep neural networks · Machine learning · Training algorithm

1 Introduction

Machine Learning (ML) is a fundamental topic in Artificial Intelligence. Its growth in the research community has been followed by a huge rise in the number of projects in the industry leveraging this technology.

Deep learning is a subset of ML, based on learning data representation through the use of neural network architectures, specifically Deep Neural Networks (DNNs). Inspired by human processing behavior, DNNs have set new state-of-art results in speech recognition, visual object recognition, object detection, and many other domains.

Work supported by MiUR, Italy (project PRIN). We gratefully acknowledge the support of NVIDIA Corporation with the donation of the Titan Xp GPU used for this research.

B. Dorronsoro et al. (Eds.): OLA 2021, CCIS 1443, pp. 3–13, 2021.
https://doi.org/10.1007/978-3-030-85672-4_1

Stochastic Gradient Descent (SGD) is *de facto* the standard algorithm for training Deep Neural Networks (DNNs). Leveraging the gradient, SGD allows one to rapidly find a good solution in the very high dimensional space of weights associated with modern DNNs. Moreover, the use of minibatches allows one to exploit modern GPUs and to achieve a considerable computational efficiency.

In the present paper we investigate the use of an alternative training method, namely, the Simulated Annealing (SA) algorithm [2]. The use of SA for training is not new, but previous proposals are mainly intended to be applied for non-differentiable objective functions for which SGD is not applied due to the lack of gradients; see, e.g., [4,7]. Instead, our SA method requires differentiability of (a proxy of) the loss function, and leverages on the availability of a gradient direction to define local moves that have a large probability to improve the current solution.

Our approach is computationally evaluated in an implementation leveraging hyper-parameters. Assume some hyper-parameter values (e.g., learning rates for SGD) are collected in a discrete set H. At each SGD iteration, we randomly pick one hyper-parameter from H, temporarily implement the corresponding *move* as in the classical SGD method (using the gradient information) and evaluate the new point on the current minibatch. If the loss function does not deteriorate too much, we *accept* the move as in the classical SGD method, otherwise we *reject* it: we step back to the previous point, change the minibatch, randomly pick another hyper-parameter from H, and repeat. The decision of accepting/rejecting a move is based on the classical SA criterion, and depends of the amount of loss-function worsening and on a certain parameter (the *temperature*) which is modified over time to lower the probability of accepting worsening moves.

A distinctive feature of our scheme is that hyper-parameters are modified within a *single* SGD execution (and not in an external loop, as customary) and evaluated on the fly on the current minibatch, i.e., their tuning is fully embedded within the SGD algorithm.

Computational results are reported, showing that the proposed approach leads to an improvement of the final validation accuracy for modern DNN architectures (ResNet34 and VGG16 on CIFAR-10).

2 Simulated Annealing

The basic SA algorithm for a generic optimization problem can be outlined as follows. Let S be the set of all possible feasible solutions, and $f : S \to \mathbb{R}$ be the objective function to be minimized. An optimal solution s^* is a solution in S such that $f(s^*) \leq f(s)$ holds for all $s \in S$.

SA is an iterative method that constructs a trajectory of solutions $s^{(0)}, \cdots, s^{(k)}$ in S. At each iteration, SA considers moving from the current feasible solution $s^{(i)}$ (say) to a candidate new feasible solution s_{new} (say). Let $\Delta(s^{(i)}, s_{new}) = f(s_{new}) - f(s^{(i)})$ be the objective function *worsening* when moving from $s^{(i)}$ to s_{new}—positive if s_{new} is strictly worse than $s^{(i)}$. The hallmark of SA is that worsening moves are not forbidden but accepted with a certain *acceptance probability* $p(s^{(i)}, s_{new}, T)$ that depends on the amount of worsening $\Delta(s^{(i)}, s_{new})$ and on a parameter $T > 0$ called *temperature*. A typical way to compute the acceptance probability is through *Metropolis' formula* [5]:

$$p(s, s_{new}, T) = \begin{cases} e^{-\Delta(s^{(i)}, s_{new})/T} & \text{if} \quad \Delta(s^{(i)}, s_{new}) > 0 \\ 1 & \text{if} \quad \Delta(s^{(i)}, s_{new}) \leq 0 \ . \end{cases} \tag{1}$$

Thus, the probability of accepting a worsening move is large if the amount of worsening $\Delta(s^{(i)}, s') > 0$ is small and the temperature T is large. Note that the probability is 1 when $\Delta(s^{(i)}, s') \leq 0$, meaning that improving moves are always accepted by the SA method.

Temperature T is a crucial parameter: it is initialized to a certain value T_0 (say), and iteratively decreased during the SA execution so as to make worsening moves less and less likely in the final iterations. A simple update formula for T is $T = \alpha \cdot T$, where $\alpha \in (0, 1)$ is called *cooling factor*. Typical ranges for this parameter are 0.95–0.99 (if cooling is applied at each SA iteration) or $0.7 - 0.8$ (if cooling is only applied at the end of a "computational epoch", i.e., after several SA iterations with a constant temperature).

The basic SA scheme is outlined in Algorithm 1; more advanced implementations are possible, e.g., the temperature can be restored multiple times to the initial value.

Algorithm 1 : SA

Input: function f to be minimized, initial temperature $T_0 > 0$, cooling factor $\alpha \in (0, 1)$, number of iterations $nIter$
Output: the very last solution $s^{(nIter)}$

1: Compute an initial solution $s^{(0)}$ and initialize $T = T_0$
2: **for** $i = 0, \ldots, nIter - 1$ **do**
3: Pick a new tentative solution s_{new} in a convenient neighborhood $\mathcal{N}(s^{(i)})$ of $s^{(i)}$
4: $worsening = f(s_{new}) - f(s^{(i)})$
5: $prob = e^{-worsening/T}$
6: **if** $random(0, 1) < prob$ **then**
7: $s^{(i+1)} = s_{new}$
8: **else**
9: $s^{(i+1)} = s^{(i)}$
10: **end if**
11: $T = \alpha \cdot T$
12: **end for**

At Step 6, $random(0, 1)$ is a pseudo-random value uniformly distributed in [0,1]. Note that, at Step 5, the acceptance probability *prob* becomes larger than 1 in case $worsening < 0$, meaning that improving moves are always accepted (as required).

2.1 A Naive Implementation for Training Without Gradients

In the context of training, one is interested in minimizing a *loss function* $L(w)$ with respect to a large-dimensional vector $w \in \Re^M$ of so-called *weights*. If $L(w)$ is differentiable (which is not required by the SA algorithm), there exists a gradient $\nabla(w)$ giving the steepest increasing direction of L when moving from a given point w.

Here is a very first attempt to use SA in this setting. Given the current solution (i.e., set of weights) w, we generate a random move $\Delta(w) \in \Re^M$ and then we evaluate the loss function in the nearby point $w' := w - \epsilon\Delta(w)$, where ϵ is a small positive real number. If the norm of $\epsilon\Delta(w)$ is small enough and L is differentiable, due to Taylor's approximation we know that

$$L(w') \simeq L(w) - \epsilon\ \nabla^T(w)\Delta(w). \tag{2}$$

Thus the objective function improves if $\nabla(w)^T\Delta(w) > 0$. As we work in the continuous space, in the attempt of improving the objective function we can also try to move in the opposite direction and move to $w'' := w + \epsilon\ \Delta(w)$. Thus, our actual move from the current w consists of picking the best (in terms of objective function) point w_{new}, say, between the two nearby points w' and w'': if w_{new} improves $L(w)$, then we surely accept this move; otherwise we accept it according to the Metropolis' formula (1). Note that the above SA approach is completely derivative free: as a matter of fact, SA could optimize directly over discrete functions such as the accuracy in the context of classification.

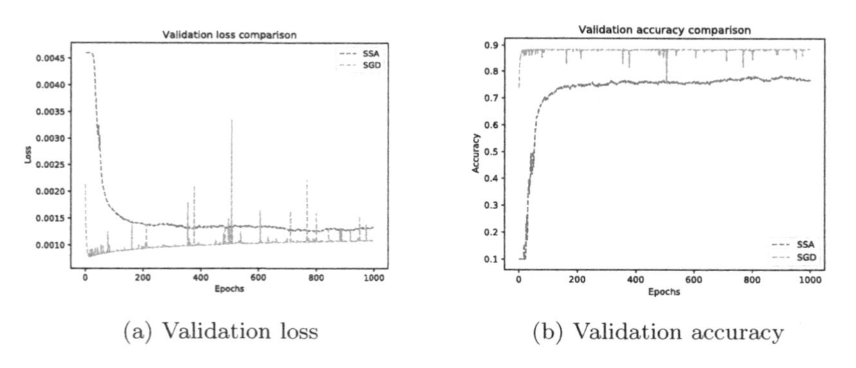

(a) Validation loss (b) Validation accuracy

Fig. 1. Performance on the validation set of our naive SA implementation (SSA) for VGG16 on Fashion-MNIST. SGD: learning rate $\eta = 0.001$, no momentum/Nesterov acceleration. `SSA`: $\epsilon = 0.01, \alpha = 0.97, T_0 = 1$.

In a preliminary phase of our work we implemented the simple scheme above in a stochastic manner, using minibatches when evaluating $L(w')$ and $L(w'')$, very much in the spirit of the SGD algorithm. Figures 1–2, compare the performance of the resulting Stochastic SA algorithm, called `SSA`, with that of a

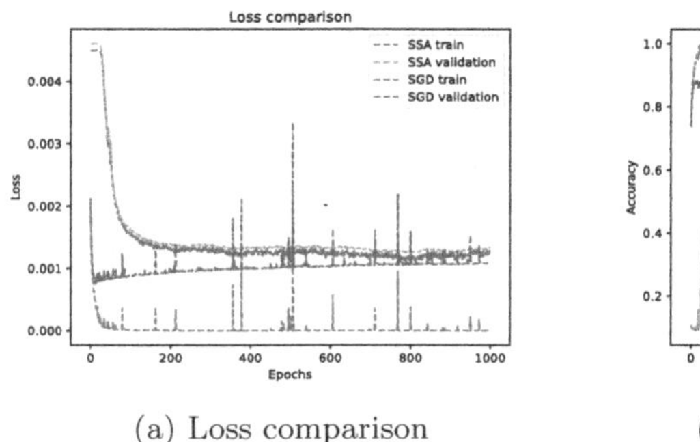

(a) Loss comparison

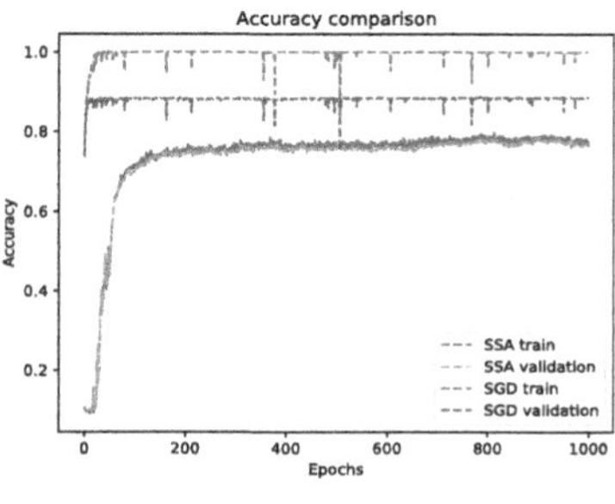

(b) Accuracy comparison

Fig. 2. Comparison of our naive SA implementation (SSA) vs SGD for VGG16 on Fashion-MNIST. SGD: learning rate $\eta = 0.001$, no momentum/Nesterov acceleration. SSA: $\epsilon = 0.01, \alpha = 0.97$, $T_0 = 1$. Subfigure (b) clearly shows that SSA has no overfitting but is not able to exploit the full capacity of VGG16, resulting into an unsatisfactory final accuracy.

straightforward SGD implementation with constant learning rate and no momentum [9] nor Nesterov [6] acceleration, using the Fashion-MNIST [10] dataset and the VGG16 [8] architecture. Figure 2(b) reports accuracy on both the training and the validation sets, showing that SSA does not suffer from overfitting as the accuracy on the training and validation sets are almost identical—a benefit deriving from the derivative-free nature of SSA. However, SSA is clearly unsatisfactory in terms of validation accuracy (which is much worse than the SGD one) in that it does not exploit well the VGG16 capacity.

We are confident that the above results could be improved by a more advanced implementation. E.g., one could vary the value of ϵ during the algorithm, and/or replace the loss function by (one minus) the accuracy evaluated on the current minibatch—recall that SSA does not require the objective function be differentiable. However, even an improved SSA implementation is unlikely to be competitive with SGD. In our view, the main drawback of the SSA algorithm (as stated) is that, due the very large dimensional space, the random direction $\pm\Delta(w)$ is very unlikely to lead to a substantial improvement of the objective function as the effect of its components tend to cancel out randomly. Thus, a more clever definition of the basic move is needed to drive SSA in an effective way.

3 Improved SGD Training by SA

We next introduce an unconventional way of using SA in the context of training. We assume the function $L(w)$ to be minimized be differentiable, so we can compute its gradient $\nabla(w)$. From SGD we borrow the idea of moving in the anti-gradient direction $-\nabla(w)$, possibly corrected using momentum/Nesterov acceleration techniques. Instead of using a certain *a priori* learning rate η, however, we randomly pick one from a discrete set H (say) of possible candidates.

In other words, at each SA iteration the move is selected randomly in a discrete neighborhood $\mathcal{N}(w^{(i)})$ whose elements correspond to SGD iterations with different learning rates. An important feature of our method is that H can (actually, should) contain unusually large learning rates, as the corresponding moves can be discarded by the Metropolis' criterion if they deteriorate the objective function too much.

A possible interpretation of our approach is in the context of SGD hyper-parameter tuning. According to our proposal, hyper-parameters are collected in a discrete set H and sampled *within a single SGD execution*: in our tests, H just contains a number of possible learning rates, but it could involve other parameters/decisions as well, e.g., applying momentum, or Nesterov (or none of the two) at the current SGD iteration, or alike. The key property here is that any element in H corresponds to a reasonable (non completely random) move, so picking one of them at random has a significant probability of improving the objective function. As usual, moves are accepted according to the Metropolis' criterion, so the set H can also contain "risky choices" that would be highly inefficient if applied systematically within a whole training epoch.

Algorithm 2 : `SGD-SA`

Parameters: A set of learning rates H, initial temperature $T_0 > 0$
Input: Differentiable loss function L to be minimized, cooling factor $\alpha \in (0, 1)$, number of epochs $nEpochs$, number of minibatches N
Output: the best performing $\mathrm{w}^{(i)}$ on the validation set at the end of each epoch

1: Divide the training dataset into N minibatches
2: Initialize $i = 0,\ \ T = T_0,\ \ w^{(0)} =$ random_initialization()
3: **for** $t = 1, \ldots, nEpochs$ **do**
4:     **for** $n = 1, \ldots, N$ **do**
5:         Extract the n-th minibatch (x, y)
6:         Compute $L(w^{(i)}, x, y)$ and its gradient $v =$ backpropagation$(w^{(i)}, x, y)$
7:         Randomly pick a learning rate η from H
8:         $w_{new} = w^{(i)} - \eta\ v$
9:         Compute $L(w_{new}, x, y)$
10:         $worsening = L(w_{new}, x, y) - L(w^{(i)}, x, y)$
11:         $prob = e^{-worsening/T}$
12:         **if** $random(0, 1) < prob$ **then**
13:             $w^{(i+1)} = w_{new}$
14:         **else**
15:             $w^{(i+1)} = w^{(i)}$
16:         **end if**
17:         $i = i + 1$
18:     **end for**
19:     $T = \alpha \cdot T$
20: **end for**

Our basic approach is formalized in Algorithm 2, and will be later referred to as `SGD-SA`. More elaborated versions using momentum/Nesterov are also possible but not investigated in the present paper, as we aim at keeping the overall computational setting as simple and clean as possible.

4 Computational Analysis of SGD-SA

We next report a computational comparison of SGD and SGD-SA for a classical image classification task involving the CIFAR-10 [3] dataset. As customary, the dataset was shuffled and partitioned into 50,000 examples for the training set, and the remaining 10,000 for the test set. As to the DNN architecture, we tested two well-known proposals from the literature: VGG16 [8] and ResNet34 [1]. Training was performed for 100 epochs using PyTorch, with minibatch size 512. Tests have been performed using a single NVIDIA TITAN Xp GPU.

Our Scheduled-SGD implementation of SGD is quite basic but still rather effective on our dataset: it uses no momentum/Nesterov acceleration, and the learning rate is set according the following schedule: $\eta = 0.1$ for first 30 epochs, 0.01 for the next 40 epochs, and 0.001 for the final 30 epochs. As to SGD-SA, we used $\alpha = 0.8$, initial temperature $T_0 = 1$, and learning-rate set $H = \{0.9, 0.8, 0.7, 0.6, 0.5, 0.4, 0.3, 0.2, 0.1, 0.09, 0.08, 0.07, 0.06, 0.05\}$.

Both Scheduled-SGD and SGD-SA use pseudo-random numbers generated from an initial random seed, which therefore has some effects of the search path in the weight space and hence on the final solution found. Due to the very large number of weights that lead to statistical compensation effects, the impact of the seed on the initialization of the very first solution $w^{(0)}$ is very limited—a property already known for SGD that is inherited by SGD-SA as well. However, random numbers are used by SGD-SA also when taking some crucial "discrete" decisions, namely: the selection of the learning rate $\eta \in H$ (Step 7) and the acceptance test (Step 12). As a result, as shown next, the search path of SGD-SA is very dependent on the initial seed. Therefore, for both Scheduled-SGD and SGD-SA we decided to repeat each run 10 times, starting with 10 random seeds, and to report results for each seed. In our view, this dependency on the seed is in fact a *positive* feature of SGD-SA, in that it allows one to treat the seed as a single (quite powerful) hyper-parameter to be randomly tuned in an external loop.

Our first order of business is to evaluate the convergence property of SGD-SA on the training set—after all, this is the optimization task that SA faces directly. In Fig. 3 we plot the average probability *prob* (clipped to 1) of accepting a move at Step 12, as well as the training-set accuracy as a function of the epochs. Subfig. 3a shows that the probability of accepting a move is almost one in the first epochs, even if the amount of worsening is typically quite large in this phase. Later on, the probability becomes smaller and smaller, and only very small worsenings are more likely to be accepted. As a result, large learning rates are automatically discarded in the last iterations. Subfig. 3b is quite interesting: even in our simple implementation, Scheduled-SGD quickly converges to the best-possible value of one for accuracy, and the plots for the various seeds (gray lines) are almost overlapping—thus confirming that the random seed has negligible effects

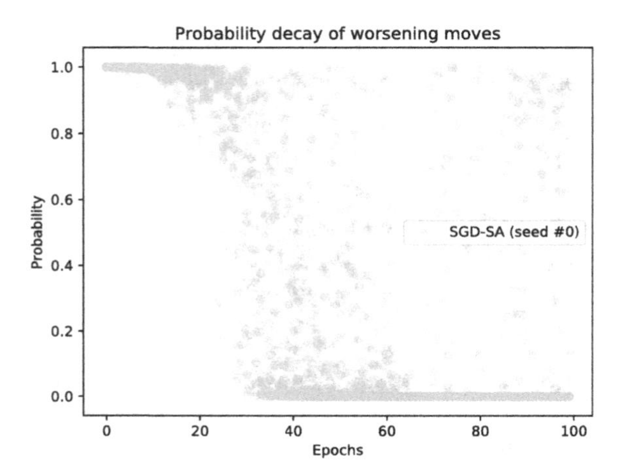

(a) Probability of accepting worsening moves

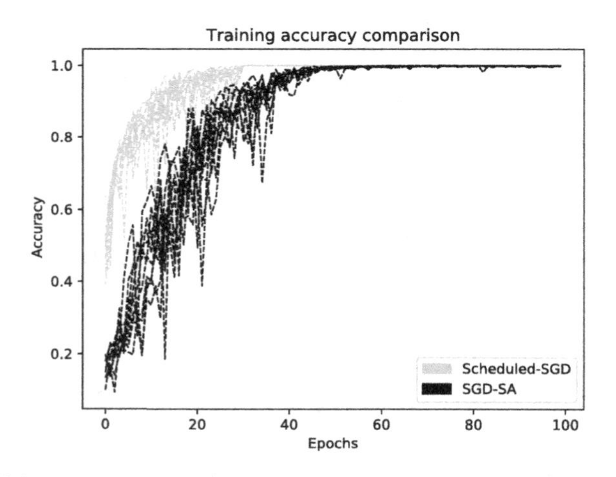

(b) Training accuracy (10 runs with different random seeds)

Fig. 3. Optimization efficiency over the training set (VGG16 on CIFAR-10)

of `Scheduled-SGD`. As to `SGD-SA` (black lines), its convergence to accuracy one is slower than `Scheduled-SGD`, and different seeds lead to substantially different curves—a consequence of the discrete random decisions taken along the search path.

Figure 4 shows the performance on the validation set of `Scheduled-SGD` and `SGD-SA` (both with 10 runs with different random seeds) when using the ResNet34 architecture—results with VGG16 are very similar, hence they are not reported. As expected, the search path of `SGD-SA` is more diversified (leading to accuracy drops in the first epochs) but the final solutions tend to generalize better than Scheduled-SGD, as witnessed by the better performance on the validation set.

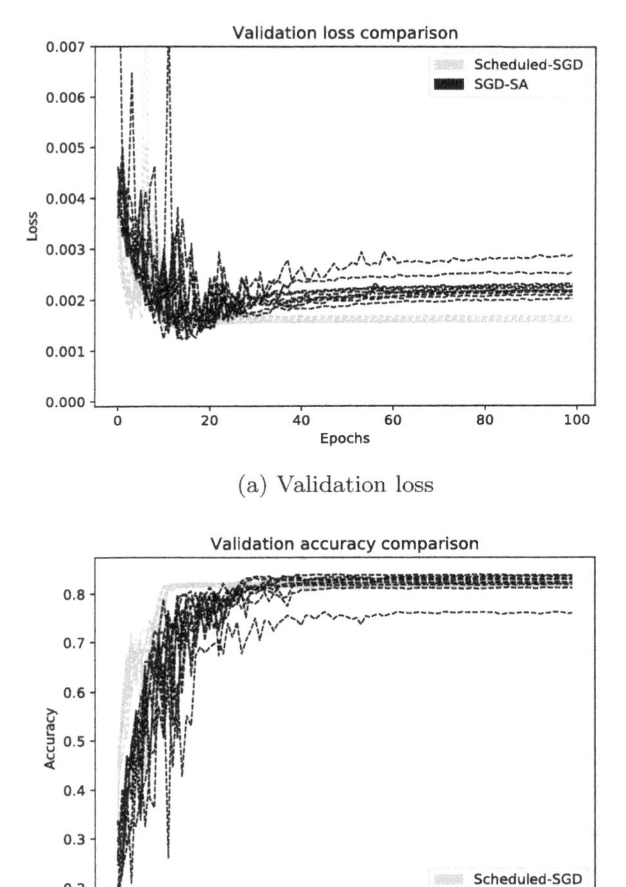

(a) Validation loss

(b) Validation accuracy

Fig. 4. ResNet34 on CIFAR-10 (validation set)

Table 1 gives more detailed results for each random seed, and reports the final validation accuracy and loss reached by Scheduled-SGD and SGD-SA. The results show that, for all seeds, SGD-SA always produces a significantly better (lower) validation loss than Scheduled-SGD. As to validation accuracy, SGD-SA outperforms Scheduled-SGD for all seeds but seeds 3, 4 and 6 for ResNet34.

Table 1. Best validation accuracy and loss, seed by seed.

Method	Seed	VGG16		ResNet34	
		Loss	Accuracy	Loss	Accuracy
Scheduled-SGD	0	0.001640	85.27	0.001519	82.18
	1	0.001564	84.94	0.001472	82.58
	2	0.001642	84.84	0.001467	82.27
	3	0.001662	84.93	0.001468	82.37
	4	0.001628	84.92	0.001602	81.69
	5	0.001677	85.37	0.001558	81.80
	6	0.001505	84.91	0.001480	82.24
	7	0.001480	85.28	0.001532	82.07
	8	0.001623	85.26	0.001574	81.52
	9	0.001680	85.41	0.001499	82.41
`SGD-SA`	0	0.001127	86.44	0.001306	82.55
	1	0.001206	86.18	0.001231	84.11
	2	0.001121	86.04	0.001238	83.32
	3	0.001133	86.76	0.001457	81.39
	4	0.001278	85.17	0.001585	76.31
	5	0.001112	86.30	0.001276	83.74
	6	0.001233	85.71	0.001405	82.07
	7	0.001130	86.59	0.001261	82.57
	8	0.001167	86.14	0.001407	83.12
	9	0.001084	86.28	0.001240	83.19
Best `Scheduled-SGD`		0.001480	85.41	0.001467	82.58
Best `SGD-SA`		0.001084	86.76	0.001240	84.11

In particular, `SGD-SA` leads to a significantly better (1–2%) validation accuracy than `Scheduled-SGD` if the best run for the 10 seeds is considered.

5 Conclusions and Future Work

We have proposed a new metaheuristic training scheme that combines Stochastic Gradient Descent and Discrete Optimization in an unconventional way.

Our idea is to define a discrete neighborhood of the current solution containing a number of "potentially good moves" that exploit gradient information, and to search this neighborhood by using a classical metaheuristic scheme borrowed from Discrete Optimization. In the present paper, we have investigated the use of a simple Simulated Annealing metaheuristic that accepts/rejects a candidate new solution in the neighborhood with a probability that depends both on the new solution quality and on a parameter (the temperature) which is varied over time. We have used this scheme as an automatic way to perform hyper-parameter tuning within a single training execution, and have shown its potentials on a classical test problem (CIFAR-10 image classification using VGG16/ResNet34 deep neural networks).

In a follow-up research we plan to investigate the use of two different objective functions at training time: one differentiable to compute the gradient (and hence a set of potentially good moves), and one completely generic (possibly black-box) for the Simulated Annealing acceptance/rejection test—the latter intended to favor simple/robust solutions that are likely to generalize well.

Replacing Simulated Annealing with other Discrete Optimization metaheuristics (tabu search, variable neighborhood search, genetic algorithms, etc.) is also an interesting topic that deserves future research.

References

1. He, K., Zhang, X., Ren, S., Sun, J.: Deep Residual Learning for Image Recognition. arXiv e-prints, December 2015
2. Kirkpatrick, S., Gelatt, C.D., Vecchi, M.P.: Optimization by simulated annealing. Science **220**(4598), 671–80 (1983)
3. Krizhevsky, A., Nair, V., Hinton, G.: CIFAR-10 (Canadian Institute for Advanced Research). http://www.cs.toronto.edu/~kriz/cifar.html
4. Ledesma, S., Torres, M., Hernández, D., Aviña, G., García, G.: Temperature cycling on simulated annealing for neural network learning. In: Gelbukh, A., Kuri Morales, Á.F. (eds.) MICAI 2007. LNCS (LNAI), vol. 4827, pp. 161–171. Springer, Heidelberg (2007). https://doi.org/10.1007/978-3-540-76631-5_16
5. Metropolis, N.C., Rosenbluth, A.W., Rosenbluth, M.N., Teller, A.H., Teller, E.: Equation of state calculation by fast computing machines. J. Chem. Phys. **21**, 1087–1092 (1953)
6. Nesterov, Y.: A method of solving a convex programming problem with convergence rate O(1/sqr(k)). Soviet Math. Doklady **27**, 372–376 (1983). http://www.core.ucl.ac.be/~nesterov/Research/Papers/DAN83.pdf
7. Sexton, R., Dorsey, R., Johnson, J.: Beyond backpropagation: using simulated annealing for training neural networks. J. End User Comput. **11** (1999)
8. Simonyan, K., Zisserman, A.: Very Deep Convolutional Networks for Large-Scale Image Recognition. arXiv e-prints, September 2014
9. Sutskever, I., Martens, J., Dahl, G., Hinton, G.: On the importance of initialization and momentum in deep learning. In: Proceedings of the 30th International Conference on International Conference on Machine Learning, ICML 2013, vol. 28, pp. III-1139–III-1147. JMLR.org (2013). http://dl.acm.org/citation.cfm?id=3042817.3043064
10. Xiao, H., Rasul, K., Vollgraf, R.: Fashion-MNIST: a novel image dataset for benchmarking machine learning algorithms (2017)

Comparing Local Search Initialization for K-Means and K-Medoids Clustering in a Planar Pareto Front, a Computational Study

Jiangnan Huang, Zixi Chen, and Nicolas Dupin(✉)

Université Paris-Saclay, LISN, 91405 Orsay, France
{jiangnan.huang,zixi.chen,nicolas.dupin}@universite-paris-saclay.fr

Abstract. Having N points in a planar Pareto Front (2D PF), k-means and k-medoids are solvable in $O(N^3)$ time by dynamic programming algorithms. Standard local search approaches, PAM and Lloyd's heuristics, are investigated in the 2D PF case to solve faster large instances. Specific initialization strategies related to 2D PF cases are implemented with the generic ones (Forgy's, Hartigans, k-means++). Applying PAM and Lloyd's local search iterations, the quality of local minimums are compared with optimal values. Numerical results are computed using generated instances, which were made public. This study highlights that local minimums of a poor quality exist for 2D PF cases. A parallel or multi-start heuristic using four initialization strategies improves the accuracy to avoid poor local optimums. Perspectives are still open to improve local search heuristics for the specific 2D PF cases.

Keywords: Clustering algorithms · K-means · K-medoids · Heuristics · Local search · Bi-objective optimization · Pareto Front

1 Introduction

K-means clustering is one of the most famous unsupervised learning problem, and is widely studied in the literature [16,18]. K-medoids clustering, the discrete variant of the k-means problem, maximizes the dissimilarity around a representative solution [17]. If k-medoids clustering is more combinatorial than k-means clustering, it is known to be more robust on noises and outliers [16]. Both k-medoids and k-means problems are NP hard in the general and planar cases [1,15,19]. One dimensional (1D) cases of k-means and k-medoids are solvable in polynomial time, with Dynamic Programming (DP) algorithms [12,14,22].

Facing the NP-hard complexity, a seminal heuristic to solve k-means problems is a steepest descent heuristic converging to local minimums, provided by Lloyd [18]. A careful initialization is prominent for this local search algorithm, many initialization strategies were proposed and discussed [2,11,13,16]. A comparative analysis was provided for general instances of k-means to analyze the

B. Dorronsoro et al. (Eds.): OLA 2021, CCIS 1443, pp. 14–28, 2021.
https://doi.org/10.1007/978-3-030-85672-4_2

impact and efficiency of initialization strategies [3]. Lloyd's local search was extended to solve heuristically k-medoids problems, this adaptation is named PAM (Partitioning Around Medoids) [17]. Comparing empirically the efficiency of initialization strategies for PAM local search is also of interest for k-medoids.

The specific cases of k-means and k-medoids clustering in a planar Pareto Front (2D PF) hold for an application to bi-objective optimization problems and algorithms [5,9]. In both cases, there is a polynomial complexity with DP algorithms running in $O(N^3)$ time [5,9]. However, this cubic complexity may be a bottleneck to solve quickly large instances, which may be required for the real-world application. PAM and Lloyd's local search algorithms are still useful. Similarly with [3,21], this paper aims to compare the impact and efficiency of local search initialization for 2D PF instances. Contrary to [3,21], local minimums can be compared to known optimal values thanks to the DP exact algorithm. Another open question is to determine if the 2D PF case induce also good properties for PAM and Lloyd's local search heuristics.

This paper is organized as follows. In Sect. 2, we define formally the problem and fix the notation. In Sect. 3, initialization heuristics are presented. In Sect. 4, the experimental conditions are presented before describing and analyzing the computational results. In Sect. 5, our contributions are summarized, discussing also future directions of research.

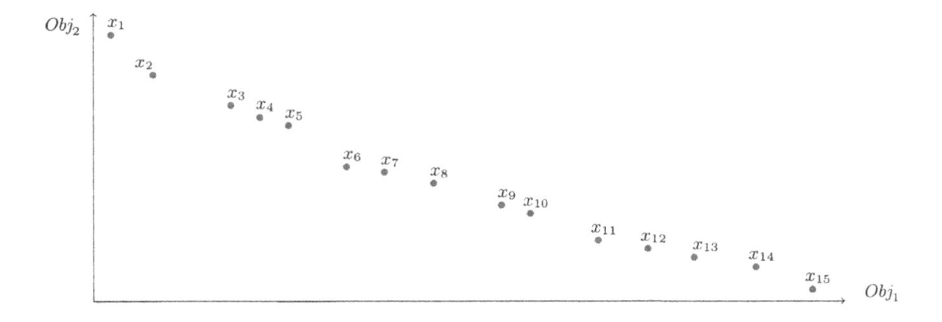

Fig. 1. Illustration of a 2D Pareto Front and its indexation

2 Problem Statement and Notations

Let E a 2D PF of size of N, $E = \{x_1, \dots, x_N\}$ is a set of N elements of $\mathbb{R}^2$ in a 2D PF. As illustrated in Fig. 1, the 2D PF E can be re-indexed such that $E = \{x_k = (y_k, z_k)\}_{k \in [\![1,N]\!]}$ such that $k \in [\![1,N]\!] \mapsto x_k$ is increasing and $k \in [\![1,N]\!] \mapsto y_k$ is decreasing, it is proven formally in [7]. Defining the binary relation $y = (y^1, y^2) \prec z = (z^1, z^2)$ with $y^1 < z^1$ and $y^2 > z^2$, $\prec$ is a total order in E and $x_1 \prec \dots \prec x_N$.

We define $\Pi_K(E)$, as the set of all the possible partitions of E in K subsets. Defining a cost function f for each subset of E to measure the dissimilarity,

K-sum-clustering problems are combinatorial optimization problems, minimizing the sum of the measure f for all the K partitioning clusters:

$$\min_{\pi \in \Pi_K(E)} \sum_{P \in \pi} f(P) \tag{1}$$

K-medoids and *K-means* problems are in the shape of (1). K-medoids problem considers the minimal sum of the squared distances from one point of P, the *medoid*, to the other points of P:

$$\forall P \subset E, \quad f_{medoids}(P) = \min_{c \in P} \sum_{x \in P} \|x - c\|^2 \tag{2}$$

Note that $\|$ denotes the standard Euclidean norm in this paper, $|$ is used to denote the cardinal of a subset of E. K-means clustering considers the sum of the squared distances from any point of P to the *centroid*:

$$\forall P \subset E, \quad f_{means}(P) = \min_{c \in \mathbb{R}^2} \sum_{x \in P} \|x - c\|^2 = \sum_{x \in p} \left\| x - \frac{1}{|p|} \sum_{y \in p} y \right\|^2 \tag{3}$$

The first equality in (3) shows the similarity with k-medoids, k-medoids is the discrete version of k-medoids. The second equality in (3), proven with convex optimization in [20], is used in this paper to compute $f_{means}(P)$ in $O(|P|)$ time, whereas $f_{medoids}(P)$ is computed in $O(|P|^2)$ time using (2). This difference is crucial in the computational efficiency to solve k-means and k-medoids.

PAM and Lloyd's algorithm are similar local search algorithms iterating over solutions encoded as partitioning subsets $P_1, \ldots, P_K$ and their respective centroids $c_1, \ldots, c_K$ (or medoids for PAM). Having a current solution encoded like previously, two steps are processed to improve the solution:

- The partitioning subsets $P_1, \ldots, P_K$ are modified, assigning each point to the closest centroid or medoid c_i. This step runs in $O(NK)$ time.
- The centroids or medoids c_i are recomputed with formulas (3) or (2) considering the updated partitioning subsets $P_1, \ldots, P_K$. This step runs in $O(N)$ time for k-means in $O(N^2)$ time for k-medoids.

If a modification is operated, these steps improve the current clustering solution, it is a steepest descent algorithm. Otherwise, the current solution is a *local minimum* for PAM or Lloyd's local search, that can be different from a *global minimum*, optimal for optimization problems (1). The initial solution given to the local searches has an influence to the quality of local minimums [3]. In the following we compare the efficiency of specific and generic initialization strategies for 2D PF cases. Considering 2D PF instances induces that optimal values of k-medoids and k-means are known even for some large instances thanks to the specific DP algorithms [5,9].

3 Initialization Heuristics

In this section, we introduce the initialization heuristics which are implemented in the computational experiments. We distinguish the generic ones applying for any clustering instances, to specific strategies using properties of 2D PF cases. Initialization strategies were more studied for the k-means problem. We adapt such strategies for the k-medoids problem defining the same clusters than the k-means initialization, i.e. keeping the assignment of points into clusters, and recomputing the medoids with (2) instead of considering the centroids with (3). Hence, we only describe how to define partitioning clusters, giving the complexity results of this phase and keeping in mind that the computations of centroids and medoids run respectively in $O(N)$ and $O(N^2)$ time.

3.1 Generic Initialization Strategies

Firstly, we present generic initialization designed for the k-means problem:

- **Random selection (RAND):** it is one of the most naive algorithms. Firstly, K points are selected randomly (and uniformly) among the N points to initialize centers. Secondly, clusters are defined assigning each point to the cluster of its closest randomly selected point. Defining such clusters runs in $O(N)$ time.
- **Furthest Point (FP):** FP selects randomly the first point. Once $k < K$ points are selected, the $k+1$ next one maximizes the minimal distance from this point to the previously k selected point. As outliers may be the furthest from other points, this algorithm is easily affected by outliers. Selecting the K points runs in $O(K^2N)$ time.
- **K-means++:** K-means++ is an upgraded and randomized version of FP. K-means++ selects randomly the first point, and use different selection probabilities instead of the deterministic max-min distance [2]. If a point is far from the already selected points, its selection probability is higher. Selecting K points for K-means++ runs in $O(K^2N)$ time.
- **Forgy's method:** Forgy's initialization uniformly and randomly assigns each point to one of the K clusters, it runs in $O(N)$ time [11].
- **Hartigan's method:** Hartigan's initialization firstly sorts the points according to their distances from the centroid of the N initial points. The k -th $i \in \{1, 2, \dots, K\}$ center is then chosen to be the point with position round$\left((1 + \frac{N(k-1)}{K})\right)$ in the sorted list [13]. Hartigan's method runs in $O(N \log N)$ time to define partitioning clusters, it is defined by the complexity of the sorting algorithm.

3.2 Initialization Using 2D PF Indexation

The 2D PF indexation, illustrated in Fig. 1, allows to provide variants of the random and Hartigan's initialization, using this specific indexation. The following strategies select initial centroids or medoids in $O(N \log N)$ time, the time complexity of the sorting re-indexation:

- **N/(K + 1) − Uniform**: using the 2d PF indexation, $N/(K+1)-Uniform$ selects the K points at indexes round$\left(\frac{kN}{K+1}\right)$ for $k \in \{1, 2 \ldots, K\}$.
- **N/(2K) − Uniform**: using the 2d PF indexation, $N/(2K)-Uniform$ selects the K points at indexes round$\left(\frac{kN}{2K}\right)$ for $k \in \{1, 2 \ldots, K\}$.
- **N/K-Random selection (N/K-RAND)** N/K-RAND is a tailored variant of RAND selection, the K points for initial centroids or medoids are randomly chosen in each of the N/K-size intervals dividing uniformly the indexes of the 2D PF.

3.3 Initialization Using p-Dispersion for 2D PF

P-dispersion problems select $P \geqslant 2$ points among N initial points, maximizing diversity measures [10]. Some variants are solvable in polynomial time in a 2D PF [4]. Similarly with Hartigan's heuristic, one may select initial centroids or medoids among diversified points. We use the standard p-dispersion (Max-Min p-dispersion) problem, having a time complexity in $O(PN \log N)$, other variants running at least in $O(PN^2)$ [4]. Selecting directly K points with (Max-Min) K-dispersion selects the two extreme points [4]. Two slight adaptations are thus provided to avoid both extreme points, keeping the time complexity in $O(KN \log N)$:

- **K+2-dispersion** ($K+2$-disp): a $K+2$-dispersion is solved using the DP algorithm [4], and the two extreme points (x_1 and x_N after re-indexation) are removed.
- **2K + 1 dispersion** ($2K+1$-disp): a $2K+1$-dispersion is solved using the tailored DP algorithm [4], and we keep the even indexes reindexing the $2K+1$ points from 1 to $2K+1$.

Note that standard p-center problems and variants have also time complexity allowing quick initialization [6,7]. However, such dissimilarity measures are based on the most extreme points, that would include outliers, which is not wished for k-means and k-medoids initialization. K-center based initialization strategies are thus not in our benchmark of initialization strategies.

3.4 Initialization Using 1D Dynamic Programming

One dimensional (1D) k-means is solvable in polynomial time, with an implementation available in a R package [22]. A first DP algorithm runs in $O(KN^2)$ time and using $O(KN)$ memory space [22]. An improvement was proposed in $O(KN)$ time and using $O(N)$ memory space [12]. Such complexity result allows for our application to use this 1D DP algorithm as an initialization heuristic to define first clusters, using two heuristics to reduce the 2D PF into a 1D case:

- **1D-DP-reduc:** using the specific shape of a 2D PF illustrated in Fig. 1, one definea an approximated 1D structure, associating to each point x_i of the 2D PF the scalar $z_i = \sum_{j=0}^{i-1} \|x_j - x_{j+1}\|$. The 1D interval clustering define the indexation for the 2D PF clustering, computing the costs of clusters in the 2D PF requires respectively $O(N)$ and $O(N^2)$ time computations for k-means and k-medoids (which computes also the initial centroids or medoids).

- **1D-DP-linReg:** this initialization is similar to 1D-DP-reduc, the difference being in the reduction from a 2D PF case to 1D. A linear regression is used, 1D points z_i are now the orthogonal projection of points x_i in the linear regression, as illustrated in Fig. 2.

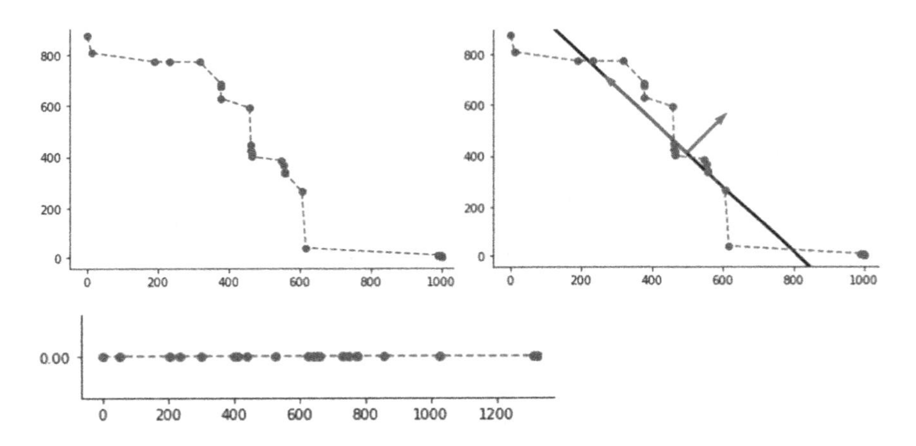

Fig. 2. From a 2dPF to a 1D case with 1D-DP-linReg approach

4 Computational Experiments and Results

4.1 Data Generation

To the best of our knowledge, no specific datasets for 2D PF are available for our study. Starting from any decreasing function $f : [0, 1] \to \mathbb{R}$, one may generate 2D PF with N random values in $(y_n) \in [0, 1]^N$ and considering the 2D points $(y_n, f(y_n))$ for $n \in [\![1; N]\!]$. With a uniform distribution in the random generation and a convex function f, such generation would be too regular, and the naive $N/(2K) - Uniform$ initialization would be very close to the trivial optimal solutions. Hence, a new data generator, described in Algorithm 1, was provided to furnish 2D PF without regularity, like the one given in Fig. 2:

In Algorithm 1, there is a loop invariant: list $\mathcal{L}$ contains non-dominated points, and $\mathcal{B}$ contains the couple of neighbor points in $\mathcal{L}$ with the indexation of Fig. 1. Starting from a point $(x_1, y_1, x_2, y_2) \in \mathcal{B}$ and having $x_3 \in]x_1, x_2[$ and $y_3 \in]y_1, y_2[$, we have $(x_1, y_1) \prec (x_3, y_3) \prec (x_2, y_2)$. Hence, adding (x_3, y_3) in $\mathcal{L}$ keep the two by two incomparability relations in $\mathcal{L}$, and the neighboring properties in $\mathcal{B}$.

Algorithm 1: random generation of a 2d-PF

Input: N the size of the wished 2D PF
  initialize $n = 2$, $\mathcal{L} = \{(0, 1000); (1000, 0)\}$, $\mathcal{B} = \{(0, 1000, 1000, 0)\}$
  for $i = 3$ to N
    select randomly $(x_1, y_1, x_2, y_2) \in \mathcal{B}$ and remove it from $\mathcal{B}$
    select randomly $x_3 \in]x_1, x_2[$ and $y_3 \in]y_1, y_2[$
    add (x_3, y_3) in $\mathcal{L}$ and add (x_1, y_1, x_3, y_3) and (x_3, y_3, x_2, y_2) in $\mathcal{B}$
  end for
return $\mathcal{L}$

4.2 Computational Experiments and Conditions

In our computational results, we used instances generated by Algorithm 1 with five values of N, $N \in \{50; 100; 500; 1000; 5000\}$, generating 10 instances for each value of N. For each generated 2D PF, we experimented K-clustering with five values of K, $K \in \{2; 3; 5; 10; 20\}$. This defines a total of $250 = 5 \times 5 \times 10$ instances for k-medoids and k-means problems. For the reproducibility of the results, these instances and the results instance by instance are available in https://github.com/ndupin/Pareto2d. For these 250 instances, following results are provided for each initialization strategy of k-means and k-medoids:

- the initial value of the heuristic solution;
- the value of the local minimum using PAM or Lloyd's local search
- the number of iterations to converge to a local minimum

Having an exact DP algorithm for K-medoids, it allows to present the quality of solutions in terms of over-cost gap related to the optimal solutions [5]. For k-means, the similar DP algorithm is only a heuristic [9]. Solutions of the DP algorithm are guaranteed optimal if a conjectured lemma is proven [9]. Our experiments did not show any counter-example where a better solution is found, the previous conjecture is not experimentally proven to be false, a perspective is still to prove this conjecture.

Note that we do not provide computation times. The heuristics converge very quickly, in order of second, which is visible in the number of iterations. Comparing the solving time with heuristics and the exact DP could be interesting. Optimal values with DP algorithms were provided using $K = 20$ computations, and storing the values in the DP matrix. Note that RAND, N/K-RAND, Forgy, Hartigan, k-means++ are randomized initialization strategies, the results are given in average using 25 runs with different seeds. Other initialization heuristics are deterministic, one run is enough to provide the results.

Similarly to [8], a parallel local search uses a portfolio of several initialization strategies, and computes independently the local search algorithms for each initial solution. It allows to have the best local minimum among the selected initialization strategies. Having two threads, the 1D DP initialization were selected, the results are reported as **Paral_1D_DP**. Allowing four threads, both previous initialization were selected with k-means++ and FP initialization, this is denoted **Parall4**. These choices were motivated by Tables 1, 2, 3 and 4.

Table 1. Comparison of statistical indicators for the relative over-costs of local minimums induced by the different initialization strategies for the k-means problem, comparison to the solutions from [9].

Init.	Time	Average	Q1	Median	Q3	Variance
RAND	$O(N)$	964,64 %	15,35 %	135,15 %	400,80 %	–
Furthest P	$O(K^2N)$	41,75 %	1,90 %	18,80 %	50,43 %	57,21 %
Kmeans++	$O(K^2N)$	74,67 %	14,03 %	48,05 %	93,10 %	94,38 %
Forgy	$O(N)$	3224,63 %	31,48 %	330,15 %	2245,43 %	–
Hartigan	$O(N \log N)$	1783,30 %	0,00 %	75,70 %	322,90 %	–
N/(K+1)-Unif	$O(N \log N)$	1280,03 %	0,00 %	43,95 %	253,60 %	–
N/2K-Unif	$O(N \log N)$	1221,94 %	0,00 %	42,40 %	249,25 %	–
N/K-RAND	$O(N \log N)$	1142,18 %	6,83 %	50,30 %	236,43 %	–
K+2 disp	$O(KN \log N)$	99,69 %	0,00 %	22,40 %	93,33 %	857,72 %
2K+1 disp	$O(KN \log N)$	399,72 %	0,00 %	12,35 %	72,20 %	–
1D-DP-reduc	$O(KN)$	336,50 %	0,00 %	0,15 %	115,18 %	–
1D-DP-proj	$O(KN)$	158,45 %	0,00 %	1,90 %	38,78 %	3677,36 %
Best Parall		2,90 %	0,00 %	0,00 %	0,90 %	0,83 %
Paral_1D_DP		154,77 %	0,00 %	0,00 %	10,95 %	3685,79 %
Parall4		3,90 %	0,00 %	0,00 %	1,90 %	1,70 %

Tables 1 and 2 present statistical indicators for the quality of local minimums induced by the different initialization strategies. For each initialization strategy, the 250 instances provide a sequence of 250 local minimums, that can be compared to the values of the DP algorithms [5,9]. Statistics are provided based on the over-cost percentages, with average values and variance and quartiles to analyze the dispersion.

Tables 3 and 4 present for k-means and k-medoids clustering the average results for the initial solution and the induced local minimums for some given values of K and N. It illustrates the improving rates of the local search for the different initialization heuristics, but also the influence of K and N in the quality of the primal heuristics.

4.3 Analyses of Computational Results

The average number of iterations for the convergence to a local minimum is illustrated when N is increasing for k-means and k-medoids clustering in Figs. 3 and 4 respectively. The number of iteration is very small, and is also slightly increasing. Keeping in mind that each iteration runs in $O(N)$ or $O(N^2)$ for k-means and k-medoids respectively, the initialization strategy is prominent in the total computation time to apply PAM and Lloyd's local searches. This property is specific for 2D PF cases, this does not hold for general instances.

Generally, local minimums exist with a very poor quality, few such local minimums degrade dramatically the average values, and the medians are much better

Table 2. Comparison of statistical indicators for the relative over-costs of local minimums induced by the different initialization strategies for the k-medoids problem, comparison to the optimal solution from [5].

Init	Time	Average	Q1	Median	Q3	Var
RAND	$O(N^2)$	1661,16 %	27,70 %	162,85 %	499,75 %	–
Furthest P	$O((N+K^2)N)$	43,38 %	3,73 %	19,60 %	54,38 %	66,13 %
Kmeans++	$O((N+K^2)N)$	79,54 %	15,78 %	53,25 %	109,48 %	118,93 %
Forgy	$O(N^2)$	3687,94 %	39,25 %	336,15 %	2304,70 %	–
Hartigan	$O(N^2)$	2214,39 %	0,15 %	102,70 %	387,18 %	–
N/(K+1)-Unif.	$O(N \log N)$	1921,86 %	0,00 %	72,55 %	358,65 %	–
N/2K-Unif.	$O(N \log N)$	1723,85 %	0,00 %	83,20 %	355,08 %	–
N/K-RAND	$O(N \log N)$	1526,19 %	11,38 %	84,20 %	330,25 %	–
K+2 disp	$O(KN \log N)$	340,92 %	0,00 %	31,85 %	131,43 %	–
2K+1 disp	$O(KN \log N)$	820,99 %	0,00 %	21,95 %	90,68 %	–
1D-DP-reduc	$O(N^2)$	677,31 %	0,00 %	0,15 %	74,45 %	–
1D-DP-proj	$O(N^2)$	617,97 %	0,00 %	0,60 %	23,40 %	–
Best Parall		5,20 %	0,00 %	0,00 %	1,20 %	2,39 %
Paral_1D_DP		615,29 %	0,00 %	0,00 %	16,53 %	–
Parall4		5,87 %	0,00 %	0,00 %	2,78 %	2,58 %

than the average. Variance computations are in such cases non significative. One must keep this in mind reading Tables 1 and 2. It is also visible in Tables 3 and 4 is some subsets. Note that Tables 1 and 2 provide results which are very similar considering k-medoids or k-means clustering in a 2D PF.

RAND and Forgy's initialization are the worst strategies in terms of quality of local minimums. N/K-RAND improves the results obtained by the RAND initialization, but the results are still unsatisfactory. FP, k-means++ are the best generic approaches, and also the best approaches to avoid local minimums of a poor quality. These initialization approaches are also the most time consuming, as given by the complexities underlined in Tables 1 and 2.

Naive initialization using 2D PF indexation, N/(K+1)-Uniform and N/(2K)-Uniform have very similar results, with more than a quartile of optimal solutions found, and a degradation of the results for the other quartiles, with a quartile of poor solutions. Initialization using p-dispersion improves the previous uniform ones, with more than a half of excellent solutions, but the last quartile is mainly composed of poor solutions. Initialization based in the 1D DP algorithm may be seen as the best individual approaches to find optimal solutions, with more than an half of excellent solutions. However, very poor local minimums still exist for the resulting local search heuristic.

Table 3. Comparison of average values on the datasets with $N = 100, 1000, 5000$ and $K = 3, 10$ of the initial overcost gap to the optimal solutions after initialization and after local search convergence for K-medoids. Bold values indicate the best average values for a given size of instances.

For $K = 3$:

	N=100 init	N=100 init+LS	N=1000 init	N=1000 init+LS	N=5000 init	N=5000 init+LS
RAND	642,6%	39,5%	785,6%	34,7%	616,7%	50,7%
N/K-RAND	273,4%	29,1%	237,6%	28,5%	283,8%	30,5%
Furthest P	181,9%	37,4%	202,5%	**23,0%**	**129,9%**	18,4%
Kmeans++	228,7%	55,5%	327,4%	56,3%	206,0%	15,0%
Hartigan's	457,6%	54,3%	497,2%	24,2%	485,9%	84,1%
Forgy's	1040,8%	97,0%	1247,0%	67,2%	1048,8%	73,6%
K+2 disp	372,8%	16,4%	562,8%	78,2%	262,4%	21,2%
2K+1 disp	165,1%	32,2%	321,8%	29,2%	161,1%	**12,7%**
N/(K+1)-Unif	307,2%	39,9%	259,2%	38,9%	373,0%	33,9%
N/2K-Unif	233,6%	41,1%	156,1%	39,1%	254,0%	36,6%
1D-DP-reduc	**11,2%**	**0,2%**	594,8%	134,8%	737,0%	154,6%
1D-DP-proj	13,1%	1,3%	**136,2%**	84,6%	737,0%	154,6%

For $K = 10$:

	N=100 init	N=100 init+LS	N=1000 init	N=1000 init+LS	N=5000 init	N=5000 init+LS
RAND	2624,14 %	501,14 %	3386,84 %	375,79 %	3633,05 %	418,77 %
N/K-RAND	822,59 %	298,63 %	1306,63 %	182,64 %	927,22 %	314,39 %
Furthest P	206,07 %	34,10 %	355,41 %	63,35 %	348,31 %	53,23 %
Kmeans++	300,26 %	100,13 %	554,78 %	103,51 %	592,68 %	137,74 %
Hartigan's	851,03 %	378,84 %	2646,14 %	249,99 %	1903,80 %	347,85 %
K+2 disp	1108,43 %	162,79 %	1170,57 %	125,64 %	1185,58 %	130,94 %
2K+1 disp	128,35 %	55,56 %	481,00 %	76,65 %	565,40 %	92,58 %
N/(K+1)-Unif	873,48 %	475,54 %	2122,56 %	212,33 %	1449,22 %	366,43 %
N/2K-Unif	733,31 %	271,75 %	1799,85 %	185,13 %	577,15 %	300,36 %
1D-DP-reduc	47,56 %	**39,55 %**	**13,67 %**	**5,63 %**	**11,76 %**	**1,08 %**
1D-DP-proj	**47,17 %**	41,58 %	23,47 %	6,87 %	19,89 %	5,06 %

Combining in parallel both 1D DP initialization significantly improves the quality of quartiles, with more than a half of optimal solutions, but this is still not enough to avoid the bad local minimums. Tables 3 and 4 shows that difficulty occurs with small values of K and high values of N, especially with $K = 3$ and $N = 5000$. This induces to consider in the pool of initialization FP and k-means++, these approaches having less dispersed results. Tables 3 and 4 illustrate the complementarity of FP and k-means++ with the 1D DP initialization. The resulting Parall4 heuristic provides accurately solutions of an excellent quality, close to the best reachable ones with local search as shown in the row

Table 4. Comparison of average values on the datasets with $N = 100, 1000, 5000$ and $K = 3, 10$ of the initial overcost gap to the optimal solutions after initialization and after local search convergence for K-means. Bold values indicate the best average values for a given size of instances.

For $K = 3$:

	N=100 init	N=100 init+LS	N=1000 init	N=1000 init+LS	N=5000 init	N=5000 init+LS
RAND	630,57 %	28,84 %	771,83 %	30,91 %	651,72 %	38,05 %
N/K-RAND	323,33 %	37,22 %	272,00 %	18,09 %	389,23 %	16,20 %
Furthest P	256,02 %	29,10 %	291,39 %	22,30 %	282,99 %	16,88 %
Kmeans++	178,97 %	27,10 %	242,87 %	22,79 %	**130,54 %**	**11,73 %**
Forgy	289,11 %	37,69 %	347,81 %	57,05 %	274,44 %	22,03 %
Hartigan	1088,12 %	79,35 %	1285,98 %	50,26 %	1093,43 %	75,43 %
N/(K+1)-Unif	391,51 %	11,76 %	586,49 %	9,43 %	274,81 %	22,00 %
N/2K-Unif	175,60 %	10,60 %	336,87 %	**7,42 %**	170,10 %	12,90 %
K+2 disp	246,78 %	38,16 %	**165,30 %**	18,09 %	266,10 %	19,06 %
2K+1 disp	479,67 %	31,57 %	518,54 %	18,09 %	506,01 %	63,30 %
1d-DP-reduc	**1,00 %**	**0,00 %**	1034,98 %	766,00 %	1113,41 %	814,74 %
1d-DP-proj	2,40 %	1,17 %	173,47 %	173,02 %	1113,41 %	814,74 %

For $K = 10$:

	N=100 init	N=100 init+LS	N=1000 init	N=1000 init+LS	N=5000 init	N=5000 init+LS
RAND	3315,60%	383,01%	2904,06%	310,10%	3826,40%	368,51%
N/K-RAND	999,86%	249,86%	2184,31%	222,75%	1487,10%	348,64%
Furthest P	973,94%	189,94%	1553,72%	151,53%	957,58%	284,20%
Kmeans++	249,06%	39,92%	369,80%	54,48%	351,74%	50,75%
Forgy	378,58%	100,34%	593,75%	112,25%	649,23%	146,26%
Hartigan	14083,05%	1591,61%	25774,46%	1281,40%	30135,19%	1516,02%
N/(K+1)-Unif	1265,31%	153,61%	1205,87%	122,00%	1217,02%	127,50%
N/2K-Unif	158,00%	38,58%	497,14%	73,00%	581,67%	90,01%
K+2 disp	841,50%	233,27%	1852,64%	137,71%	593,71%	280,50%
2K+1 disp	974,50%	224,33%	2722,44%	246,10%	1952,80%	328,46%
1d-DP-reduc	**57,04%**	**6,95%**	**5,70%**	**2,12%**	**1,28%**	**1,00%**
1d-DP-proj	59,70%	9,82%	12,08%	5,07%	7,01%	5,20%

BestParall of Tables 1 and 2. Parall4 heuristic does not require a lot of computation times as in [5,9], and requires only 4 threads for parallel computations, or one thread in a multi-start local search with four sequential local searches, which runs in small computation times.

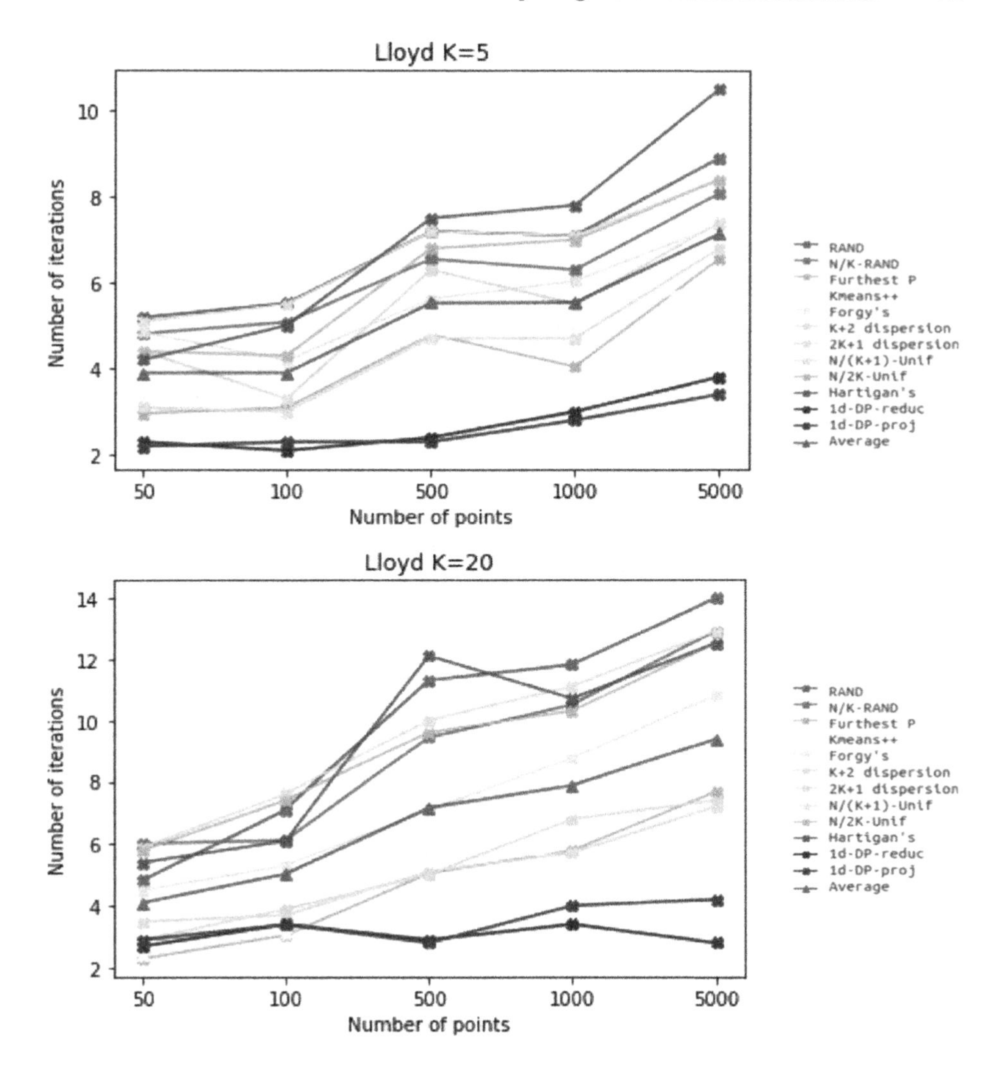

Fig. 3. Number of iterations for K = 5,20 and Lloyd's local search applied to the initialization strategies, with N varying from 50 to 5000

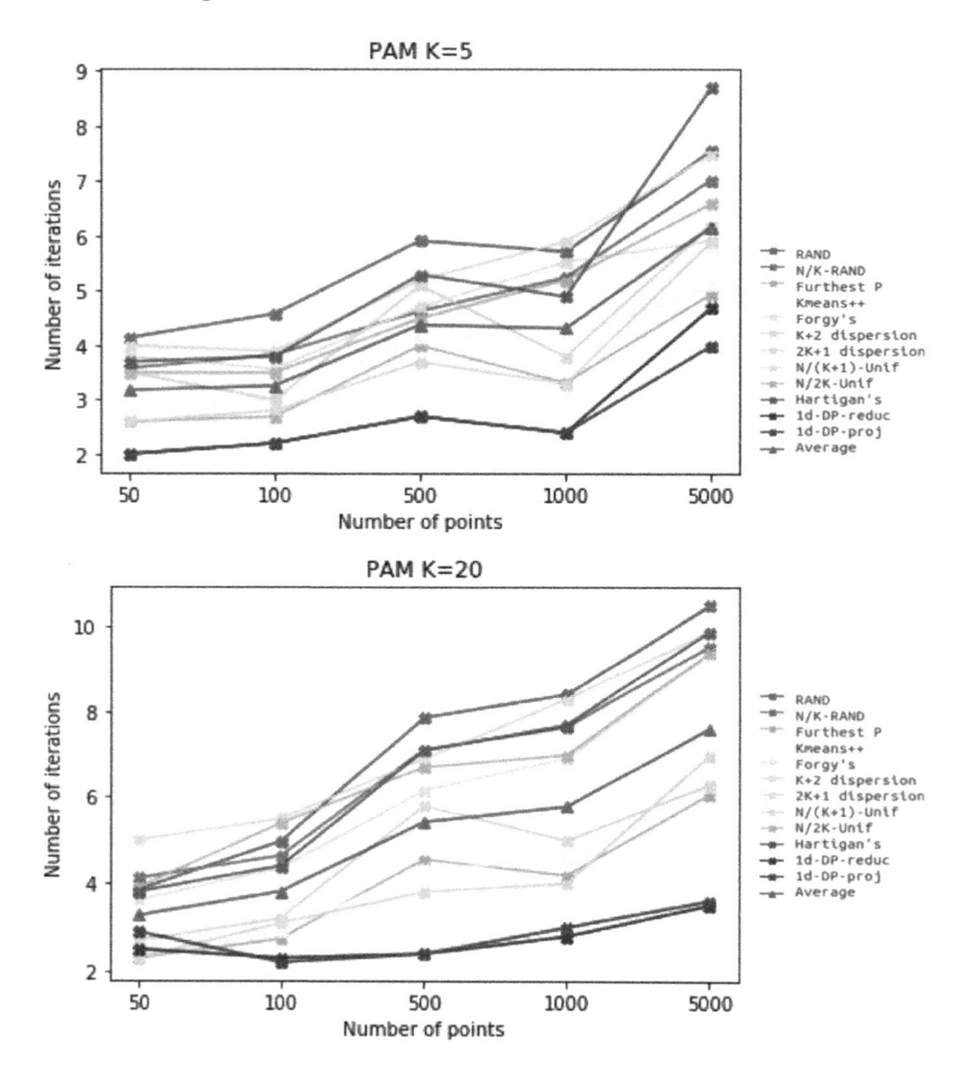

Fig. 4. Number of iterations for K = 5,20 and PAM local search applied to the initialization strategies, with N varying from 50 to 5000

5 Conclusions and Perspectives

Standard local search approaches, PAM and Lloyd's heuristics, are investigated in the 2D PF case to solve faster large instances of k-means and k-medoids clustering than the DP algorithms. Generally, local minimums with a poor quality are found using any initialization heuristic. Two heuristic reductions to the 1D cases, allowing to use the specific 1D DP algorithm provided in average the solutions with the best quality. The generic initialization methods, k-means++

and Furthest Point, were the best to avoid local minimums of a poor quality. Combining these four approaches in a parallel or multi-start local search heuristic allows to have more accurately solutions of a good quality. Such approach is much faster than the exact DP algorithms [5,9], for an application to large instance sizes, this is due to the complexity of the initialization heuristics and that a small number of iterations is required to converge to local optimums.

This numerical study offers several research perspectives. In the applications to evolutionary algorithms of 2D PF clustering already discussed by [4,6,9], it allows to have faster solutions of a very good quality. Perspective are also to prove the global optimality of the solutions provided for k-means clustering in a 2D PF by the DP algorithm [9]. Perspectives may be to improve the local search for the specific 2D PF case instead of the generic PAM and Lloyd's local search algorithms, to improve the accuracy and avoid local minimums of a poor quality. Lastly, one may try to generalize such study and results to dimension 3 and more. A challenging issue would be to extend an efficient projection to the 1D PF case, it was easier considering the 2D PF case.

References

1. Aloise, D., Deshpande, A., Hansen, P., Popat, P.: NP-hardness of Euclidean sum-of-squares clustering. Mach. Learn. **75**(2), 245–248 (2009)
2. Arthur, D., Vassilvitskii, S.: K-means++: the advantages of careful seeding. In: Proceedings of the Eighteenth Annual ACM-SIAM Symposium on Discrete Algorithms, pp. 1027–1035. Society for Industrial and Applied Mathematics (2007)
3. Celebi, M., Kingravi, H., Vela, P.: A comparative study of efficient initialization methods for the k-means clustering algorithm. Expert Syst. Appl. **40**, 200–210 (2013). https://doi.org/10.1016/j.eswa.2012.07.021
4. Dupin, N.: Polynomial algorithms for p-dispersion problems in a 2D Pareto Front. arXiv preprint arXiv:2002.11830 (2020)
5. Dupin, N., Nielsen, F., Talbi, E.-G.: K-medoids clustering is solvable in polynomial time for a 2D Pareto Front. In: Le Thi, H.A., Le, H.M., Pham Dinh, T. (eds.) WCGO 2019. AISC, vol. 991, pp. 790–799. Springer, Cham (2020). https://doi.org/10.1007/978-3-030-21803-4_79
6. Dupin, N., Nielsen, F., Talbi, E.-G.: Clustering a 2D Pareto Front: P-center problems are solvable in polynomial time. In: Dorronsoro, B., Ruiz, P., de la Torre, J.C., Urda, D., Talbi, E.-G. (eds.) OLA 2020. CCIS, vol. 1173, pp. 179–191. Springer, Cham (2020). https://doi.org/10.1007/978-3-030-41913-4_15
7. Dupin, N., Nielsen, F., Talbi, E.: Unified polynomial Dynamic Programming algorithms for p-center variants in a 2D Pareto Front. Mathematics **9**(4), 453 (2021)
8. Dupin, N., Talbi, E.: Parallel matheuristics for the discrete unit commitment problem with min-stop ramping constraints. Int. Trans. Oper. Res. **27**(1), 219–244 (2020)
9. Dupin, N., Talbi, E., Nielsen, F.: Dynamic programming heuristic for k-means clustering among a 2-dimensional pareto frontier. In: 7th International Conference on Metaheuristics and Nature Inspired Computing, META 2018 (2018)
10. Erkut, E., Neuman, S.: Comparison of four models for dispersing facilities. INFOR: Inf. Syst. Oper. Res. **29**(2), 68–86 (1991)

11. Forgy, E.: Cluster analysis of multivariate data: efficiency vs. interpretability of classification. Biometrics **21**(3), 768–769 (1965)
12. Grønlund, A., et al.: Fast exact k-means, k-medians and Bregman divergence clustering in 1D. arXiv preprint arXiv:1701.07204 (2017)
13. Hartigan, J., Wong, M.: Algorithm AS 136: a k-means clustering algorithm. J. R. Stat. Soc. Ser. C (Appl. Stat.) **28**(1), 100–108 (1979)
14. Hassin, R., Tamir, A.: Improved complexity bounds for location problems on the real line. Oper. Res. Lett. **10**(7), 395–402 (1991)
15. Hsu, W., Nemhauser, G.: Easy and hard bottleneck location problems. Discret. Appl. Math. **1**(3), 209–215 (1979)
16. Jain, A.: Data clustering: 50 years beyond k-means. Pattern Recogn. Lett. **31**(8), 651–666 (2010)
17. Kaufman, L., Rousseeuw, P.: Clustering by Means of Medoids. North-Holland (1987)
18. Lloyd, S.: Least squares quantization in PCM. IEEE Trans. Inf. Theory **28**(2), 129–137 (1982)
19. Mahajan, M., Nimbhorkar, P., Varadarajan, K.: The planar K-means problem is NP-hard. Theor. Comput. Sci. **442**, 13–21 (2012)
20. Nielsen, F.: Introduction to HPC with MPI for Data Science. Springer, Heidelberg (2016)
21. Pena, J., Lozano, J., Larranaga, P.: An empirical comparison of four initialization methods for the k-means algorithm. Pattern Recogn. Lett. **20**(10), 1027–1040 (1999)
22. Wang, H., Song, M.: Ckmeans.1d.dp: optimal k-means clustering in one dimension by dynamic programming. R J. **3**(2), 29–33 (2011)

Reinforcement Learning-Based Adaptive Operator Selection

Rafet Durgut[1(✉)] and Mehmet Emin Aydin[2]

[1] Engineering Faculty, Computer Engineering Department, Karabuk University, Karabük, Turkey
rafetdurgut@karabuk.edu.tr
[2] UWE Bristol, Department of Computer Science and Creative Technologies, Bristol, UK
mehmet.aydin@uwe.ac.uk

Abstract. Metaheuristic and swarm intelligence approaches require devising optimisation algorithms with operators to let produce neighbouring solutions to conduct a move. The efficiency of algorithms using single operator remains recessive in comparison with those with multiple operators. However, use of multiple operators require a selection mechanism, which may not be always as productive as expected; therefore an adaptive selection scheme is always needed. In this study, an experience-based, reinforcement learning algorithm has been used to build an adaptive selection scheme implemented to work with a binary artificial bee colony algorithm in which the selection mechanism learns when and subject to which circumstances an operator can help produce better and worse neighbours. The implementations have been tested with commonly used benchmarks of uncapacitated facility location problem. The results demonstrates that the selection scheme developed based on reinforcement learning, which can also be named as smart selection scheme, performs much better that state-of-art adaptive selection schemes.

Keywords: Adaptive operator selection · Reinforcement learning · Artificial bee colony · Uncapacitated Facility Location Problem (UFLP)

1 Introduction

Metaheuristic and swarm intelligence algorithms have gained a deserved popularity through the success accomplished over last few decades. Although they do not guarantee globally optimal solutions within a reasonable time, the success in offering useful near-optimum solutions within an affordable time has helped gain such credit. This does not mean that metaheuristic and swarm intelligence algorithms can be seamlessly implemented for a productive algorithmic solution. The main shortcoming arises in handling local optima capabilities, which enforces researchers to build a balance in exploration for new and fresh solutions while exploiting the gained success level within the search space. That is known

B. Dorronsoro et al. (Eds.): OLA 2021, CCIS 1443, pp. 29–41, 2021.
https://doi.org/10.1007/978-3-030-85672-4_3

as Exploration versus Exploitation (*EvE*) rate in the literature [5]. *EvE* rate guides to search through as many neighbourhoods as possible while retaining exploitation of achieved success and gained experience for a better performance, where weaker exploration causes falling in local optima while weaker exploitation would cause higher fluctuations in performance [12].

Metaheuristic approaches, especially population-based ones, use neighbourhood functions, also known as operators, to let the search process identify next solutions to move to. It is conceivable that search with single operators would have higher likelihood to stick in a local optima than multiple operators. Many hybridisation approaches and memetic algorithms have been designed to help diversify the search through a balanced *EvE*, which usually appear in the form of using multiple operators subject to a selection scheme. The idea an operator to apply after another would prevent the search falling in local optima contributing to diversification of the search. It appears that the nature of the operators to be applied in an order and the order managed in use play very important role in the success level of the algorithms. Adaptive operator selection schemes have been studied for a while to achieve a useful balance in *EvE* and level of diversification in search [11].

Adaptive operator selection is a process of two phases; (i) *credit assignment* in which the selected operators are credited based on the level of success measured, or (ii) *operator selection* in which an operator is identified to run based on the credit level in order to produce a neighbour [10]. The amount of credit to assign is decided using either the positive difference achieved in fitness values or the categories of success or fail [9]. Credit assignment phase also covers the calculation of the time window in which the amount of credit to assign to selected operators is estimated [4]. On the other hand, operator selection phase imposes prioritisation/rank of operators within a pool of functions/operators. *Probability Matching* (PM), *Adaptive Pursuit* (AP) and *Upper Confidence Bound* (UCB) are known to be among state-of-art operators selection schemes [4].

Adaptive operator selection schemes have been used in the literature with evolutionary algorithms and swarm intelligence. Failho et al. [8] uses a multi-armed bandits approach with genetic algorithms, while Durgut and Aydin [7] comparatively studied the success of PM, AP, and UCB schemes to supply a binary artificial bee colony algorithm. Yue et al. [18] proposes a self-adaptive particle swarm optimisation algorithm adaptively selecting among 5 operators to solve large scale feature selection problems.

Adaptive operator selection schemes estimate likelihood of each operator within the pool relying on credits gained to the time. The selection happens through the estimated likelihoods irrespective of the problem state in hand. It is clear that the success of selected operator is not sensitive to the problem state; whether it is in a harsh neighbourhood or trapped in a difficult local optima or not. Reinforcement learning (RL) gains more and more popularity day-by-day to solve dynamic problems progressively, gaining experiences through problems solving process [3,16]. There are renown powerful RL algorithms let map input sets to outputs through experiencing the the problems states and collecting environmental responses to the actions taken [19].

In this study, an artificial bee colony (ABC) algorithm has been implemented for solving uncapacitated facility location problems (UFLP) represented in binary form. ABC algorithms have been implemented to solve many real-world engineering problems. Among them are combinatorial optimisation problems, which formulated as binary optimisation problems. ABC can be viewed as multi-start hill-climbing algorithms in optimisation, where new neighbouring solutions are generated with operators as discussed above. In this study, the ABC algorithm is furnished with multiple operators selected with reinforcement learning-based selection scheme.

The rest of this paper is organised as follows; Adaptive operator selection schemes are introduced in Sect. 2, the operator selection scheme developed based on reinforcement algorithm is explained in Sect. 3. Experimentation and results are presented and discussed in Sect. 4 while conclusions are briefed in Sect. 5.

2 Adaptive Operator Selection

One of the common problem of heuristic-based optimisation algorithms is that search is inevitably driven into local optima, which sometimes remains as the offered final solution. The aim of use multiple operator is to help rescue the search from local optima by the means of diversifying search using different neighbourhood functions/operators interchangeably or systematically. Operator selection schemes are used for this purpose.

Operator selection is not necessarily to be adaptive by nature, but, most of recent studies have been developed as adaptive to insert smartness in the process of selection. Metaheuristic and evolutionary approaches can come up with self-imposing operator selection. Evolutionary algorithms such as genetic algorithms and genetic programming have self-contained probabilistic operator selection while metaheuristics such as variable neighbourhood search imposes a systematic count-based operator change mechanism to achieve diversity in search and manage neighbourhood change. Operator selection built-in algorithms do not offer much flexibility in working with multiple operators, while memetic algorithms, hill-climbing style heuristic algorithms and modern swarm intelligence algorithms allow customising operator selection mechanism to engineer bespoke efficient optimisation algorithms.

Adaptive operator selection is the process of prioritisation of the operators based on merits, which can be imported in the algorithms via crediting each operator based on achievements gained. Although there are a number of adaptive operator selection schemes studied, the general mechanism is depicted in Fig. 1 in which a two phase process is run; (i) *operator selection* and (ii) *credit assignment.* As suggested, the pool of operators holds a finite number of operators to select an operator from in order to produce neighbours to move to, while the selected operators is credited upon its action and success level it achieves in producing new solutions. The credit level to assign to the selected operator is estimated based on preferred rules.

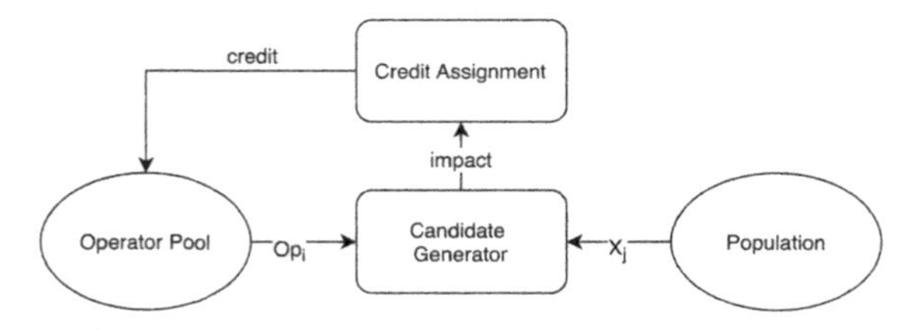

Fig. 1. General overview of adaptive operator selection process with support of population and pool of operators

2.1 Operator Selection

The first phase of operator selection process is to execute the selection rule imposed by operator selection scheme in order to produce neighbouring solutions to move to. The main aim is to keep a *EvE* rate as balanced as possible so that the search to be intensified within the neighbourhood as long as it produces positively and to be diversified as soon as it turns to negative productivity. Literature reports a number of operator selection schemes; random selection, merit-based selection, probability matching, adaptive pursuit and multi-arm bandit approaches, e.g. upper confidence bound (UCB). Random selection chooses an operator from the pool completely randomly, Roulette-wheel takes the success counts of each operator into account to calculate a probability-based prioritisation, while probability matching (PM) approach accounts the success as merits and lets to increase the selectability of non-chosen operators using the following rule:

$$p_{i,t} = p_{min} + (1 - Kp_{min})\frac{q_{i,t}}{\sum_{j=1}^{K} q_{j,t}}, \quad i = 1, 2..K \tag{1}$$

where K is the number of operators in the pool, $p_{min} \in [0, 1]$ represents the minimum probability of being selected, and $q_{i,t}$ is the credit level/value of operation i at time t. Both PM and AP use p_{min} to set a base probability for each operator, which would help address the *EvE* dilemma with allocating a minimum chance to every operators to be selected. PM imposes to calculate the probabilities of being selected per operation, while AP uses the strategy of "*winner takes all*" approach that credits more to promising options. adaptive pursuit (AP) calculates the probabilities with Eq. 2.

$$p_{i,t} = \begin{cases} p_{i,t-1} + \beta(p_{max} - p_{i,t-1}), & \text{if } i = i_t* \\ p_{i,t-1} + \beta(p_{min} - p_{i,t-1}), & \text{otherwise} \end{cases} \tag{2}$$

Both of PM and AP impose higher dominance for exploitation, which is aimed to decrease by UCB using the following rule, which selects the operator with highest probability.

$$p_{i,t} = \begin{cases} 1 - p_{min} * (K - 1) & \text{if } i = i_t* \\ p_{min}, & \text{otherwise} \end{cases} \tag{3}$$

where K is the number of operators in the pool, $p_{min} \in [0, 1]$ represents the minimum (base) probability for being selected, i_t* is calculated with 4.

$$i_t* = \arg \max_{i=1,..,K} \{q_{i,t} + C \times \sqrt{\frac{2 \log \sum_{j=1}^{K} n_{j,t}}{n_{i,t}}}\} \quad (4)$$

where op_t represents the selected operator, C works as a scaling factor, n is number of times the operator selected while $q_{i,t}$ and $n_{i,t}$ on the right-hand-side of equation help control *EvE* dilemma, respectively.

2.2 Credit Assignment

The next phase of adaptive operator selection process is to estimate a credit to be assigned to the operator just used. This involves how to estimate the amount of reward to assign and what to be the base for estimate of a credit. Literature suggests that mainly two classes of approaches have been implemented; whether a success has been achieved or not, or how much positive difference accomplished. The former approach considers if the result is "success" or "fail", while the latter processes the amount of achievement in quantity to estimate the level of reward to assign.

The process of credit assignment entails clarifying the time window with which the reward level is to be estimated. The time window can span from last single step to a pre-defined number of previous steps in which the credit level and/or the achievement level can be averaged. This reveals that a credit can be decided as *instant* credit, an *averaged* credit or the *maximum* credit.

3 Proposed Approach: Adaptive Selection with Reinforced-Clusters

Operator selection adaptively developed and used for higher efficiency in diversification of search process. The operator selection schemes, even the adaptive ones, propose choosing an operator based on credits gained over the success counts through out the search, but, regardless of the input sets, the problem state, and search circumstances. The merit-based schemes usually select operators through a blind process, where the total gained credit is relied on regardless of the status of search etc. It is known that operators do not always produce success due to their limitations; each performs better under some circumstances, while does worse in other circumstances. Once the fruitful circumstances are ascertained for each operator, a complementary policy can be customised for deliberative selection to achieve success.

This study aims to propose a more conscious selection process developed based on reinforcement learning approach implemented into a distance-based clustering algorithm in which the distance in between the input set and the fine-tuned cluster centres is estimated and made reference index in operator selection. The idea of setting up a selection scheme based on clusters is discussed and implemented in machine learning studies. Reinforcement learning is known to be very useful in handling dynamically changing environment and for solving dynamic problems, particularly for operating within unknown dynamic environments. One of earlier studies proposes embedding reinforcement learning in a distance-based clustering algorithm, namely hard-c-means algorithm, to train agents to select the best scheduling operator subject to dynamic production environments to solve dynamic scheduling problems [2]. Inspiring of this study, a reinforced-clustering algorithm is put together to optimise the cluster centres so that the problem states can be classified with optimised clusters, where each cluster will correspond to an operator. The algorithm will impose selecting the cluster centre, operator, closer to the input set in distance. This will facilitates a selection scheme conscious with problem state.

Operators are selected based on probabilities, $p_{i,t}$, calculated as in Eq. 3, where the best operator is determined using Eq. 5. The other operators are also prioritised based on the distance in between the problem state at time t, $\mathbf{x_t}$, and the cluster centres, $\mathbf{c_t}$ - corresponding to the operators. Here, the distance metric used in this study is *hamming* distance due to the binary representation of the problem and the operators.

$$i_t* = \arg \min_{i=1,..,K} \{\beta q_{i,t} + \gamma e_i(x_t)\} \tag{5}$$

where $q_{i,t}$ is the credit level/value of operation i at time t, while $e_i(x_t) = \|\mathbf{x_t} - c_i\|$, the estimated distance between an input set and cluster c_i, β and γ are coefficients to balance between credit and distance metrics. Note that unlike other methods, the reward value of good solutions is reflected as negative.

4 Experimental Results

The reinforced-clustering-based operator selection scheme has been tested with a binary ABC algorithm to solve uncapacitated facility location problem (UFLP) instances, which is one of well-known NP-Hard combinatorial problem. The details of UFLP benchmarking instances taken form OR-Library can be found in many articles [1,7].

The problem solving algorithm to use reinforced-clustering-based operator selection scheme is chosen as the standard artificial bee colony (ABC) algorithm reported in [13]. The standard ABC is designed for continuous numerical optimisation problems, while UFLP is a combinatorial optimisation problem represented in binary form [17]. The algorithm has been rearranged to work with state-of-art binary operators; *binABC* [15] and *ibinABC* [6] work on the basis of XOR logical operator and *disABC* [14] uses a hamming distance-based binary logic.

Algorithm 1 presents a pseudo code of ABC algorithm embedded with reinforced-cluster-based operator selection scheme implemented for UFL problems. As seen, ABC imposes a three-phase process to evolve a swarm (population) of solutions. The first phase exploits *employed* bees to generate new

Algorithm 1. The pseudo code of binary ABC embedded with reinforced-cluster based operator selection scheme

```
1: Initialisation phase:
2: Set algorithm parameters
3: Create initial population
4: while Termination criteria is not met do
5:     Employed bee phase:
6:     Select operators and assign to bees
7:     for i=1 to N do
8:         Select neighbour, apply operator and obtain candidate solution (v_i)
9:         if f(v_i) is better than f(x_i) then
10:            Replace v_i with x_i
11:            Get reward and add to r_{op,t} and update centroid of c_{op,t}
12:            Reset trial counter
13:        else
14:            Increment trial counter
15:        end if
16:    end for
17:    Onlooker bee phase:
18:    Calculate probabilities for food sources
19:    Select operators and assign to bees
20:    Increment operator counter, t=0
21:    for i=1 to N do
22:        Determine current solution according to probability
23:        Select neighbour food source
24:        Apply operator and obtain candidate solution (v_c)
25:        if f(v_c) is better than f(x_c) then
26:            Replace v_c with x_c
27:            Get reward and add to r_{op,t} and update centroid of c_{op,t}
28:            Reset trial counter
29:        else
30:            Increment trial counter
31:        end if
32:    end for
33:    Update Phase:
34:    Credit assignment
35:    Memorisation
36:    Scout bee phase:
37:    if Limit is exceed for any bee then
38:        Create random solution for the first exceeding bee and evaluate it
39:    end if
40: end while
```

solutions with selected binary operators applying to the materials taken from a selected solutions and one of its neighbours. The generated solution is added to the swarm if it is better than the parents, the amount of reward to allocate to the operators is estimated and the position of centre for selected and used operator is updated. If the the generated new solution is not better than the parent solution no reward is generated and the trail counter is incremented.

The *onlooker* bees conduct the second phase of ABC in which the solutions are selected with a probabilistic approach to let randomness contribute the diversity of the swarm. Similar to the first phase, the operator selection, the reward estimation and crediting are performed and the corresponding cluster centres are updated. The *scout* bees follow up the *onlookers* to replace from non-improvable solutions with randomly generated ones to keep the swarm further divers.

The experimentation has started with parametric study to fine-tune parameters used in both the algorithm and within the mechanics of the operator selection scheme. The experimentation for parametric study has been conducted using the hardest benchmarking instance of UFL problem, which is known as CapC. The parameters configured for best fit are tabulated in Table 1 and averaged over 30 repetitions.

Table 1. Parameter configurations tested

Parameter	Values		
Reward	Inst	Avrg	Max
P_{min}	0.10	0.20	0.30
W	10.00	25.00	50.00
β	0.01	0.05	0.10
γ	0.10	0.50	0.90

Table 2 presents the *hit* metric, which is the number of trails attained the optimum. The best performance so far is 25 hits out of 30 trails, where $\gamma = 0.5$, $\beta = 0.01$ and $P_{min} = 0.1$ are found and setup. Next, the reward estimation across a time/iteration window is fine-tuned, where the parametric study results obtained for *average* and *extreme* rewards are tabulated in Table 3. The best hit values are obtained *25* and *27* out of *30* trails for *average* and *extreme* reward cases. respectively.

Table 2. Parameter tuning for Instant reward measured with *hit* metric

γ	P_{min}								
	0.1			0.2			0.3		
	β			β			β		
	0.01	0.05	0.1	0.01	0.05	0.1	0.01	0.05	0.1
0.1	24	16	20	18	24	24	24	24	24
0.5	25	21	19	19	19	21	24	19	14
0.9	16	21	14	21	21	14	17	21	17

The window size (W) of *25* and *50* produce best results, while all trails are tested with $P_{min} = 0.1$, $\beta = 0.05$ and $\gamma = 0.1$. The averaged achievements conclude that $W = 25$ produces the best configuration.

The best configuration concluded out of parametric study has been run with hardest benchmark instances, CapC, to trace the operator selection through timeline, where the progress of operation selection is plotted in Fig. 2. The plot demonstrates that *disABC* operates best over the first 200 iterations and then *ibinABC* takes over the best delivery. *binABC* doesn't perform well in comparison to other two as suggested in the plot.

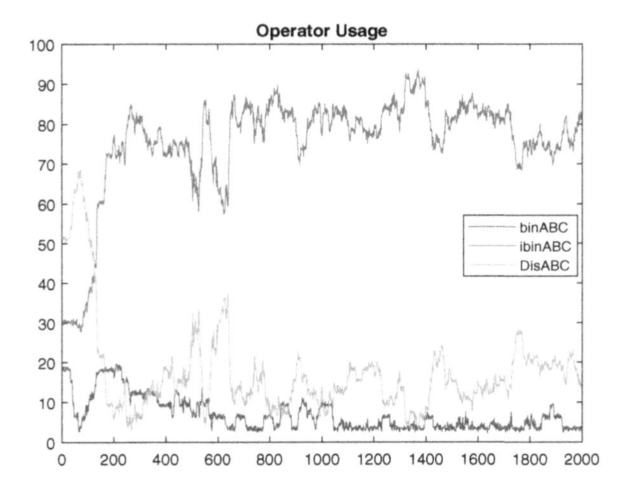

Fig. 2. Operator usage rates through search process

The results by the proposed approach have been tabulated in Table 4 alongside of other adaptive operator selection methods explained above for comparative purposes. As seen, all adaptive methods embedded in binary ABC algorithm have assisted solve all UFLP benchmark instances with 100% success except CapC, where the *Gap* and *St. Dev* metrics are *0* and the *hit* measure is 30 out of 30 for all instances except CapC. It is paramount to define the *gap* as the

Table 3. Parametric fine-tuning results in *hit* metric for both *average* and *extreme* rewards

W	P_{min}	β	Average Reward			Extreme Reward			W	P_{min}	β	Average Reward			Extreme Reward		
			γ			γ						γ			γ		
			0.1	0.3	0.9	0.1	0.3	0.9				0.1	0.3	0.9	0.1	0.3	0.9
5	0.1	0.01	16	21	16	24	25	24	25	0.1	0.01	22	20	22	22	23	24
		0.05	23	19	17	23	24	23			0.05	25	17	19	27	21	23
		0.1	21	19	18	23	24	19			0.1	24	21	18	22	19	21
	0.2	0.01	24	21	23	19	19	23		0.2	0.01	23	21	21	17	25	23
		0.05	22	20	18	21	19	17			0.05	15	19	21	21	22	22
		0.1	20	21	19	19	18	19			0.1	22	20	21	17	25	23
	0.3	0.01	20	21	23	19	18	20		0.3	0.01	21	20	20	20	16	26
		0.05	20	23	20	21	20	19			0.05	20	18	18	22	23	17
		0.1	22	15	18	21	16	19			0.1	21	18	19	25	22	16
10	0.1	0.01	25	21	24	22	16	25	50	0.1	0.01	23	19	19	21	18	19
		0.05	21	23	20	20	18	22			0.05	21	22	19	27	21	19
		0.1	24	15	23	22	19	22			0.1	20	20	18	21	18	23
	0.2	0.01	25	19	20	21	18	13		0.2	0.01	18	20	25	21	22	19
		0.05	21	22	20	14	23	17			0.05	19	21	19	21	18	18
		0.1	24	21	20	15	21	21			0.1	23	14	22	22	17	19
	0.3	0.01	24	20	20	24	20	19		0.3	0.01	16	25	20	21	18	20
		0.05	21	14	16	20	24	19			0.05	22	17	21	22	16	20
		0.1	23	21	19	20	20	22			0.1	16	19	16	21	14	18

Table 4. The comparative results obtained; the proposed operator selection scheme vs alternatives

Benchmarks	PM-ABC			AP-BABC			UCB-BABC			C-BABC		
	Gap	Std. Dev.	Hit	Gap	Std. Dev.	Hit	Gap	Std. Dev.	Hit	Gap	Std. Dev.	Hit
Cap71	0	0	30	0	0	30	0	0	30	0	0	30
Cap72	0	0	30	0	0	30	0	0	30	0	0	30
Cap73	0	0	30	0	0	30	0	0	30	0	0	30
Cap74	0	0	30	0	0	30	0	0	30	0	0	30
Cap101	0	0	30	0	0	30	0	0	30	0	0	30
Cap102	0	0	30	0	0	30	0	0	30	0	0	30
Cap103	0	0	30	0	0	30	0	0	30	0	0	30
Cap104	0	0	30	0	0	30	0	0	30	0	0	30
Cap131	0	0	30	0	0	30	0	0	30	0	0	30
Cap132	0	0	30	0	0	30	0	0	30	0	0	30
Cap133	0	0	30	0	0	30	0	0	30	0	0	30
Cap134	0	0	30	0	0	30	0	0	30	0	0	30
CapA	0	0	30	0	0	30	0	0	30	0	0	30
CapB	0	0	30	0	0	30	0	0	30	0	0	30
CapC	0.0055	1428.003	25	0.0043	1302.539	26	0.0087	1694.457	22	0.0033	1149.5	27

average difference in between the optimum value and the fitness/cost value found, while *St. Dev.* is the standard deviation calculated over 30 repeated trails. CapC seems to be the hardest benchmark instance, which helps fine-tuning the hyper parameters and comparing the results produced by each rival approaches. The proposed method, labelled as "C-BABC" in the tables, produces the lowest *gap* and *st. dev* and the highest *hit* in comparisons to "PM-BABC", "AP-BABC" and "UCB-BABC", which are the binary ABC algorithms embedded with *PM*, *AP* and *UCB* as explained above.

The success of proposed method has been comparatively tested with a number of recently published studies, which can be considered as state-of-art works. The comparative results have been picked up form corresponding articles [1] and tabulated with the results produced by the proposed approach. As clearly seen on Table 5, the proposed method, C-BABC, outperforms all the algorithms known to be the state-of-the-art with a 100% success of solving all benchmark instances except CapC, which is solved with the highest score, while binAAA and JayaX solve all instances except CapB and CapC. Due to level of hardness in solving CapB and CapC approaches are tested with, so is the proposed approach in comparative way. The difference between the results by the proposed approach and other competitor algorithms have been tested statistically with Wilcoxon signed rank and the results are presented in Table 6, where C-BABC, the proposed method is significantly performed better.

Table 5. Comparative results; The proposed method (C-BABC) versus some state-of-art approaches

Benchmark	GA-SP			BPSO			binAAA			JayaX			C-BABC		
	Gap	Std. Dev.	Hit	Gap	Std. Dev.	Hit	Gap	Std. Dev.	Hit	Gap	Std. Dev.	Hit	Gap	Std. Dev.	Hit
Cap71	0	0	30	0	0	30	0	0	30	0	0	30	0	0	30
Cap72	0	0	30	0	0	30	0	0	30	0	0	30	0	0	30
Cap73	0.066	899.65	19	0.024	634.625	26	0	0	30	0	0	30	0	0	30
Cap74	0	0	30	0.0088	500.272	29	0	0	30	0	0	30	0	0	30
Cap101	0.068	421.655	11	0.0432	428.658	18	0	0	30	0	0	30	0	0	30
Cap102	0	0	30	0.00989	321.588	28	0	0	30	0	0	30	0	0	30
Cap103	0.063	505.036	6	0.04939	521.237	14	0	0	30	0	0	30	0	0	30
Cap104	0	0	30	0.040	1432.239	28	0	0	30	0	0	30	0	0	30
Cap131	0.068	720.877	16	0.171	1505.749	10	0	0	30	0	0	30	0	0	30
Cap132	0	0	30	0.058	1055.238	21	0	0	30	0	0	30	0	0	30
Cap133	0.091	685.076	10	0.082	690.192	10	0	0	30	0	0	30	0	0	30
Cap134	0	0	30	0.195	2594.211	18	0	0	30	0	0	30	0	0	30
CapA	0.046	22451.21	24	1.69	319855.4	8	0	0	30	0	0	30	0	0	30
CapB	0.58	66658.65	9	1.40	135326.7	5	0.24	39224.74	15	0.07	27033.02	26	0	0	30
CapC	0.70	51848.28	2	1.62	115156.4	1	0.29	29766.31	1	0.021	5455.94	17	0.0033	1149.5	27

Table 6. Statistical test results for state-of-art methods compared with proposed approach

Benchmarks	binAAA		JayaX		BPSO		GA-SP	
	p-value	H	p-value	H	p-value	H	p-value	H
Cap71	1	0	1	0	1	0	1	0
Cap72	1	0	1	0	1	0	1	0
Cap73	1	0	1	0	1.E-01	0	1.E-03	1
Cap74	1	0	1	0	3.E-06	1	4.E-08	1
Cap101	1	0	1	0	2.E-01	0	4.E-04	1
Cap102	1	0	1	0	5.E-01	0	1	0
Cap103	1	0	1	0	1.E-06	1	1.E-06	1
Cap104	1	0	1	0	5.E-01	0	1	0
Cap131	1	0	1	0	1.E-06	1	1.E-06	1
Cap132	1	0	1	0	1.E+00	0	4.E-08	1
Cap133	1	0	1	0	2.E-06	1	1.E-06	1
Cap134	1	0	1	0	5.E-04	1	1	0
CapA	1	0	1	0	5.E-05	1	1.E-01	0
CapB	6.E-05	1	2.E-07	1	2.E-06	1	2.E-06	1
CapC	4.E-06	1	1.E-04	1	3.E-06	1	4.E-06	1

5 Conclusion

This study has been done to investigate how machine learning can help adapt a dynamically updating scheme for operator selection within ABC algorithms as one of recently developed swarm intelligence approaches in solving binary problems. The research has been done embedding an online learning mechanism into binary ABC to learn which operator performs better in given circumstances. The main contribution of this research is that the adaptive operator selection has been achieved through reinforcement learning which is implemented with Hard-C-means clustering algorithm converted its unsupervised nature into reinforcement learning. Unlike the previously suggested adaptive selection schemes, this approach maps the binary input set into corresponding operators, hence, each time the hamming distance between both binary sets is used to make the selection, while the centres of the clusters are optimised/fine-tuned with estimated rewards per operator selection. The optimised cluster centres remain as the basis of operator selection. The proposed algorithm is tested with solving UFL problems, and statistically verified that the proposed approach significantly outperforms the state-of-art approaches in solving the same benchmark instances. It is also demonstrated that other existing adaptive approaches are also outperformed.

References

1. Aslan, M., Gunduz, M., Kiran, M.S.: Jayax: jaya algorithm with xor operator for binary optimization. Appl. Soft Comput. **82**, 105576 (2019)
2. Aydin, M.E., Öztemel, E.: Dynamic job-shop scheduling using reinforcement learning agents. Robot. Auton. Syst. **33**(2–3), 169–178 (2000)
3. Coronato, A., Naeem, M., De Pietro, G., Paragliola, G.: Reinforcement learning for intelligent healthcare applications: a survey. Artif. Intell. Med. **109**, 101964 (2020)
4. DaCosta, L., Fialho, A., Schoenauer, M., Sebag, M.: Adaptive operator selection with dynamic multi-armed bandits. In: Proceedings of the 10th annual conference on Genetic and evolutionary computation, pp. 913–920 (2008)
5. Dokeroglu, T., Sevinc, E., Kucukyilmaz, T., Cosar, A.: A survey on new generation metaheuristic algorithms. Comput. Ind. Eng. **137**, 106040 (2019)
6. Durgut, R.: Improved binary artificial bee colony algorithm. Frontiers of Information Technology & Electronic Engineering (in press) (2020)
7. Durgut, R., Aydin, M.E.: Adaptive binary artificial bee colony algorithm. Appl. Soft Comput. **101**, 107054 (2021)
8. Fialho, Á.: Adaptive operator selection for optimization. Ph.D. thesis, Université Paris Sud-Paris XI (2010)
9. Fialho, Á., Da Costa, L., Schoenauer, M., Sebag, M.: Extreme value based adaptive operator selection. In: Rudolph, G., Jansen, T., Beume, N., Lucas, S., Poloni, C. (eds.) PPSN 2008. LNCS, vol. 5199, pp. 175–184. Springer, Heidelberg (2008). https://doi.org/10.1007/978-3-540-87700-4_18
10. Fialho, Á., Da Costa, L., Schoenauer, M., Sebag, M.: Analyzing bandit-based adaptive operator selection mechanisms. Ann. Math. Artif. Intell. **60**(1–2), 25–64 (2010)
11. Hussain, A., Muhammad, Y.S.: Trade-off between exploration and exploitation with genetic algorithm using a novel selection operator. Complex Intell. Syst. **6**, 1–14 (2019)
12. Hussain, K., Salleh, M.N.M., Cheng, S., Shi, Y.: On the exploration and exploitation in popular swarm-based metaheuristic algorithms. Neural Comput. Appl. **31**(11), 7665–7683 (2019)
13. Karaboga, D., Basturk, B.: On the performance of artificial bee colony (ABC) algorithm. Appl. Soft Comput. **8**(1), 687–697 (2008)

14. Kashan, M.H., Nahavandi, N., Kashan, A.H.: DisABC: a new artificial bee colony algorithm for binary optimization. Appl. Soft Comput. **12**(1), 342–352 (2012)
15. Kiran, M.S., Gündüz, M.: Xor-based artificial bee colony algorithm for binary optimization. Turkish J. Electr. Eng. Comput. Sci. **21**(Sup. 2), 2307–2328 (2013)
16. Moerland, T.M., Broekens, J., Jonker, C.M.: Model-based reinforcement learning: A survey. arXiv preprint arXiv:2006.16712 (2020)
17. Ozturk, C., Hancer, E., Karaboga, D.: Dynamic clustering with improved binary artificial bee colony algorithm. Appl. Soft Comput. **28**, 69–80 (2015)
18. Xue, Y., Xue, B., Zhang, M.: Self-adaptive particle swarm optimization for large-scale feature selection in classification. ACM Trans. Knowl. Discov. Data (TKDD) **13**(5), 1–27 (2019)
19. Yang, T., Zhao, L., Li, W., Zomaya, A.Y.: Reinforcement learning in sustainable energy and electric systems: a survey. Ann. Rev. Control **49**, 145–163 (2020)

Learning for Optimization

A Learning-Based Iterated Local Search Algorithm for Solving the Traveling Salesman Problem

Maryam Karimi-Mamaghan(✉), Bastien Pasdeloup, Mehrdad Mohammadi, and Patrick Meyer

IMT Atlantique, Lab-STICC, UMR CNRS 6285, 29238 Brest, France
{maryam.karimi,bastien.pasdeloup,mehrdad.mohammadi, patrick.meyer}@imt-atlantique.fr

Abstract. In this paper, we study the use of reinforcement learning in adaptive operator selection within the Iterated Local Search metaheuristic for solving the well-known NP-Hard Traveling Salesman Problem. This metaheuristic basically employs single local search and perturbation operators for finding the (near-) optimal solution. In this paper, by incorporating multiple local search and perturbation operators, we explore the use of reinforcement learning, and more specifically Q-learning as a machine learning technique, to intelligently select the most appropriate search operator(s) at each stage of the search process. The Q-learning is separately used for both local search operator selection and perturbation operator selection. The performance of the proposed algorithms is tested through a comparative analysis against a set of benchmark algorithms. Finally, we show that intelligently selecting the search operators not only provides better solutions with lower optimality gaps but also accelerates the convergence of the algorithms toward promising solutions.

Keywords: Adaptive operator selection · Iterated local search · Reinforcement learning · Q-learning · Traveling salesman problem

1 Introduction

Combinatorial Optimization Problems (COPs) are a complex class of optimization problems with discrete decision variables and a finite search space. Many COPs are NP-hard for which no polynomial-time algorithm exists. Metaheuristics (MHs) can solve these problems in a reasonable time and provide them with acceptable solutions; however, they do not guarantee the optimality [25]. MHs employ either single or multiple search operators to evolve a single or a population of solutions toward (near-) optimal solutions. When using multiple search operators, the problem of *operator selection* arises.

Individual search operators may be effective in particular stages of the search process and not throughout the search process. The reason is that the search space of COPs is a non-stationary environment that includes different search

B. Dorronsoro et al. (Eds.): OLA 2021, CCIS 1443, pp. 45–61, 2021.
https://doi.org/10.1007/978-3-030-85672-4_4

regions with dissimilar characteristics. Therefore, different search operators act differently in different regions of the search space [7]. Accordingly, solving COPs with single search operators does not necessarily lead to the highest performance of the search process. Intuitively, employing multiple search operators selected in an appropriate way during the search process not only leads to a more robust behavior of a MH with respect to the process of finding the optimal solution [22], but also significantly affects the exploration (i.e., explore undiscovered regions) and exploitation (i.e., intensify the search in promising regions) abilities of a MH, and provides an Exploration-Exploitation balance during the search process. The main question in this regard is in which order the search operators should be employed such that the MH can go toward the global optimum. One efficient way is to dynamically select and apply the most appropriate operators based on their history of performance during the search process. This is referred to as *Adaptive Operator Selection* (AOS) [7]. Adaptive selection strategies may differ from very simple strategies to more advanced ones. In simple strategies, such as score-based selection strategy [19], an initial score is assigned to each search operator and the scores are updated based on the performance of each operator at each step of the search process. In this strategy, the selection chance of each search operators is then proportional to its accumulated score. Regardless of the neglectable overhead that they impose to the search process, the added-value of simple strategies may not be necessarily significant [27]. Hence, more advanced adaptive strategies should be embedded into the AOS.

In this regard, Machine Learning (ML) techniques can be used in AOS to provide a more intelligent adaptive strategy when selecting the search operators during the search process. The integration of ML techniques into MHs is an emerging research field that has attracted numerous researchers in recent years [3,5,8,10,11,18,23,26]. In particular, ML techniques help the AOS to use feedback information on the performance of the search operators during the search process. In this situation, operators are selected based on a credit assigned to each operator (i.e., feedback from their historical performance). Considering the nature of the feedback, the learning can be *offline* or *online*. In *offline* learning, knowledge is extracted from a set of training instance with the aim to solve new problem instances. In *online* learning, the knowledge is extracted and incorporated into the resolution process dynamically while solving a problem instance [4,26].

In this paper, we study the use of reinforcement learning (RL), particularly Q-learning as a ML technique, in AOS within the Iterated Local Search (ILS) meta-heuristic [13] for solving the well-known NP-hard Traveling Salesman Problem (TSP). The ILS basically employs single local search and perturbation operators for finding the (near-) optimal solutions. However, there are several specific and efficient local search and perturbation operators for TSP in the literature (e.g., 2-opt, 3-opt, insertion, etc. as local search operators and double-bridge, shuffle-sequence, etc. as perturbation operators) [25] that can be employed simultaneously. In this paper, we incorporate multiple local search and perturbation operators into the ILS and use Q-learning to adaptively select among them during the search process. Indeed, Q-learning is integrated into ILS to adaptively select its operators during the search process. This integrated algorithm is called Q-ILS hereafter. In this paper, two variants of Q-ILS are proposed: in the first algorithm called Q-ILS-1, Q-learning is used to select appropriate local search operators at each stage of the search process, and in the second algorithm called Q-ILS-2, Q-learning is used for selecting appropriate perturbation operators. We will show that both Q-ILS-1 and Q-ILS-2 are able to find good solutions and outperform the ILS with single operator and also ILS with multiple randomly selected operators.

The rest of the paper is organized as follows. Section 2 reviews the recent relevant papers studying Q-learning for AOS in solving different COPs. Section 3 explains the preliminaries and main concepts of this paper. The two Q-ILS algorithms (Q-ILS-1 and Q-ILS-2) are proposed in Sect. 4. The performances of the proposed algorithms are investigated in Sect. 5. Finally, the conclusion is given in Sect. 6.

2 Literature Review

AOS has been widely studied within different MHs for adaptively selecting the search operators [7,15,16,30]. Most of the studies use simple score-based methods that select operators based on their accumulated score [7]. Besides simple score-based mechanisms for AOS, RL techniques, in particular Q-learning algorithm, have been used for AOS in recent years [2,9,17,21,22]. In the following, the studies on the use of Q-learning algorithm for AOS for solving different COPs are elaborated.

In [22], Q-learning has been integrated into a Variable Neighborhood Search algorithm to solve the symmetric TSP. The role of Q-learning is to select appropriate local search operators during the search process, where both the states and actions are a set of local search operators (i.e., interchange, insertion, 2-opt, and double-bridge). The authors show that using Q-learning to intelligently select the local search operators achieves satisfactory results for small-sized instances of the TSP. In [21], the Q-learning algorithm is used to select the search operators of a Genetic algorithm, namely mutation and crossover operators, during the search process for solving TSP. The authors discuss that adaptive operator selection based on the immediate performance of the operators might lead to a short-sighted optimization. Therefore, to overcome this shortcoming, they recommended using RL that can learn a policy to maximize the expected reward in a long-term prospect. In [17], the authors have used Q-learning algorithm to select appropriate local search operators of a Simulated Annealing algorithm. The proposed algorithm is applied to mixed-model sequencing problem to select among exchange, shift, and knowledge sharing operators. The states are defined as the number of successful neighbor moves (i.e., moves that improve the objective function) occurred during an episode, and actions are a set of triplet local search operators. They show that the integration of Q-learning into Simulated Annealing significantly improves its performance comparing to other Simulated

Annealing-based algorithms. In [2], the authors have employed Q-learning to select the order of applying mutation and crossover operators in each generation of the Genetic algorithm. In their algorithm, five states are defined depending on the number of chromosomes within the population that are replaced by executing an action, and there are two possible actions; apply crossover first and mutation next or apply mutation first and crossover next. To show the performance of the proposed method, it is applied to job sequencing and tool switching problem. The authors show that the proposed algorithm is competitive and even superior to the state-of-the-art algorithms for solving some instances of the problem.

As shown by the reviewed papers, the use of Q-learning in AOS has provided promising results in solving different COPs, and even in some cases it has been superior to some state-of-the-art algorithms. Motivated by the good performance of Q-learning, this paper aims at investigating the integration of Q-learning into AOS to select local search (Q-ILS-1 algorithm) and perturbation (Q-ILS-2 algorithm) operators of the ILS for solving the TSP.

The main contributions of this paper compared to the literature are threefold: 1) for the first time, this paper investigates the use of Q-learning in ILS for intelligently selecting the search operators throughout the search process, 2) the Q-learning is integrated into ILS in two levels for selecting local search and perturbation operators, with the aim of investigating the effect of intelligent AOS in each level, and 3) a new design of Q-learning is proposed where a set of appropriate states and actions are defined according to the level of integration. In Q-ILS-1 the states are defined as the sequence of last k local search operators and the actions are the local search operators. In Q-ILS-2, we define two states; 0 if there is no improvement in the best found solution during an episode and 1; otherwise, and the actions are a set of perturbation operators.

3 Preliminaries

In this section, first a short introduction to the TSP is provided. Next, the basics of the ILS algorithm and the Q-learning algorithm are explained.

3.1 Traveling Salesman Problem

TSP is a classical NP-hard COP, which requires exponential time to be solved to optimality [12]. TSP can be formally defined by means of a weighted graph $G = (V, A)$ where V is the set of vertices representing cities and A is the set of edges that connect the vertices of V. The edge that connects cities i and j has a weight of d_{ij}, which represents the distance between cities i and j; $i, j \in V$. In TSP, the aim is to find the Hamiltonian cycle of minimum total travel distance such that all vertices are visited exactly once.

3.2 Iterated Local Search

Iterated Local Search (ILS) is a well-known MH for its effectiveness in both exploration and exploitation and its simplicity in practice. When the search gets trapped in a local optimum, ILS attempts to escape from the trap without losing many of the good properties of the current solution [13]. Considering s_{best} as the best solution found in the history of ILS, the general pseudo code of ILS is given in Algorithm 1. For a given initial solution s_0, a `LocalSearch(.)` function is performed on solution s_0 to search its neighborhood with the hope to find better solutions, particularly the local optimal solution s^*. Subsequently, s^* is archived as the current best solution s_{best}. Then, the main loop of the ILS starts by performing a `Perturbation(.)` function over the current local optimum solution s^* to help the search process to escape from the local optimum; whereby an intermediate solution s' is generated. The `LocalSearch(.)` function is performed on the intermediate solution s' to obtain a new local optimal solution $s^{*'}$. Next, the $\texttt{Acceptance}(s^*, s^{*'}, s_{best})$ function is employed to check whether the new local optimal solution $s^{*'}$ is accepted. The `Acceptance(.)` function can only accept better solution (i.e., *Only Improvement* strategy) or it can even accept worse solution with a small gap (i.e., *Metropolis acceptance* strategy [14]). Finally, the best solution s_{best} is updated. The algorithm terminates when the termination criterion is satisfied.

Algorithm 1. Pseudo code of the ILS

1 **get** an initial solution s_0
2 $s^* := \texttt{LocalSearch}(s_0)$
3 $s_{best} := s^*$
4 **while** *termination criterion not reached* **do**
5 $\quad s' := \texttt{Perturbation}(s^*)$
6 $\quad s^{*'} := \texttt{LocalSearch}(s')$
7 $\quad s^* := \texttt{Acceptance}(s^*, s^{*'}, s_{best})$
8 **end**
9 **return** the best found solution s_{best}

3.3 The Q-Learning Algorithm

In RL, an agent interacts with the environment and aims to iteratively learn which action to take at a given state of the environment to achieve a goal. At each interaction depending on the state s $(s \in S)$ the agent takes an action a $(a \in A(s))$ and receives a numerical feedback from the environment. Through this process, the agent attempts to iteratively maximize the cumulative received reward. Classical RL methods need the complete model of the environment (i.e., all possible states of the system, the set of possible actions per state, and the matrix of transition probabilities as well as the expected values of the feedback). However, in most problems including COPs, it is not possible to have a complete model of the environment [29]. In such cases, Monte Carlo and Temporal Difference algorithms can be used [24]. The Q-learning algorithm [28] is a model-free RL algorithm based on temporal differences. In Q-learning, a Q-value is associated with each state-action pair (s, a) that represents the expected gain of the choice of action a at state s. The Q-value of each state-action pair (s, a) is updated using Expression (1).

$$Q(s,a) \leftarrow Q(s,a) + \alpha[r + \gamma \max_{a'} Q(s',a') - Q(s,a)] \quad (1)$$

where r is the reward (punishment) received after performing action a in state s and γ $(0 \leq \gamma < 1)$ and α $(0 \leq \alpha < 1)$ are the discount factor and the learning rate, respectively.

One strategy to select the actions in Q-learning is to always select the action with the maximum Q-value. In this strategy, the best state-action pairs with the maximum Q-values are exploited sufficiently, while other state-action pairs remain unexplored. To cope with this issue and to make a balance between exploration and exploitation, the ϵ-greedy strategy (Expression (2)) [24] is an efficient strategy that assigns an ϵ selection probability to other actions to give them a chance to be explored.

$$a = \begin{cases} \operatorname{argmax}_a Q(s,a) & \text{with probability } 1-\epsilon \\ \text{any other action} & \text{with probability } \epsilon \end{cases} \quad (2)$$

To move from exploration of new actions toward exploitation of the best actions, the value of ϵ gradually degrades throughout the search process using a parameter β called ϵ-decay.

4 Proposed Q-ILS Algorithms

This section proposes two Q-ILS-1 and Q-ILS-2 algorithms and explains their corresponding operators and properties.

4.1 Q-ILS-1 Algorithm

The novelty of the proposed Q-ILS-1 algorithm is development of a new local search procedure for ILS based on the ideas from AOS and Q-learning. The proposed local search procedure adaptively selects appropriate operators during the search process based on the currently employed operator and operators' history of performance. In the first step, a pool of local search operators are incorporated into the algorithm. Then, the proposed Q-learning algorithm is integrated into AOS to select local search operators.

General Framework. In Q-ILS-1, the local search operators perform a descent-based search and continue until no more improvements are found. As the perturbation operator, we employ *double-bridge* operator wherein four edges are removed from the route of the cities and sub-routes are reconnected in another way to explore a new route [25]. The `Acceptance(.)` function in Q-ILS-1 applies a *Metropolis acceptance* strategy [14] that accepts all improved solutions and even non-improved solutions with a probability of $\exp \frac{\Delta f}{T}$, where Δf is the difference between the objective function before and after applying the local search operator, and parameter T denotes the temperature. The higher the value of T, the higher the chance to accept worse moves and vice versa.

Local Search Operators. In Q-ILS-1, three efficient local search operators are used; the *basic 2-opt* [25], *a new 2-opt*, and *a new insertion* operators. The basic 2-opt removes two edges from the route of the cities and reconnects the sub-routes with new edges.

In this paper, we propose a new 2-opt operator based on the idea of best-move 2-opt presented in [6]. In the best-move 2-opt, in each iteration of the local search, all the improving moves are identified and sorted based on their improvement value, and only the best improving move is performed. In this way, the information gathered about other improving moves is neglected and remain unused. However, in the proposed 2-opt, the main idea is to use the gathered information about the improving moves and to perform all possible moves simultaneously as long as they can be done independently (i.e., they do not share any segment of the route). In this way, in an iteration of the local search, a greater value of improvement achieves. We call this new 2opt, the best-independent-moves 2-opt. To explain the procedure of the proposed 2-opt, consider a simple example of Fig. 1. In the first step, all improving moves are identified (moves 1, 2, 3, and 4 with improving values in parenthesis). Then, the improving moves are sorted based on their improvement values in a descending order (moves 2, 4, 1, 3). Finally, starting from the first move, all the independent moves are performed simultaneously (moves 2, 4, and 3). Indeed, move 1 cannot be applied immediately after move 4 since they share the same segment "Q-R-A".

In addition, we propose a new insertion operator in this paper. In the new proposed insertion operator, four types of moves are employed: *forward-left*, *forward-right*, *backward-left*, and *backward-right*. Let's consider two $k \rightarrow i \rightarrow l$ and $m \rightarrow j \rightarrow n$ segments of the route, where the first segment is visited before the second segment. In addition, consider that two cities i and j undergo the insertion operator. The four above-mentioned insertion moves produce $m \rightarrow i \rightarrow j \rightarrow n$, $m \rightarrow j \rightarrow i \rightarrow n$, $k \rightarrow j \rightarrow i \rightarrow l$, and $k \rightarrow i \rightarrow j \rightarrow l$, respectively. Finally, the best insertion move among all four moves are applied to the solution.

Action, State, and Reward. In Q-ILS-1, the actions are the set of local search operators to be selected and applied at each iteration and the states are the sequence of last k local search operators (k is equal to 1). At the end of each iteration, the performance of the employed perturbation operator is evaluated. Then, a reward or punishment is assigned to the employed operator. If the operator has been able to improve the best found solution, it receives a reward equal to the proportional improvement of the objective function; otherwise, it receives a punishment and is deleted from the set of available actions for the next iteration. In some cases, where no operator is able to improve the solution and set of available operators is empty, one operator is selected and applied randomly.

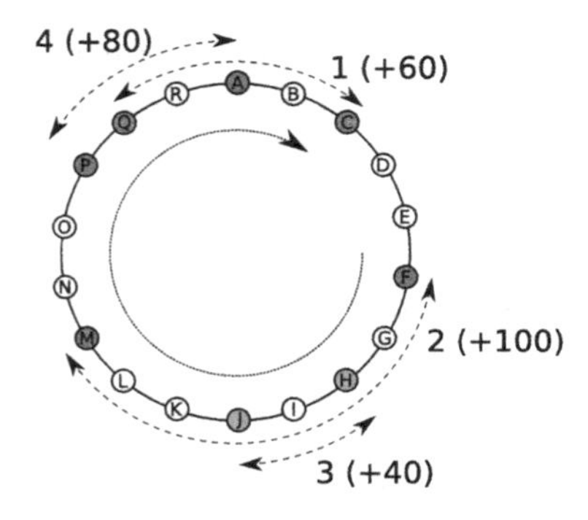

Fig. 1. Independent improving moves in the best-independent-moves 2-opt operator

4.2 Q-ILS-2 Algorithm

The novelty of the proposed Q-ILS-2 algorithm is development of a new perturbation procedure for ILS based on the ideas from AOS and Q-learning. In this algorithm, the type of perturbation operators and the number of times to apply them are adaptively selected based on the status of the search using the Q-learning algorithm. The aim of the proposed perturbation procedure is to adapt the exploration level to the status of the search. The general framework of Q-ILS-2 is the same as Q-ILS-1 except that Q-ILS-2, employs *the best-independent-moves 2-opt* operator as its single local search operator.

Perturbation Operators. In Q-ILS-2, a pool of three different perturbation operators are employed; the *Double-bridge* operator, the *Shuffle-sequence* operator that perturbs the solution by re-ordering a randomly selected sequence at random, and the *Reversion-sequence* operator that perturbs the solution by reversing a randomly selected sequence from the solution.

Action, State, and Reward. In Q-ILS-2, the actions are tuples (P, R), where P is the type of the perturbation operator and R is the repetition number of the perturbation operator P. Each action is given a chance of one episode equal to a fixed number of iterations to help the solution to escape from the local optimum. Accordingly, the states are the set of $S = \{0, 1\}$. $s = 1$ if the current perturbation operator P with R number of repetition followed by the local search has been able to improve the best found solution in an episode and $s = 0$, otherwise. After evaluation of the current action at the end of each episode, a reward (punishment) is assigned to the corresponding action. If the operator has been able to improve the best found solution, it receives a reward equal to the proportional improvement of the objective function; otherwise, it receives a punishment.

5 Results and Discussion

In this section, the performance of the two proposed algorithms, Q-ILS-1 and Q-ILS-2 are validated through a set of experimental results. For this aim, the experiments are designed in Sect. 5.1. Next, the numerical results are presented in Sect. 5.2.

5.1 Experimental Design

The performance of the proposed algorithms are validated through a set of 24 randomly selected symmetric TSP instances from the TSPLIB library [1] with different number of cities ranging from 50 to 2150. Two different experiments are done in this paper. First, the performance of the proposed algorithms in finding the optimal solution is investigated. Second, a comparative study is done to assess the efficiency of employing Q-learning in AOS. For this aim, first, to show the effectiveness of intelligent operator selection, Q-ILS-1 and Q-ILS-2 are compared to their corresponding Random ILS (R-ILS) with the same set of operators selected randomly. Second, in order to show the effectiveness of incorporating multiple operators into ILS, Q-ILS-1 and Q-ILS-2 are compared to their corresponding S-ILS algorithms, each employing single local search and perturbation operators.

For the Q-ILS-2, the maximum number of repetitions R of *double-bride*, *shuffle-sequence* and *reversion-sequence* are considered equal to 3, 1, and 1, respectively. The input parameters of the proposed algorithms are tuned using Design of Experiments [20] where $\epsilon = 0.8, \alpha = 0.6, \gamma = 0.5, \beta = 0.999$, and $episode = 3$. Each algorithm has been executed 30 times on each instance and is stopped after $0.2N$ number of iterations without improvement, where N is the number of cities in each instance. All algorithms have been coded in Python 3 and executed on an Intel Core i5 with 2.7 GHz CPU and 16G of RAM.

The performance of the algorithms is measured using two main criteria [25]:

- The *solution quality* represented as the Relative Percentage Deviation (RPD). The RPD is calculated as $RPD = \frac{OF-OF^*}{OF^*} \times 100$, where OF is the objective function (i.e., tour length) of the best found solution by each algorithm and OF^* is the objective function of the optimal solution.
- The *convergence behavior* of the algorithms that measures how fast (i.e., when/ at which iteration) an algorithm converges to the best found (optimal) solution.

5.2 Numerical Results

The performance of Q-ILS-1 and Q-ILS-2 in achieving the (near-) optimal solution are investigated through Tables 1 and 2. In these tables, the columns "Best RPD" and "Best time" are the gap of the best found solution and its corresponding CPU time, respectively and the columns "Average RPD" and "Average time" are the average gap and average CPU time over 30 executions.

Table 1. Result of the proposed Q-ILS-1 in comparison to the optimal solutions

Instance	Optimal	Best RPD (%)	Best time (s)	Average RPD (%)	Average time (s)
berlin52	7542	0	0.1	0	0.7
st70	675	0	2.4	0.037	9.9
kroA100	21282	0	0.3	0.005	10.2
rd100	7910	0	9.3	0.173	25.9
lin105	14379	0	0.5	0	15.2
pr124	59030	0	4.7	0.004	34.1
ch130	6110	0.262	42.9	0.546	73.4
ch150	6528	0.077	50.5	0.465	50.8
u159	42080	0	5.8	0	55.4
d198	15780	0.165	286.8	0.263	220.9
kroA200	29368	0.051	97.3	0.352	237.5
ts225	126643	0	1.1	0	48.0
pr264	49135	0	303.5	0.402	207.1
a280	2579	0	451.8	0.587	396.8
pr299	48191	0.151	734.7	0.805	704.3
lin318	42029	0.895	737.2	1.559	688.7
fl417	11861	0.430	605.5	0.696	796.1
pr439	107217	0.755	840.1	2.308	778.2
pcb442	50778	1.061	747.7	1.568	758.4
d493	35002	1.451	16.5	2.135	643.4
vm1084	239297	3.025	1461.4	4.217	2012.2
d1291	50801	3.173	782.4	4.295	1835.3
u1817	57201	4.142	254.0	4.858	664.4
u2152	64253	3.836	235.3	4.841	919.8

Table 1 indicates that Q-ILS-1 is able to find optimal solution in both small- and medium-sized instances and it is able to find near-optimal solutions with an optimality gap of 3.83% for the largest instance with 2152 cities. By looking at the "Best RPD" and the "Average RPD" results, it can be seen that Q-ILS-1 has produced small gaps over all 30 executions. In terms of the CPU time, the higher the size of the instance, the higher the CPU time of the algorithm. By looking at the "Best Time" and the "Average Time" results, it can be seen how expensive certain instances are in terms of CPU time. For example, instance "ts225" with 225 cities is much less expensive comparing to instance "ch130" with 130 cities. Accordingly, the number of cities is not the only factor that affects the computational complexity of the instance, but the geographical distribution of the cities is also an important factor.

Table 2. Result of the proposed Q-ILS-2 in comparison to the optimal solutions

Instance	Optimal	Best RPD (%)	Best time (s)	Average RPD (%)	Average time (s)
berlin52	7542	0	0	0	0.2
st70	675	0	0.7	0.109	4.5
kroA100	21282	0	1.3	0	7.3
rd100	7910	0	3.6	0.201	14.2
lin105	14379	0	0.4	0	4.5
pr124	59030	0	1.8	0	8.7
ch130	6110	0	17.4	0.359	40.4
ch150	6528	0	19.8	0.374	33.2
u159	42080	0	9.6	0.112	42.9
d198	15780	0.057	149.4	0.180	147.5
kroA200	29368	0	117.3	0.150	194.8
ts225	126643	0	56.5	0.002	108.9
pr264	49135	0	61.4	0.172	135.8
a280	2579	0	246.4	0.498	276.5
pr299	48191	0	453.1	0.274	440.7
lin318	42029	0.302	502.4	0.795	479.6
fl417	11861	0.211	574.5	0.339	553.0
pr439	107217	0.438	573.9	1.392	528.2
pcb442	50778	0.640	518.4	1.294	521.6
d493	35002	0.923	571.5	1.499	587.8
vm1084	239297	3.061	2729.6	3.717	2608.3
d1291	50801	2.281	1985.9	3.418	2060.1
u1817	57201	3.449	412.1	4.366	1416
u2152	64253	3.947	1967.5	4.531	1022.3

Some of the observations from Table 1 can be also generalized to the results of Table 2. Besides the zero optimality gap for small- and medium-sized instances, Q-ILS-2 is even able to find optimal solution for some large-sized instances up to 300 cities. For larger instances, small gaps have been also reported with an optimality gap of 3.94% for the largest instance with 2152 cities. Similar to Q-ILS-1, the results of "Best RPD" and "Average RPD" show that Q-ILS-2 produces small gaps over all 30 executions for almost all instances.

Table 3 shows the comparative results of Q-ILS-1 and Q-ILS-2 against R-ILS and S-ILS algorithms. Considering Q-ILS-1 with three local search operators, there are three S-ILS; S-ILS-1 to S-ILS-3 that stand for the use of *basic 2-opt*, *best-independent-moves 2-opt* and *insertion* local search operators, respectively.

Considering Q-ILS-2 with three perturbation operators, S-ILS-1 to S-ILS-3 stand for the use of *double-bridge*, *shuffle-sequence*, and *reversion-sequence*

Table 3. The RPD (%) of R-ILS and S-ILS comparing to Q-ILS-1 and Q-ILS-2

Instance	Q-ILS-1				Q-ILS-2			
	R-ILS	S-ILS			R-ILS	S-ILS		
		1	2	3		1	2	3
berlin52	2.04	1.34	2.09	3.01	3.34	3.48	3.17	2.10
st70	1.05	1.03	1.21	1.00	1.19	1.25	0.74	0.83
kroA100	0.34	0.76	0.43	2.17	0.97	1.23	0.58	0.64
rd100	1.23	1.53	1.63	3.00	1.43	1.79	1.20	1.62
lin105	1.39	1.07	1.19	1.02	0.83	1.84	1.24	0.85
pr124	0.60	0.72	0.93	1.87	1.06	1.21	0.65	0.70
ch130	0.99	1.51	1.12	2.60	1.68	2.18	1.12	1.23
ch150	0.95	1.19	1.29	3.20	1.31	2.36	1.01	1.17
u159	1.78	1.44	1.26	2.28	2.28	2.99	1.28	1.74
d198	0.43	0.39	0.51	2.65	0.73	1.92	0.71	0.96
kroA200	0.53	0.31	0.37	1.07	1.33	1.91	0.60	1.02
ts225	0.72	0.81	0.97	3.67	0.72	1.54	0.74	0.80
pr264	1.00	1.15	1.22	5.05	1.67	3.39	1.12	1.14
a280	1.96	1.84	2.27	4.23	2.23	4.17	1.53	2.00
pr299	1.63	2.07	1.56	5.06	2.22	4.23	1.65	2.18
lin318	0.58	0.48	0.58	0.97	1.81	3.34	1.44	1.79
fl417	1.67	2.00	2.19	5.11	1.14	3.88	2.19	1.99
pr439	1.51	1.22	1.68	3.16	2.78	3.65	1.75	1.45
pcb442	0.77	1.34	1.40	3.36	1.26	3.16	1.66	1.97
d493	0.85	0.87	1.49	3.28	1.26	3.58	2.18	2.06
vm1084	0.67	1.01	1.44	3.58	1.33	2.74	1.40	1.89
d1291	1.15	1.08	1.74	2.82	2.06	3.33	2.30	2.62
u1817	2.19	1.35	2.82	5.17	1.57	3.18	3.14	3.24
u2152	2.45	1.26	2.46	5.59	1.75	2.72	2.43	2.52

operators perturbation operators repeated only once, respectively. The values in Table 3 represent the RPD of other algorithms (i.e., R-ILS and S-ILS) comparing to Q-ILS-1 and Q-ILS-2 which is calculated as Eq. 3.

$$RPD^{R(S)} = \frac{OF^{R(S)} - OF^{Q}}{OF^{Q}} \times 100 \tag{3}$$

where $OF^{R(S)}$ is the average tour length obtained by R-ILS (S-ILS) for each instance and OF^{Q} is the average tour length obtained by Q-ILS-1/Q-ILS-2. A positive RPD value for an algorithm represents that the corresponding algorithm has a positive gap compared to the Q-ILS-1/Q-ILS-2. The RPD values equal to 0 shows both the two algorithms have led to the same solution. It can be

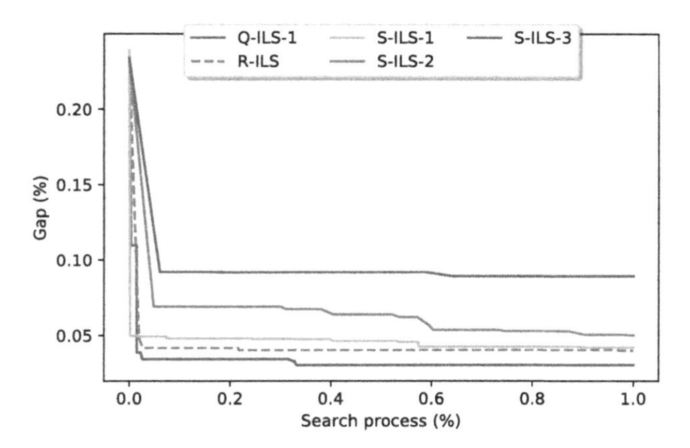

Fig. 2. Convergence behavior of Q-ILS-1 comparing to its benchmarks for instance d493

seen that both R-ILS and S-ILS for almost all the instances have positive gap compared to Q-ILS-1 and Q-ILS-2. This highlights the outperformance of the proposed Q-ILS algorithms over R-ILS and S-ILS in terms of the optimality gap. Investigating the results of R-ILS with positive gaps illustrates the efficiency of integrating the knowledge from the Q-learning algorithm into the operator selection mechanism of the ILS algorithm. Furthermore, the performance of Q-ILS algorithms are better than the S-ILS algorithms with single local search or perturbation operators. It shows the efficiency of employing different operators when solving TSP instances. Based on the obtained results, it can be concluded that intelligent selection of the operators at each stage of the search process using Q-learning provides promising results when solving TSP instances.

In addition to investigating the performance of Q-ILS-1 and Q-ILS-2 over R-ILS and S-ILS in finding (near-) optimal solutions, the algorithms are also compared based on their convergence behavior. In this regard, the gap to the optimal solution for the instance "d493" for different algorithms at different stages of the search are depicted in Fig. 2 for Q-ILS-1 and Fig. 3 for Q-ILS-2.

As Figs. 2 and 3 illustrate, the convergence of Q-ILS-1 and Q-ILS-2 happens at earlier stages of the search, about 40% of the search process for Q-ILS-1 and 60% of the search process for Q-ILS-2, which leads to solutions with higher quality in both algorithms. Considering Fig. 3, although R-ILS converges at earlier stages, it is a premature convergence which cannot be improved by the end of the search process. Considering both Figs. 2 and 3, all algorithms are competitive but the Q-ILS-1 and Q-ILS-2 always converge faster to the good solutions. The faster convergence of Q-ILS-1 and Q-ILS-2 is also observed for all TSP instances.

Based on the obtained results, the integration of Q-learning into ILS in both levels provides promising results. Comparing the performance of the proposed Q-ILS-1 and Q-ILS-2 algorithms, it can be seen that Q-ILS-2 outperforms the Q-ILS-1 in all selected instances. It can be concluded that incorporating

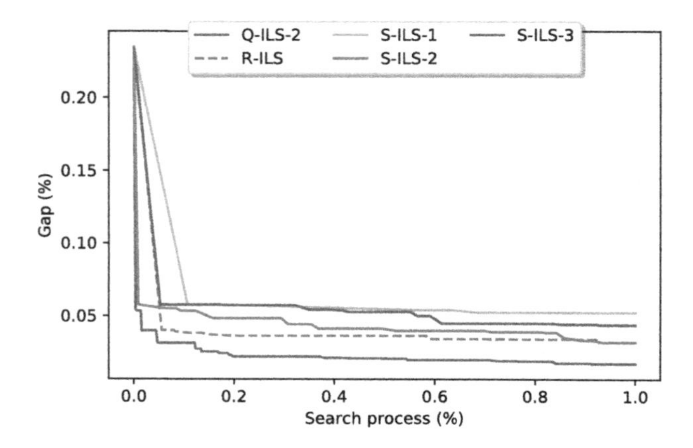

Fig. 3. Convergence behavior of Q-ILS-2 comparing to its benchmarks for instance d493

multiple efficient perturbation operators with different characteristics into ILS and intelligently selecting among them significantly enhances the exploration ability of the ILS.

6 Conclusion

In this paper, we have integrated the Q-learning algorithm as a machine learning technique to select the most appropriate search operators in the ILS algorithm for solving TSP. For this aim, the Q-learning has been integrated into the ILS algorithm in two levels: 1) selecting the appropriate local search operators and 2) selecting the appropriate perturbation operators at each stage of the search process. In the first integration level, a set of three local search operators including the basic 2-opt, a new 2opt, and a new insertion operator are considered. In the second integration level, the selection is done among three perturbation operators including double-bridge, shuffle-sequence and reversion-sequence.

The performance of the proposed algorithms has been tested on a set of 24 symmetric TSP instances from the TSPLIB library. In addition, a comparative study has been conducted to investigate the efficiency of intelligently selecting search operators using Q-learning algorithm. The results showed that the proposed algorithms are able to find optimal solutions for small- and medium-sized instances and near-optimal solutions for large-sized instances with small gaps. Through the comparative analysis, it was observed that employing several search operators provides better performance for the ILS when solving the TSP instances. Furthermore, the impact of the Q-learning for intelligently selecting the appropriate search operators at each stage of the search process was significant.

Finally, it was concluded that employing different perturbation operators provides better results in comparison to employing different local search operators.

Indeed, the ILS algorithm is inherently powerful in exploitation while it gets trapped easily in local optimum. Accordingly, considering different perturbation operators and selecting the most appropriate one at each stage of the search process helps the ILS to escape from the local optimum. ILS with multiple perturbation operators becomes more and more efficient when the knowledge obtained from the Q-learning algorithm is injected into its operator selection mechanism.

Testing the performance of the proposed algorithms on TSP instances with larger sizes could be an interesting future research direction. In addition, considering other types of local search and perturbation operators and testing their performance is another future research direction that is worth of further investigation. Finally, comparing the performance of the proposed algorithms against the benchmark algorithms in the literature and statistically checking their differences could be another future research work.

References

1. www.elib.zib.de/pub/mp-testdata/tsp/tsplib/tsplib.html
2. Ahmadi, E., Goldengorin, B., Süer, G.A., Mosadegh, H.: A hybrid method of 2-TSP and novel learning-based GA for job sequencing and tool switching problem. Appl. Soft Comput. **65**, 214–229 (2018)
3. Bengio, Y., Lodi, A., Prouvost, A.: Machine learning for combinatorial optimization: a methodological tour d'horizon. Eur. J. Oper. Res. **290**(2), 405–421 (2021)
4. Burke, E.K., Hyde, M.R., Kendall, G., Ochoa, G., Özcan, E., Woodward, J.R.: A classification of hyper-heuristic approaches: revisited. In: Gendreau, M., Potvin, J.-Y. (eds.) Handbook of Metaheuristics. ISORMS, vol. 272, pp. 453–477. Springer, Cham (2019). https://doi.org/10.1007/978-3-319-91086-4_14
5. Calvet, L., de Armas, J., Masip, D., Juan, A.A.: Learnheuristics: hybridizing metaheuristics with machine learning for optimization with dynamic inputs. Open Math. **15**(1), 261–280 (2017)
6. El Krari, M., El Benani, B., et al.: Breakout local search for the travelling salesman problem. Comput. Inform. **37**(3), 656–672 (2018)
7. Fialho, Á.: Adaptive operator selection for optimization. Ph.D. thesis, Université Paris Sud - Paris XI (2010)
8. Karimi-Mamaghan, M., Mohammadi, M., Jula, P., Pirayesh, A., Ahmadi, H.: A learning-based metaheuristic for a multi-objective agile inspection planning model under uncertainty. Eur. J. Oper. Res. **285**(2), 513–537 (2020)
9. Karimi-Mamaghan, M., Mohammadi, M., Pasdeloup, B., Billot, R., Meyer, P.: An online learning-based metaheuristic for solving combinatorial optimization problems. In: 21ème congrès annuel de la société Française de Recherche Opérationnelle et d'Aide à la Décision (ROADEF) (2020)
10. Karimi-Mamaghan, M., Mohammadi, M., Pirayesh, A., Karimi-Mamaghan, A.M., Irani, H.: Hub-and-spoke network design under congestion: a learning based metaheuristic. Transp. Res. Part E: Logist. Transp. Rev. **142**, 102069 (2020)
11. Karimi-Mamaghan, M., Mohammadi, M., Meyer, P., Karimi-Mamaghan, A.M., Talbi, E.G.: Machine Learning at the service of Meta-heuristics for solving Combinatorial Optimization Problems: A state-of-the-art. Eur. J. Oper. Res. (2021)

12. Karp, R.M.: On the computational complexity of combinatorial problems. Networks **5**(1), 45–68 (1975)
13. Lourenço, H.R., Martin, O.C., Stützle, T.: Iterated local search. In: Glover, F., Kochenberger, G.A. (eds.) Handbook of Metaheuristics. ISOR, vol. 57, pp. 320–353. Springer, Heidelberg (2003). https://doi.org/10.1007/0-306-48056-5_11
14. Metropolis, N., Rosenbluth, A.W., Rosenbluth, M.N., Teller, A.H., Teller, E.: Equation of state calculations by fast computing machines. J. Chem. Phys. **21**(6), 1087–1092 (1953)
15. Mohammadi, M., Jula, P., Tavakkoli-Moghaddam, R.: Reliable single-allocation hub location problem with disruptions. Transp. Res. Part E: Logist. Transp. Rev. **123**, 90–120 (2019)
16. Mohammadi, M., Tavakkoli-Moghaddam, R., Siadat, A., Dantan, J.Y.: Design of a reliable logistics network with hub disruption under uncertainty. Appl. Math. Model. **40**(9–10), 5621–5642 (2016)
17. Mosadegh, H., Ghomi, S.F., Süer, G.A.: Stochastic mixed-model assembly line sequencing problem: mathematical modeling and q-learning based simulated annealing hyper-heuristics. Eur. J. Oper. Res. **282**(2), 530–544 (2020)
18. Pasdeloup, B., Karimi-Mamaghan, M., Mohammadi, M., Meyer, P.: Autoencoder-based generation of individuals in population-based metaheuristics. In: ROADEF 2020: 21ème Congrès Annuel de la Société Française de Recherche Opérationnelle et d'Aide à la Décision (2020)
19. Peng, B., Zhang, Y., Gajpal, Y., Chen, X.: A memetic algorithm for the green vehicle routing problem. Sustainability **11**(21), 6055 (2019)
20. Ridge, E., Kudenko, D.: Tuning an algorithm using design of experiments. In: Bartz-Beielstein, T., Chiarandini, M., Paquete, L., Preuss, M. (eds.) Experimental Methods for the Analysis of Optimization Algorithms, pp. 265–286. Springer, Heidelberg (2010). https://doi.org/10.1007/978-3-642-02538-9_11
21. Sakurai, Y., Takada, K., Kawabe, T., Tsuruta, S.: A method to control parameters of evolutionary algorithms by using reinforcement learning. In: 2010 Sixth International Conference on Signal-Image Technology and Internet Based Systems, pp. 74–79. IEEE (2010)
22. dos Santos, J.P.Q., de Melo, J.D., Neto, A.D.D., Aloise, D.: Reactive search strategies using reinforcement learning, local search algorithms and variable neighborhood search. Expert Syst. Appl. **41**(10), 4939–4949 (2014)
23. Song, H., Triguero, I., Özcan, E.: A review on the self and dual interactions between machine learning and optimisation. Progr. Artif. Intell. **8**(2), 143–165 (2019). https://doi.org/10.1007/s13748-019-00185-z
24. Sutton, R.S., Barto, A.G.: Reinforcement Learning: An Introduction. MIT Press, Cambridge (2018)
25. Talbi, E.G.: Metaheuristics: From Design to Implementation, vol. 74. Wiley, Hoboken (2009)
26. Talbi, E.G.: Combining metaheuristics with mathematical programming, constraint programming and machine learning. Ann. Oper. Res. **240**(1), 171–215 (2016). https://doi.org/10.1007/s10479-015-2034-y
27. Turkeš, R., Sörensen, K., Hvattum, L.M.: Meta-analysis of metaheuristics: quantifying the effect of adaptiveness in adaptive large neighborhood search. Eur. J. Oper. Res. **292**(2), 423–442 (2021)
28. Watkins, C.J.C.H.: Learning from delayed rewards (1989)

29. Wauters, T., Verbeeck, K., De Causmaecker, P., Berghe, G.V.: Boosting metaheuristic search using reinforcement learning. In: Talbi, E.G. (ed.) Hybrid Metaheuristics. Studies in Computational Intelligence, vol. 434, pp. 433–452. Springer, Heidelberg (2013). https://doi.org/10.1007/978-3-642-30671-6_17
30. Zhalechian, M., Tavakkoli-Moghaddam, R., Zahiri, B., Mohammadi, M.: Sustainable design of a closed-loop location-routing-inventory supply chain network under mixed uncertainty. Transp. Res. Part E: Logist. Transp. Rev. **89**, 182–214 (2016)

A Hybrid Approach for Data-Based Models Using a Least-Squares Regression

Malin Lachmann(✉) and Christof Büskens

Center for Industrial Mathematics, University of Bremen, Bibliothekstraße 5, 28359 Bremen, Germany
mlachmann@uni-bremen.de, bueskens@math.uni-bremen.de

Abstract. An increased use of renewable energy could significantly reduce greenhouse gas emissions but is difficult to realize since most renewable energy sources underlie volatile availability. Making use of storage devices and scheduling consumers to times when energy is available allows to increase the amount of renewable energy that can be used. For this purpose, adequate models for forecasting the energy generated and consumed as well as for the behavior of storage devices are essential. Many data-based modeling approaches are computationally costly and therefore difficult to apply in real-world systems. Hence we present a computationally efficient modeling approach using a least-squares regression. Besides, we propose to use a hybrid model approach and evaluate it on real-world data at the examples of modeling the state of charge of a battery storage and the temperature inside a milk cooling tank. The experiments indicate that the hybrid approach leads to better forecasting results, especially for modeling more complicated behavior. Also, it is investigated if the behavior of the models is qualitatively realistic and we find that the battery model fulfills this requirement and is thus suitable for the application in a smart energy management system. Even though forecasts for the hybrid milk cooling model have even lower error values than the ones for the battery storage, further steps need to be taken to avoid undesired effects when using this model in such a sophisticated system.

Keywords: Data-based modeling · Least-squares regression · Hybrid models · Multiple models

1 Introduction

Even though its greenhouse gas emissions are decreasing, the energy supply sector is still the sector causing most of these environmentally hazardous emissions [1]. Increased use of renewable energy sources could reduce those emissions and thus allow climate change to decelerate. However, most of these sources

This research is based on a project funded by the Federal Ministry for Economic Affairs and Energy of Germany (project title SmartFarm, project number 0325927).

B. Dorronsoro et al. (Eds.): OLA 2021, CCIS 1443, pp. 62–73, 2021.
https://doi.org/10.1007/978-3-030-85672-4_5

have volatile availability. On the one hand, there are times where the demand is higher than the availability, on the other hand, if the demand is low at times of high availability, the grid stability might be in danger. To still allow increasing the use of renewable energy sources, the installation of storage devices can help to absorb this undesired behavior. In this context, storage devices are not limited to electrical storages but could also include devices that can be used as thermal storage such as cooling systems or heat pumps. Locally installed smart energy management systems can now allow using such storage devices in an optimal way. For this purpose, models that forecast the local energy generation and consumption as well as the behavior of storage devices are essential.

Modeling approaches are usually classified into two groups: Physics-based and data-based modeling [2,3]. The first aims at finding a model by analyzing the underlying physical laws and requires a deep understanding of the dependencies in the system while the latter determines a model by data for input and output values that is recorded during a training horizon and highly depends on the quality of this data. Their biggest advantage is that they are transferable to many different devices. For data-based approaches, it is often distinguished between models based on statistical methods and techniques using artificial intelligence (AI) [4]. AI techniques include fuzzy regression models, artificial neural networks and support vector machines while examples for statistical methods are (linear) regression models and autoregressive and moving average models. All these techniques are often used to model energy generation, consumption or storage behavior. In [5] and [6], for instance, forecasts for the energy generation are made by applying regression methods and a neural network, respectively. In [7–9], models for batteries' states of charge are determined with a neural network together with a Kalman filter, using a neuro-fuzzy inference system and a resistor-capacitor model.

In this work, we extend the data-based technique used in [10–12] such that the model can forecast complex behavior better. To achieve that, a data-based method based on a least-squares regression is combined with a hybrid model approach as introduced in [13] allowing multiple models for one device, each identified on and valid for a subset of the data. This paper investigates the application of these approaches on real-world data and evaluates if hybrid models are likely to be more plausible, i.e. to show more realistic qualitative model behavior.

In Sect. 2 the modeling approach is presented and the extension to hybrid models is explained. The subsequent section deals with numerical results of applying this modeling approach to real-world data using the examples of a battery storage and the temperature inside a milk cooling tank. In Sect. 4 this work is closed with a conclusion and possible future work is outlined.

2 Data-Based Modeling Approach

One approach to data-based modeling is to fit the data by applying a least-squares regression. This method has two advantages over other data-based techniques. First, it allows fast computation of models even for large data sets which is very convenient in case calculations need to be done on locally installed hardware with little computational power. Second, the method can be extended to include an even more efficient adaptation of the models to new data without recalculating the model on the entire data set, but only on the newly acquired data. This method will only be sketched here since it is frequently applied e.g. in [14] where it is also explained in more detail. In [11] and [10], this method is applied to a similar problem where it is extended to determining probabilistic forecasts and analyzing the capability of adapting to new data as well as improving forecasts by taking very recent data into account. In contrast to that, the focus in this paper is on a hybrid model approach that is explained in this section.

2.1 Data-Based Modeling with Least-Squares Regression

When identifying data-based models using a least-squares regression, we want to find a model $f : \mathbb{R}^m \to \mathbb{R}$ that best fits to a given data set with output data $y_i \in \mathbb{R}$, $i \in \{1, \dots, n\}$, measured at n different points in time, and input data $x_i \in \mathbb{R}^m$, $i \in \{1, \dots, n\}$ measured at the same time points from m different inputs $x^1, \dots, x^m$.

According to Taylor's Theorem, such a function f can be approximated around a point $x_0 \in \mathbb{R}^m$ by its derivatives at that point if it is sufficiently smooth. Since the function is not given, the derivatives which determine the coefficients of the polynomial are not available. Nevertheless, assuming a normally distributed error, the coefficients can be determined as minimizers of the mean-square deviation between the model $f(x_i)$ and the measured output y_i at all times. This problem can be reformulated as a linear least-squares problem which can be solved by QR-decomposition very efficiently even for large data sets [15].

2.2 Introducing Hybrid Models

In [10], it is found that high polynomial degrees often result in an overfitted model, i.e. one that fits very well to the training data but does not generalize well. It is also mentioned that if the model might be extrapolated to a bigger data range, a polynomial degree of one is best to use. However, a low polynomial degree does often not allow to model complex behavior. Using *hybrid models* is a possible approach to reduce the chance of overfitting but still allow modeling complex dependencies. For more details on this, we refer to [13].

Since the behavior of the device to be modeled might be very different in different phases, i.e. in $\kappa \in \mathbb{N}$ different subsets of the data set, we want to identify κ different models on the respective subsets, each valid only for these data subsets. If we denote the κ disjoint data subsets as $X_1, \dots, X_\kappa$, then we can write our model as

$$f(x_i) = \begin{cases} f_1(x_i) & \text{if } x_i \in X_{k_1} \\ f_2(x_i) & \text{if } x_i \in X_{k_2} \\ \vdots & \vdots \\ f_\kappa(x_i) & \text{if } x_i \in X_{k_\kappa}. \end{cases}$$

There exist many different approaches for choosing these subsets (see [13]) and it can be expected that the modeling results are sensitive to the data partitioning strategy, although the investigation of the effects is not part of this paper. Our choice in the following is to divide the data into subsets depending on the value of one integer input x having κ different values. This input can be measured data, obtained by a classification algorithm or be generated by hand. When calculating a forecast for time i, it is determined to which subset X_k the point x_i belongs and the corresponding function f_k is chosen for calculating the forecasted model value at time i, that means $f(x_i) = f_k(x_i)$ if $x_i \in X_k$.

3 Experimental Results Comparing Non-hybrid and Hybrid Models on Real-World Data

3.1 Setup and Data

The modeling approaches are now evaluated on real-world data at the example of the state of charge of a battery storage and the temperature of a milk cooling tank. The data was recorded by a measurement system comprised of one- and three-phase smart meters, current terminals and 1-wire temperature sensors on two demonstration sites in Lower Saxony, Germany within the scope of a research project aiming at developing an energy management system that controls storage devices and shiftable consumers such that the use of self-produced energy is maximized.

At both demonstration sites, a photovoltaic plant produces energy that can either be locally used or exported to the grid. One of the sites is a four-person household in which a lithium-ion battery storage with a capacity of 106 A h, a usable energy of 5.0 kW h and a one-phase inverter with a maximum apparent power flow of 6 kV A is installed. Internal values from the inverter can be accessed via a modbus interface and are also used. On the other site, a milk farm, a milk cooling tank with an energy consumption of up to 13 kW is installed which can be used as thermal storage by cooling the milk to a lower temperature within constraints that guarantee no quality loss.

The data used for the battery storage was recorded on 37 days in April and May 2017 and interpolated to 30 min with a moving average filter to reduce noise in the measurements. The first 20 days of data are used for training a model while all other data is used for validating that model. For modeling the milk cooling tank temperature, active power and temperature data measured between February 18, 2018 and April 10, 2018 is used and interpolated to five minutes again using a moving average filter. Here, further preprocessing was required since up to eight values per day (i.e. from 1440 values) exceeded all other values by several orders of magnitude. They were replaced by the value measured before that value since the data does often not change within a minute. The 52 days of data are again divided into a training period and a validation period where the first comprises the first 20 days of that set.

Within that setup, we now want to determine non-hybrid and hybrid models for the state of charge of the battery and the temperature inside the milk cooling tank. Both values show a dynamic behavior and depend on the state of charge or the temperature that was measured before as well as the active power which is available as a forecast in the energy management system. In the experiments, the actual measurements of the active power will be used since the quality of the power forecasts could influence the results. Also, the state of charge or temperature measured one time step before will be used as an input to the model. In addition to that, setpoints for the milk cooling tank would be determined in an energy management system indicating whether the cooling is *on* or *off*. These were reconstructed from the power data and are also used in the experiments.

Within the energy management system in which the models presented here will be applied, forecasts for 24 h are required. Since the models use the state of charge or temperature measured one step earlier which is not available at time points in the future, we evaluate the models iteratively to simulate the models' predictions within that energy management system. This means for a forecast starting at time t_0 that is one step into the future, we use the state of charge or temperature measured one step ago which is available. For the next time step t_1 we do not have a measurement at the time t_0 to use, but instead calculate the output value at time t_1 using the forecasted value at time t_0. This value is then used as an input to forecasting the output at time t_2. This procedure is continued until the end of the forecast horizon, e.g. until values for 24 h into the future are calculated.

Naturally, those forecasts will be more accurate during the beginning of the forecast horizon since small deviations from the data within the first hours can propagate and lead to huge differences in the last hours. Thus, the energy management system requests forecasts minutely and recalculates the optimal operation schedule to reduce this effect. However, in the following evaluations, we will consider forecasts for 24 h since these are of interest to the energy management system. To allow clear depictions, we will simulate that forecasts are requested only at midnight.

To evaluate the quality of a model, the deviation between the forecast by a model and the actual measurements is considered during the training and the validation period separately. To measure this deviation, we calculate the *normalized root-mean-square deviation* (nRSMD) which is the root-mean-square deviation (RSMD) (often also referred to as *root-mean-square error* (RSME)) normalized to the biggest value $y_{\max}$ measured for the output y, i.e.

$$\text{nRSMD} = \frac{1}{y_{\max}} \sqrt{\frac{\sum_{i=1}^{n} (y_i - f(x_i))^2}{n}}.$$

3.2 Comparing Non-hybrid and Hybrid Models on Real-World Data

Forecasting the State of Charge of a Battery Storage. To identify the parameters of not only a non-hybrid, but also a hybrid model, the data for the battery storage device as mentioned above needs to be divided into subsets that show similar behavior. For this division, we add an input x^{m+1} to the data that indicates whether the battery storage is charged or not, i.e. $\kappa = 2$ and

$$x_i^{m+1} = \begin{cases} 1 & \text{if the battery storage is charged at time } i \\ 0 & \text{otherwise.} \end{cases}$$

Based on this data, one non-hybrid and one hybrid model for the state of charge of a battery storage device are determined and forecasts are calculated. An excerpt of the results is depicted in Fig. 1 and shows five days at the beginning of the validation period. It can be observed that on some days the forecast calculated using the hybrid model (green) is closer to the measured data (purple) than the one with the non-hybrid model (blue) and on other days they are very similar. For both models, the forecasts do usually not reach the maximum state of charge of 90 %. This behavior can similarly be observed on the other days of the validation period indicating that the model predicts the charging process to be slower than in the measurements. The discharging, however, can be predicted better. In total, forecasts from both models are close to the measurements while the hybrid model seems to yield better predictions than the non-hybrid one.

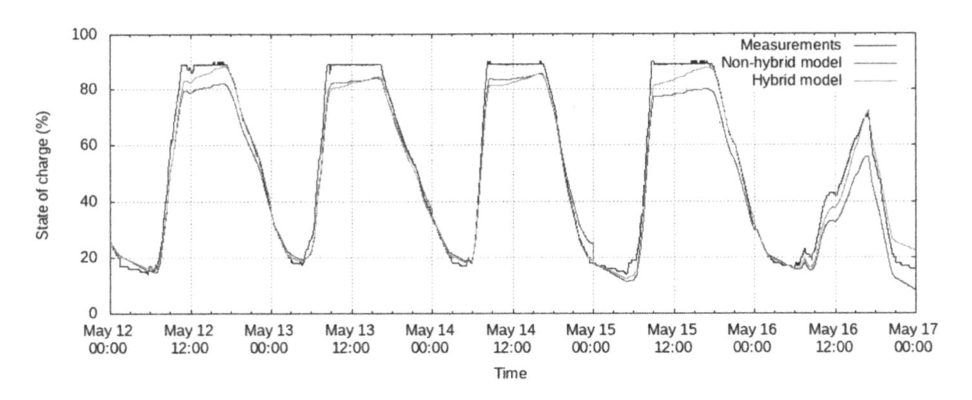

Fig. 1. Model for the state of charge of a battery storage during five days from the validation data. Measurements are depicted in purple, the results of the iteratively computed forecast by the non-hybrid model in blue and the ones obtained by the hybrid model in green. (Color figure online)

This can also be observed in the error values of both models. The nRSMD for the non-hybrid model is 8.6 % during the training period and 6.9 % during the validation period. This indicates that the model is not overfitted to the training data but generalizes well. The error values for the hybrid model are 8.0 % on the training data and 6.0 % on the validation data indicating that the hybrid model has the potential to improve forecasts even for devices in which a non-hybrid approach already leads to good models. Regarding the fact that the iterative forecast calculation is determined for a horizon of 24 h, the error values here are comparable to the results of other works, e.g. [9] in which the state of charge of batteries is forecasted with an error of less than 5 %.

Forecasting the Temperature Inside a Milk Cooling Tank. To forecast the temperature in the milk cooling tank, it is interesting to know the general behavior of the temperature. In the measurements, it can be observed that there is a pattern that repeats with a periodicity of two days. In Fig. 2, the measurements, depicted in purple, are often constantly at about 5 °C which is the temperature at which the milk is stored. In the power data, it can be seen that the isolation of the tank allows it to keep that temperature constant without cooling after it is reached. Twice a day, pre-cooled milk is added to the tank raising the temperature to about 8 °C depending on the amount of milk in the tank. Every other day the tank is emptied and cleaned with warm water resulting in temperatures of up to 53 °C. After that, no cooling is activated and the tank is left open until the next milking. The first milking after the cleaning procedure occurs while the tank still has a temperature of about 15 °C.

From this knowledge an additional input x^{m+1} is added to the data set indicating whether it is cleaned, milk is added or none of these, i.e. $\kappa = 3$ and

$$x_i^{m+1} = \begin{cases} 1 & \text{if milk is added to the tank at time } i \\ -1 & \text{if the milk tank is cleaned at time } i \\ 0 & \text{otherwise} \end{cases}$$

while the cleaning process is considered to start once the tank is cleaned with hot water and ends when the subsequent milking starts. An addition of milk occurs when a rise in temperature can be observed even though the power indicates that the cooling is active.

Based on this indicator vector for the hybrid model, again a non-hybrid and a hybrid model for the temperature inside the milk cooling tank are determined based on temperature data from one time step earlier, active power data and reconstructed setpoints. These results are depicted in Fig. 2 which shows four days from the validation period. First, it can be observed that the non-hybrid model (blue) is neither close to the measurements (purple) nor able to predict the behavior. It shows a periodicity of two days but the predicted temperature decreases when the one in the measurement increases and vice versa.

In contrast, the hybrid model is much closer to the measurements. During the cleaning process, i.e. when temperatures are above 10 °C, it fails to forecast the decrease in the data but instead predicts a constant temperature of about 25 °C. This could be since the model calculated during cleaning times is – as

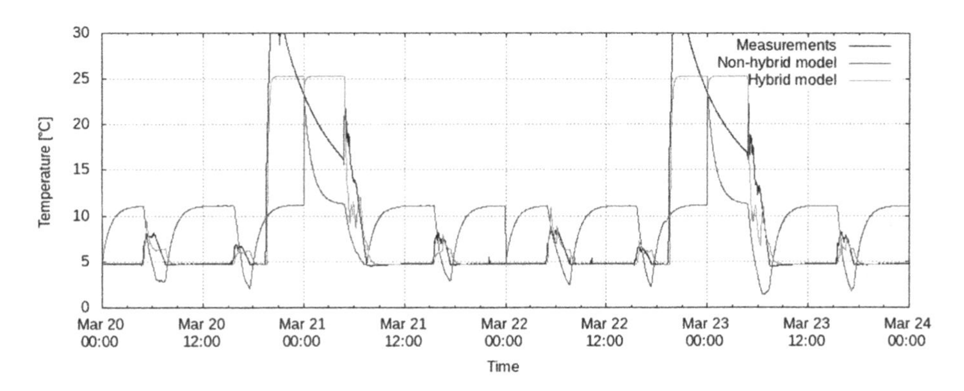

Fig. 2. Model for the temperature inside a milk cooling tank during five days of the validation data. Measurements are depicted in purple, the results of the iteratively computed forecast by the non-hybrid model in blue and the ones obtained by the hybrid model in green. (Color figure online)

all models – linear in its inputs and can thus not model the decrease in the data but chooses an average temperature. At all other times it can be observed that the forecast qualitatively behaves as the measurement: At times where the measurement is constant, the hybrid model forecasts constant behavior at the correct temperature and at times where the temperature is changing, this change is also visible in the forecasts even though the temperature is often lower than the measurements and decreases to a constant value later than the actual measurement does.

The hybrid model outperforms the non-hybrid one also when regarding the error values. The error (nRSMD) of the non-hybrid model during the training period is 10.6 % which does not seem to be very high. However, during the validation period, the error is 16.0 % which is much higher and indicates that this model does not generalize well. The error values of the hybrid model are much lower, being 4.9 % on the training data and 6.4 % on the validation data. In [12], the error for the temperature of a milk cooling model is calculated with an error of 11 % indicating that the hybrid model is not only an improvement over the non-hybrid model but also better than what other approaches have achieved.

3.3 Checking the Models' Plausibility

After finding that the non-hybrid and hybrid models are mostly close to the actual measurements, it is of interest if they could be used in a smart energy management system. That requires that the models behave logically correct even if the storage device is operated differently than in the measurements, e.g. the battery storage could be charged or discharged at different times than in the measurement. In case of the milk cooling, a smart system could decide to cool it to a slightly lower temperature at times of an energy surplus, such that during the next milking less energy needs to be used for cooling. To check whether the models identified above could be used in such a context, we now simulate their qualitative behavior on artificial data.

For the battery storage, it is checked how the non-hybrid and hybrid model forecast the state of charge during one day for given power values. For that, the power is set to 0 W during the first quarter of the forecast horizon, then charging at constant 200 W is simulated during the second quarter, followed by a quarter where the active power is −200 W which means that the battery is discharged. In the last quarter of the forecast horizon, the power is again set to 0 W. This power curve (black) is depicted in Fig. 3 together with the forecasts generated by the non-hybrid (blue) and hybrid (green) model. Discharging and charging can be observed in the forecasts as expected. Furthermore, when the simulated active power is 0 W the state of charge also decreases, but slower than at a power of −200 W. This passive discharging, i.e. the process of slowly decreasing the state of charge even though no energy is actively used, matches the actual behavior of all battery storages. The hybrid model forecasts a slightly slower passive discharging than the non-hybrid one. The biggest difference between the non-hybrid and the hybrid model can be observed during the charging in the second quarter where both models show an increase in the state of charge which is realistic. However, the hybrid model reaches much higher states of charge, but since the constant charging with 200 W for a longer time cannot be found in the measurements, it cannot be assessed which of the models is more realistic. Nevertheless, the fact that the hybrid model forecasts higher states of charge after a charging period could explain why in Fig. 1 forecasts by the hybrid model were closer to the measurements than the ones by the non-hybrid model. In summary, both models show realistic behavior here and could thus be applied in an energy management system where they would be suitable to forecast behavior that has not occurred in the data.

For the temperature inside a milk cooling tank, we simulate an additional cooling period at noon after the temperature has been constant for several hours to check if the models would predict a decrease in the temperature and stay at that lower temperature once the cooling is turned off. For simulating the additional cooling process, we set the active power to 11 850 W, the average power measured during all cooling processes, and adjust the setpoints to indicate cooling. The non-hybrid model predicts a decrease in temperature, but is, as in Fig. 2, not able to predict the constant temperature. For the hybrid model, the simulated additional cooling leads to a rise in temperature once the cooling starts and the temperature falls to the constant temperature of 5 °C once the process stops. Thus both models are not able to predict an additional cooling process qualitatively correct and also other choices of the additional input x^{m+1} do not lead to better qualitative behavior. This might be since the training data does not include temperatures below 5 °C. Another explanation is that the models learn that the temperature rises once the cooling starts since at the beginning of each cooling warm milk is added. Adding data from a flow sensor as an input could thus be interesting. Also, other data, such as the amount of milk inside the

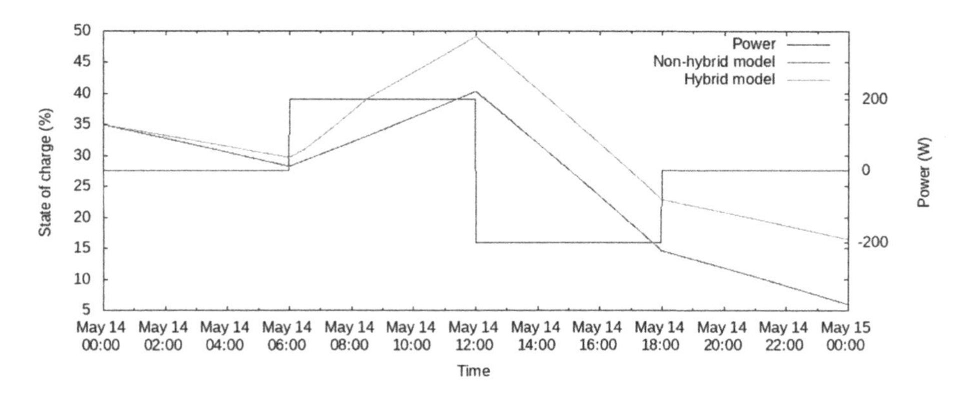

Fig. 3. Model for the state of charge of a battery storage evaluated on data simulating an active power profile as depicted by the black line. The forecast by the non-hybrid model is depicted in blue and the one obtained by the hybrid model in green. (Color figure online)

tank might improve the models. Furthermore, it would be interesting to choose the subsets for the hybrid model in a more sophisticated way, e.g. by a clustering approach, or to generate data containing additional cooling periods.

4 Conclusion

In this work it is evaluated to which extent computationally efficient data-based models can be applied to forecast the behavior of storage devices. The modeling approach uses a least-squares regression and is extended to hybrid models, where each submodel is trained on a subset of the data. The division into subsets is based on integer-valued indicator vectors that are added to the data manually. These two approaches are evaluated at the examples of the state of charge of a battery storage and the temperature inside a milk cooling tank. For both devices, two different models, a non-hybrid and a hybrid one, are calculated and compared to the actual measurements. It is found that the hybrid model is closer to the measurements in both cases. For the battery storage device, the error values of the models are not too far apart, but for the milk cooling tank, the non-hybrid model fails to forecast the temperature inside the tank while the hybrid model's prediction is close to the measurements. Additionally, it is investigated whether the models show plausible behavior which would be essential for their application in an energy management system. It is found that both battery models show realistic behavior while the hybrid model predicts a higher state of charge after a period of charging. In contrast, the models for the milk cooling tank both do not show plausible behavior. This can be explained by the fact that the data might lack information such as the flow of milk into the tank that influences the temperature heavily or the fact that the model is extended to data not contained in the training data.

To tackle this, adding further data would be interesting as well as a repetition of the experiments once data is available where the milk cooling tank is controlled and thus includes behavior that could not be predicted in the experiments in this paper. Additionally, it would be very interesting to evaluate if other divisions of the data into subsets lead to better results, e.g., if the data is divided into subsets by a clustering approach.

In summary, we show on real-world data that a hybrid data-based modeling approach can indeed improve forecasts calculated by models identified using least-squares regression. However, even though the hybrid models are much closer to the actual measurements, when applying them in a smart energy management system it must be carefully checked if their qualitative behavior is plausible.

References

1. World Resources Institute: World Greenhouse Gas Emissions (2016). https://www.wri.org/resources/data-visualizations/world-greenhouse-gas-emissions-2016. Accessed 30 Nov 2020
2. Foley, A.M., Leahy, P.G., Marvuglia, A., McKeogh, E.J.: Current methods and advances in forecasting of wind power generation. Renewable Energy **37**(1), 1–8 (2012)
3. Antonanzas, J., Osorio, N., Escobar, R., Urraca, R., Martinez-de-Pison, F.J., Antonanzas-Torres, F.: Review of photovoltaic power forecasting. Sol. Energy **136**, 78–111 (2016)
4. Pedro, H.T.C., Inman, R.H., Coimbra, C.F.M.: 4 - Mathematical Methods for Optimized Solar Forecasting. In: Kariniotakis, G. (eds.) Renewable Energy Forecasting. From Models to Applications, pp. 111–152 Woodhead Publishing (2017). https://doi.org/10.1016/B978-0-08-100504-0.00004-4
5. Tsekouras, G., Dialynas, E., Hatziargyriou, N., Kavatza, S.: A non-linear multivariable regression model for midterm energy forecasting of power systems. Electr. Power Syst. Res. **77**(12), 1560–1568 (2007)
6. Li, S., Wunsch, D.C., O'Hair, E.A., Giesselmann, M.G.: Using neural networks to estimate wind turbine power generation. IEEE Trans. Energy Convers. **16**(3), 276–282 (2001)
7. Chen, Z., Qiu, S., Masrur, M.A., Murphey, Y.L.: Battery state of charge estimation based on a combined model of extended Kalman filter and neural networks. In: Proceedings of the 2011 International Joint Conference on Neural Networks, pp. 2156–2163. IEEE, San Jose, USA (2011)
8. Cai, C.H., Du, D., Liu, Z.Y.: Battery State-of-Charge (SOC) Estimation Using Adaptive Neuro-Fuzzy Inference System (ANFIS). In: Proceedings of the 12th IEEE International Conference on Fuzzy Systems (FUZZ 2003), pp. 1068–1073. IEEE, St. Louis, USA (2003)
9. Eichi, H.R., Chow, M.: Adaptive parameter identification and state-of-charge estimation of lithium-ion batteries. In: Proceedings of the 38th Annual Conference of the IEEE Industrial Electronics Society, pp. 4012–4017. IEEE, Montreal, Canada (2012)
10. Lachmann, M., Maldonado, J., Bergmann, W., Jung, F., Weber, M., Büskens, C.: Self-learning data-based models as basis of a universally applicable energy management system. Energies **13**(8), 2084 (2020)

11. Jung, F., Büskens, C.: Probabilistic data-based models for a reliable energy management. In: Proceedings of the 2018 IEEE International Conference on Environment and Electrical Engineering and 2018 IEEE Industrial and Commercial Power Systems Europe, pp. 1–6. IEEE, Palermo, Italy (2018)
12. Lachmann, M., Jung, F., Büskens, C.: Computationally efficient identification of databased models applied to a milk cooling system. In: Conference of Computational Interdisciplinary Science, 2019, pp. 1–10. Galoa, Atlanta, USA (2020)
13. Hastie, T., Tibshirani, R., Friedman, J.: The Elements of Statistical Learning. Data Mining, Inference, and Prediction. 2nd edn. Springer Science & Business Media, New York, USA (2009)
14. Chen, S., Wassel, D., Büskens, C.: High-precision modeling and optimization of cogeneration plants. Energ. Technol. **4**, 177–186 (2016)
15. Hanke-Bourgeois, M.: Grundlagen der numerischen Mathematik und des wissenschaftlichen Rechnens. Vieweg+Teubner Verlag/GWV Fachverlage GmbH, Wiesbaden, Germany, pp. 119 (2009)

A Comparison of Learnheuristics Using Different Reward Functions to Solve the Set Covering Problem

Broderick Crawford, Ricardo Soto, Felipe Cisternas-Caneo(✉), Diego Tapia, Hanns de la Fuente-Mella, Wenceslao Palma, José Lemus-Romani, Mauricio Castillo, and Marcelo Becerra-Rozas

Pontificia Universidad Católica de Valparaíso, Valparaíso, Chile
{broderick.crawford,ricardo.soto,hanns.delafuente, wenceslao.palma}@pucv.cl, {felipe.cisternas.c,diego.tapia.r,jose.lemus.r, mauricio.castillo.d,marcelo.becerra.r}@mail.pucv.cl

Abstract. The high computational capacity that we have thanks to the new technologies allows us to communicate two great worlds such as optimization methods and machine learning. The concept behind the hybridization of both worlds is called Learnheuristics which allows to improve optimization methods through machine learning techniques where the input data for learning is the data produced by the optimization methods during the search process. Among the most outstanding machine learning techniques is Q-Learning whose learning process is based on rewarding or punishing the agents according to the consequences of their actions and this reward or punishment is carried out by means of a reward function. This work seeks to compare different Learnheuristics instances composed by Sine Cosine Algorithm and Q-Learning whose different lies in the reward function applied. Preliminary results indicate that there is an influence on the quality of the solutions based on the reward function applied.

Keywords: Learnheuristic · Sine cosine algorithm · Q-Learning · Reward function · Reinforcement learning

1 Introduction

Optimization problems are very recurrent in the real world being very complex to solve and it is necessary to obtain results in reasonable times. Approximate optimization techniques such as metaheuristics provide good results in reasonable times but when solving increasingly complex industry problems the quality of the solutions worsen.

Nowadays there is a high computational capacity thanks to new technologies allowing machine learning techniques to process large volumes of data in short times. During the optimization process a large amount of data is generated serving as input for some machine learning techniques to improve the optimization process and obtain better quality solutions.

B. Dorronsoro et al. (Eds.): OLA 2021, CCIS 1443, pp. 74–85, 2021.
https://doi.org/10.1007/978-3-030-85672-4_6

This hybridization is called Learnheuristic [1], where it is composed of an optimization module and a machine learning module. The optimization module solves an optimization problem and in each iteration generates a quantity of data that is delivered to the machine learning module so that it learns and makes decisions that affect the optimization technique in order to improve the quality of the solutions.

This work applies the concept of Learnheuristic, where the optimization module is composed of the metaheuristic Sine Cosine Algorithm (SCA) and the machine learning module is composed of Q-Learning. Q-Learning is used to learn how to select binarization schemes when SCA solves the Set Covering Problem. In particular, we seek to demonstrate the impact generated by different reward functions applied in Q-Learning.

The paper is organized as follows: Sect. 2 presents the Set Covering Problem, Sect. 3 presents the reinforcement learning, Q-Learning and the reward functions found in the literature, Sect. 4 presents Sine Cosine Algorithm, why it is necessary to use binarization schemes and how the Learnheuristic is generated, Sect. 5 presents the experimental results to finish with the conclusions in Sect. 6.

2 Set Covering Problem

The Set Covering Problem (SCP) is a classic optimization problem, which can be used to model in different applications and various domains, such as assignment problems, transport networks, among others. This problem is class NP-Hard [8] and consists of finding the set of elements with the minor cost that meets a certain amount of needs. The mathematical model of the Set Covering Problem is available at [10].

The SCP allows modeling of real-life optimization problems such as the location of gas detectors for industrial plants [22], the location of electric vehicle charging points in California [25], the optimal UAV locations for the purpose of generating wireless communication networks in disaster areas [17] and airline and bus crew scheduling [18]. These studies allow us to appreciate the importance of solving this problem with optimization techniques that guarantee good results.

3 Reinforcement Learning

Reinforcement learning is a subcategory of Machine Learning whose learning process consists of an agent performing different actions in different states and the objective of the agent is to learn what is the best action for each of the states by judging the consequence of performing each of the actions. Some of the classic examples are Q-Learning [23], Monte Carlo RL [11] and SARSA [19].

3.1 Q-Learning

Watkins et al. [23] proposed Q-Learning in 1992 and it is one of the best known reinforcement learning techniques. The main objective is to maximize the accumulated reward of an action in a particular state, in other words, to find the best action for a state.

An agent travels in different states and in each one of them an action is experienced immediately obtaining a reward or a punishment and the moment when an action is taken in a particular state is called a episode.

The agent should learn to select the best action for each one of the possible states. As the episodes pass the agent performs all possible actions for a state and the best action is the one that obtains the best accumulated reward [23].

The Q-Learning algorithm tries to learn how much accumulated reward the agent will get in the long run for each pair of action-state. The action-state function is represented as $Q(s,a)$ which returns the reward that the agent will get when performing the action a in the state s and assumes that it will follow the same policy obtained by the Q function until the end of the episode, this value is called Q-Value.

These Q-Values are stored in the Q-Table, which is a matrix where the rows correspond to the states and the columns correspond to the actions.

The Q-Value obtained for action a in state s when the n-th episode occurs is calculated as follows:

$$Q_n(s,a) = \begin{cases} (1-\alpha_n)Q_{n-1}(s,a) + \alpha_n[r_n + \gamma V_{n-1}(s_{n+1})] \ if \ \ s = s_n \ and \ a = a_n \\ Q_{n-1}(s,a) \qquad otherwise, \end{cases} \tag{1}$$

where,

$$V_{n-1}(s_{n+1}) = max \ \ Q_{n-1}(s_{n+1}, b) \tag{2}$$

s_n corresponds to the current state, a_n corresponds to the action selected and performed for the n-th episode, $max \ Q_{n-1}(s_{n+1}, b)$ corresponds to the highest Q-Value obtained for episode $n-1$ for the following state s_{n+1}, in other words, the best action for the following state s_{n+1}, r_n corresponds to the reward function that allows rewarding or punishing the action based on its consequence, α_n corresponds to the learning factor and γ corresponds to the discount factor.

3.2 Reward Function

The big question when using Q-Learning is: How to reward or punishment the consequences of carrying out an action? A good balance of reward and penalty allows an equitable variation of the selection of actions so the best action found is more reliable.

In the literature different Learnheuristics were found where metaheuristics incorporate Q-Learning as a machine learning technique. The reward function used by these Learnheuristics are diverse and adapted to the behavior of the metaheuristic.

For example, in [15] proposes 3 ways of reward or penalty for Ant Colony Optimization where all are associated to the best ant tour and, in some cases, they are calculated with respect to a predefined W constant.

On the other hand, other ways of rewarding or penalizing were found where they are oriented to the performance of Metaheuristics and can be applied to any of them. Such are the cases of [24] where the reward is 1 when the fitness is improved or -1 otherwise and as a result of this the reward or penalty visible in equation (4) is born where only the reward is given. The different reward functions are shown in Table 1.

Table 1. Reward function

Reference	Reward Function
[24]	$r_n = \begin{cases} 1, & if\, the\, current\, action\, improves\, fitness \\ -1, & otherwise. \end{cases} \quad (3)$
–	$r_n = \begin{cases} 1, & if\, the\, current\, action\, improves\, fitness \\ 0, & otherwise. \end{cases} \quad (4)$
[15]	$r_n = \begin{cases} W \cdot f_{best}, & if\, (r,s) \in tour\, done\, by\, the\, best\, agent \\ 0, & otherwise. \end{cases} \quad (5)$
[15]	$r_n = \begin{cases} \frac{W}{f_{best}}, & if\, (r,s) \in tour\, done\, by\, the\, best\, agent \\ 0, & otherwise. \end{cases} \quad (6)$
[15]	$r_n = \begin{cases} \sqrt{f_{best}}, & if\, (r,s) \in tour\, done\, by\, the\, best\, agent \\ 0, & otherwise. \end{cases} \quad (7)$

4 Sine Cosine Algorithm

Sine Cosine Aalgorithm (SCA) is a population-based metaheuristic where initial solutions are randomly generated and altered during the search process [13]. The equations of movements proposed for both phases are as follows:

$$X_{i,j}^{t+1} = \begin{cases} X_{i,j}^{t} + r_1 \cdot \sin(r_2) \cdot \left| r_3 P_j^t - X_{i,j}^t \right|, & r_4 < 0.5 \\ X_{i,j}^{t} + r_1 \cdot \cos(r_2) \cdot \left| r_3 P_j^t - X_{i,j}^t \right|, & r_4 \geq 0.5 \end{cases} \tag{8}$$

$$r_1 = a - t\frac{a}{T} \tag{9}$$

where r_1 is a parameter by Eq. 9, r_2, r_3 and r_4 are random numbers, $X_{i,j}^t$ is the position of the $i-th$ solution in the $j-th$ dimension in $t-th$ iteration and P_j^t it is the position of the best solution in the $j-th$ dimension in the $t-th$ iteration.

SCA is a metaheuristic that was built to solve problems in continuous domains, that is why to solve combinatorial optimization problems such as the Set Covering Problem it is necessary to transform the solutions from the continuous domain to the discrete domain [4].

One of the most used mechanisms to transfer solutions from the continuous domain to the binary domain is the Two-Step Technique. This technique consists of first transferring the values of the solutions from the continuous domain to the [0, 1] domain by means of transfer functions and then taking the values of the solutions in the [0, 1] domain and discretizing them by means of discretization functions.

4.1 Learnheuristic Framework

Recent studies [2,20,21] built a general Learnheuristics framework that incorporates Q-Learning as a machine learning module for operator selection. In particular, they select binarization schemes derived from combinations between transfer functions and discretization functions of the Two-Step Technique.

In the present work, modifications were made to the proposal presented by the authors in [2]. The **actions** to be taken by the agents are the **binarization schemes**, the **states** are the **phases of the metaheuristic**, i.e. exploration or exploitation, the **episodes** where an action is selected and applied in a particular state will be the **iterations** and the **agents** will be the **individuals** of the Sine Cosine Algorithm.

4.2 Balancing Exploration and Exploitation

To design a good metaheuristic is to make a proper trade-off between two forces: exploration and exploitation. It is one of the most basic dilemmas that both individuals and organizations constantly face. Exploration consists of looking for alternatives different from those already found while exploitation consists of exploiting a previously known alternative. This translates into selecting a new product for an individual, innovating or not for an organization and for metaheuristics into exploring the search space or exploiting a limited region of it. Ambidextrous Algorithms [3,12] aim at balancing exploration and exploitation oriented to decision making.

Before solving the balancing problem between exploration and exploitation, one must first have the ability to measure these indicators and then make a decision. Diversity metrics allow the measurement of exploration and exploitation because they quantify the dispersion of the individuals of the population.

There are different ways to quantify this diversity where the metrics based on central measurements and others based on frequency stand out. Metrics based on central measurements quantify diversity levels only in population algorithms. This is because the best search agents tend to attract the other solutions to them. In other words, the distance between solutions increases in exploration search processes and decreases in exploitation search processes [14].

In the literature there are different ways to quantify diversity and for this work the Dimensional-Hussain Diversity [9] was used. Dimensional-Hussain Diversity is a diversity based on central measurements and is defined as:

$$D_{dh}(X) = \frac{1}{l \cdot n} \sum_{d=1}^{l} \sum_{i=1}^{n} |mean(x^d) - x_i^d| \tag{10}$$

Where $D_dh(X)$ corresponds to the Dimensional-Hussan Diversity of the population X, $mean(x^d)$ is average of the d-th dimension, n is the number of search agents in the population X and l is the number of dimension of the optimization problem.

Morales-Castañeda et al. in [14] propose some equations which can obtain a percentage of exploration (XPL%) and a percentage of exploitation (XPT%) based on the diversity of the population. The particularity of these equations is that they are generic, that is, any diversity metric can be used, since the percentages are calculated around the diversity that the population has in a given iteration in relation to the maximum diversity obtained during the search process. These equations are as follows:

$$XPL\% = \left(\frac{Div_t}{Div_{max}}\right) \times 100, \quad XPT\% = \left(\frac{|Div_t - Div_{max}|}{Div_{max}}\right) \times 100 \tag{11}$$

Where Div_t corresponds to the current diversity in the t-th iteration and Div_{max} corresponds to the maximum diversity obtained in the entire search process.

By obtaining these percentages of exploration and exploitation we can determine the current state to be used in Q-Learning. This determination is done in the following way:

$$next\ state = \begin{cases} Exploration\ if\ XPL\% \geq XPT\% \\ Exploitation\ if\ XPL\% < XPT\% \end{cases} \tag{12}$$

All of the above are summarized in Algorithm 1.

5 Experimental Results

To evaluate the impact of the reward function in Q-Learning is that 5 different instances of BQSCA were implemented where the main change lies in the reward function to be applied. Table 2 shows the applied functions and their respective names.

Each of the instances was run 31, with a population of 40 individuals and 1000 iterations performed in each run. As mentioned in Sect. 3.1, Q-Learning has 2 additional parameters and they are the learning factor (α) and the discount factor (γ). The 5 versions of BQSCA were configured with the same learning factor whose value is $\alpha = 0.1$ proposed in [5] and the same discount factor whose value is $\gamma = 0.4$ proposed in [5]. On another hand, the value of the constant a of the parameter r_1 of Eq. 9 is 2 [13].

Table 2. Q-Learning implementation

Reference	Reward Type	Equation	Name
[24]	Penalty -1	Eq. (3)	BQSCA-QL1
–	With Out Penalty	Eq. (4)	BQSCA-QL2
[15]	Global Best	Eq. (7)	BQSCA-QL3
[15]	Root Adaptation	Eq. (6)	BQSCA-QL4
[15]	Escalating Multiplicative Adaptation	Eq. (5)	BQSCA-QL5

The Set Covering Problem instances proposed by OR-Library, which are betchmarck instances where the different authors make their comparisons, have been solved.

Table 4 shows the details obtained for the algorithms. The first column refers to each evaluated instance, the second column refers to the best known optimum for each instance, the fourth column indicates the best optimum obtained for each algorithm, the fifth column indicates the average of the results obtained for each algorithm and the sixth column indicates the Relative Percetage Deviation (RPD%) for each algorithm. These last 3 columns are repeated for each algorithm under analysis.

The RPD measures the percentage deviation of the best result obtained Z in relation to the best known result Z_{opt} for each instance. The measure is calculated as follows:

$$RPD = \frac{Z - Z_{opt}}{Z_{opt}} \times 100 \tag{13}$$

The results indicate that the 5 instances of BQSCA obtain good results, reaching optimal results in some instances. It should be noted that BQSCA-QL1 obtains preliminarily better results than the other instances.

Additionally, a statistical test was performed to validate the results obtained. Since the data does not come from nature, it does not have a normal distribution. On the other hand, since the data are not independent of each other, a nonparametric statistical test was performed. The Wilcoxon-Mann-Whitney test was applied. The hypothesis used for this statistical test is the following:

$$H_0 = \text{Algorithm A} \geq \text{Algorithm B}\,, \quad H_1 = \text{Algorithm A} < \text{Algorithm B}$$

If the result of the statistical test is obtained a p-value <0.05, we cannot assume that Algorithm A has worse performance than Algorithm B, rejecting H_0.

The results indicate that the 5 instances of BQSCA obtain good results, reaching optimal results in some instances. It should be noted that BQSCA-QL1 obtains preliminarily better results than the other instances.The results are shown in Table 3.

Table 3. p-value average BQSCA

	QL1	QL2	QL3	QL4	QL5
QL1	–	>0.005	**0.0484**	**0.0360**	>0.005
QL2	>0.005	–	>0.005	>0.005	>0.005
QL3	>0.005	>0.005	–	>0.005	>0.005
QL4	>0.005	>0.005	>0.005	–	>0.005
QL5	>0.005	>0.005	>0.005	>0.005	–

By being able to quantify the diversity of the population in a particular iteration, we can analyze the behavior of the Binary Q-Sine Cosine Algorithm in terms of exploration and exploitation. This information is shown in Figs. 1a and Fig. 1b. It can be seen that the Learnheuristics on average reach a balance close to 50% of exploration and exploitation but it is not a smooth transition as there are sharp jumps from exploration to exploitation or from exploitation to exploitation. In conclusion, the proposal perturb exploration and exploitation but fails to control it.

By analyzing the results obtained from both Table 4 and the statistical test in Table 3, we can see that BQSCA-QL1 stands out over some proposals. A peculiarity of this instance is that it is the only one that considers negative reward [6,7,16]. Experimental results show that considering negative reward "deters" exploration in reinforcement learning algorithm.

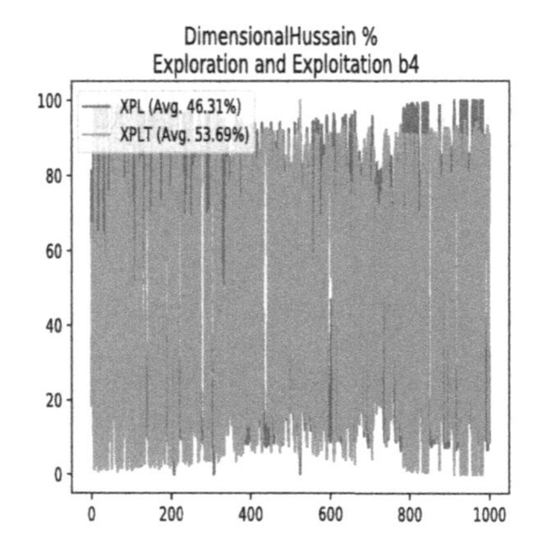

(a) BQSCA-QL1 solving instance scpb4

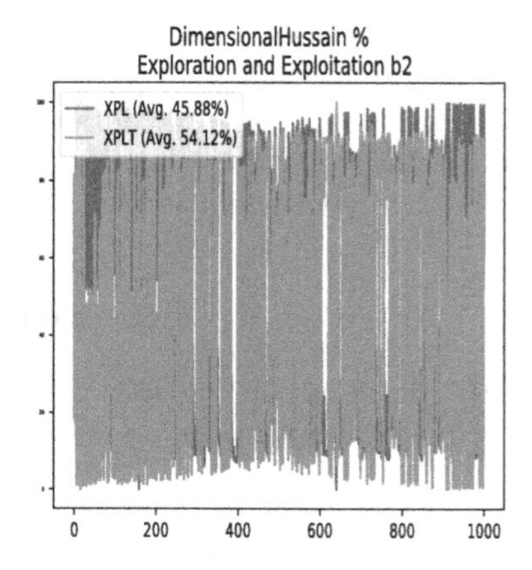

(b) BQSCA-QL3 solving instance scpb2

Fig. 1. BQSCA-QL1 Exploration-Exploitation

Algorithm 1. Binary Q-Sine Cosine Algorithm

Input: The population $X = \{X_1, X_2, ..., X_n\}$
Output: The updated population $X' = \{X'_1, X'_2, ..., X'_n\}$ and X_{best}

1: **Initialize Q-Table with** q_0
2: Initialize random population X
3: Set initial r_1
4: **Calculate Initial Population Diversity (X)**
5: **Define the initial state using equation (12)**
6: **for** *iteration* (t) **do**
7: $\quad a$ **: Select action from Q-Table**
8: $\quad$ **for** *solution* (i) **do**
9: $\quad\quad$ Evaluate solution X_i in the objective function
10: $\quad\quad$ **for** *dimension* (j) **do**
11: $\quad\quad\quad$ Update P_j^t, where $P_j^t = X_{best,j}$
12: $\quad\quad\quad$ Randomly generate the value of r_2, r_3, r_4
13: $\quad\quad\quad$ Update the position of $X_{i,j}$
14: $\quad\quad$ **end for**
15: $\quad$ **end for**
16: $\quad$ **Binarization** X **with action** a **and apply reward function**
17: $\quad$ **Calculate Population Diversity (X)**
18: $\quad$ **Define the next state using equation (12)**
19: $\quad$ **Update Q-Table using equation (1)**
20: $\quad$ Update r_1
21: $\quad$ Update X_{best}
22: **end for**
23: Return the updated population X where X_{best} is the best result

---

Table 4. Results obtained by solving instances of OR-library

Inst.	Opt.	QL1			QL2			QL3			QL4			QL5		
		Best	Avg	RPD	Best	Avg	RPD	Best	Avg	RPD	Best	Avg	RPD	Best	Avg	RPD
41	429	**431.0**	439.19	**0.47**	435.0	442.72	1.4	439.0	444.06	2.33	438.0	444.32	2.1	438.0	444.13	2.1
42	512	**535.0**	549.52	**4.49**	537.0	553.71	4.88	541.0	554.89	5.66	542.0	555.45	5.86	537.0	552.39	4.88
43	516	532.0	545.48	3.1	534.0	552.03	3.49	535.0	552.19	3.68	535.0	550.52	3.68	**527.0**	547.0	**2.13**
44	494	**510.0**	527.48	**3.24**	514.0	530.44	4.05	516.0	530.43	4.45	512.0	531.9	3.64	511.0	530.68	3.44
45	512	532.0	547.45	3.91	537.0	553.17	4.88	**527.0**	549.48	**2.93**	532.0	551.58	3.91	531.0	550.16	3.71
46	560	573.0	585.74	2.32	573.0	588.68	2.32	577.0	588.68	3.04	576.0	591.77	2.86	**568.0**	588.84	**1.43**
47	430	**437.0**	447.26	**1.63**	441.0	449.77	2.56	440.0	449.84	2.33	439.0	449.81	2.09	439.0	451.06	2.09
48	492	502.0	511.55	2.03	509.0	516.39	3.46	**499.0**	513.87	**1.42**	503.0	514.13	2.24	507.0	514.77	3.05
49	641	**672.0**	690.26	**4.84**	683.0	697.48	6.55	685.0	697.42	6.86	686.0	696.55	7.02	676.0	697.52	5.46
410	514	**521.0**	532.0	**1.36**	**521.0**	533.88	**1.36**	529.0	537.26	2.92	526.0	535.58	2.33	522.0	534.52	1.56
51	253	**263.0**	269.13	**3.95**	264.0	272.75	4.35	264.0	272.32	4.35	264.0	273.16	4.35	265.0	273.29	4.74
52	302	326.0	332.55	7.95	327.0	335.58	8.28	**325.0**	334.61	**7.62**	328.0	335.48	8.61	327.0	335.71	8.28
53	226	231.0	234.94	2.21	**230.0**	235.62	**1.77**	231.0	235.81	2.21	232.0	236.29	2.65	**230.0**	235.77	**1.77**
54	242	**250.0**	253.93	**3.31**	**250.0**	254.6	**3.31**	**250.0**	255.39	**3.31**	252.0	255.39	4.13	251.0	254.84	3.72
55	211	216.0	219.85	2.37	218.0	221.46	3.32	217.0	220.87	2.84	**214.0**	221.39	**1.42**	217.0	220.97	2.84
56	213	**217.0**	228.23	**1.88**	221.0	231.26	3.76	225.0	231.23	5.63	223.0	231.9	4.69	220.0	230.9	3.29
57	293	305.0	312.97	4.1	304.0	316.4	3.75	**303.0**	317.03	**3.41**	306.0	318.32	4.44	312.0	317.48	6.48
58	288	**295.0**	299.37	**2.43**	296.0	301.32	2.78	296.0	300.77	2.78	**295.0**	300.9	**2.43**	**295.0**	301.1	**2.43**
59	279	285.0	291.73	2.15	**284.0**	293.42	**1.79**	289.0	293.1	3.58	288.0	294.1	3.23	286.0	293.23	2.51
510	265	**271.0**	278.19	**2.26**	274.0	281.35	3.4	277.0	281.74	4.53	273.0	281.9	3.02	274.0	281.84	3.4
61	138	**141.0**	145.77	**2.17**	144.0	148.16	4.35	144.0	148.42	4.35	146.0	148.29	5.8	146.0	148.65	5.8
62	146	**151.0**	156.26	**3.42**	152.0	159.06	4.11	154.0	158.77	5.48	152.0	158.58	4.11	153.0	158.06	4.79
63	145	**149.0**	151.19	**2.76**	**149.0**	151.29	**2.76**	**149.0**	151.65	**2.76**	**149.0**	151.74	**2.76**	**149.0**	151.74	**2.76**
64	131	**133.0**	135.35	**1.53**	**133.0**	136.03	**1.53**	**133.0**	135.74	**1.53**	134.0	136.29	2.29	134.0	136.32	2.29
65	161	**169.0**	179.87	**4.97**	173.0	183.26	7.45	173.0	183.19	7.45	173.0	182.84	7.45	175.0	182.97	8.7
a1	253	**262.0**	267.24	**3.56**	266.0	269.42	5.14	263.0	268.81	3.95	263.0	268.9	3.95	265.0	269.68	4.74
a2	252	**267.0**	271.65	**5.95**	**267.0**	273.8	**5.95**	269.0	274.16	6.75	269.0	274.1	6.75	268.0	274.23	6.35
a3	232	243.0	247.85	4.74	245.0	248.87	5.6	245.0	249.45	5.6	**242.0**	249.71	**4.31**	**242.0**	248.63	**4.31**
a4	234	**245.0**	250.81	**4.7**	**245.0**	252.61	**4.7**	249.0	252.77	6.41	**245.0**	252.13	**4.7**	**245.0**	252.68	**4.7**
a5	236	245.0	248.92	3.81	247.0	251.27	4.66	245.0	250.48	3.81	245.0	251.29	3.81	**244.0**	251.26	**3.39**
b1	69	**70.0**	71.81	**1.45**	71.0	72.68	2.9	72.0	72.9	4.35	72.0	72.87	4.35	72.0	73.03	4.35
b2	76	**76.0**	80.16	**0.0**	78.0	81.35	2.63	**76.0**	81.13	**0.0**	78.0	81.06	2.63	78.0	81.23	2.63
b3	80	**81.0**	82.85	**1.25**	82.0	83.87	2.5	82.0	83.77	2.5	82.0	83.58	2.5	**81.0**	83.55	**1.25**
b4	79	82.0	83.82	3.8	83.0	84.9	5.06	82.0	84.68	3.8	83.0	84.9	5.06	**81.0**	85.1	**2.53**
b5	72	**73.0**	74.55	**1.39**	**73.0**	75.03	**1.39**	74.0	74.9	2.78	74.0	75.23	2.78	**73.0**	74.84	**1.39**
c1	227	240.0	245.55	5.73	246.0	251.85	8.37	244.0	251.03	7.49	**239.0**	251.0	**5.29**	244.0	250.94	7.49
c2	219	234.0	240.19	6.85	237.0	242.89	8.22	238.0	242.61	8.68	**231.0**	242.65	**5.48**	233.0	242.19	6.39
c3	243	**255.0**	260.97	**4.94**	259.0	263.25	6.58	257.0	262.74	5.76	256.0	263.0	5.35	256.0	263.48	5.35
c4	219	232.0	236.03	5.94	**230.0**	236.1	**5.02**	233.0	236.94	6.39	232.0	237.13	5.94	232.0	236.97	5.94
c5	215	**225.0**	231.62	**4.65**	229.0	234.2	6.51	226.0	233.35	5.12	227.0	233.42	5.58	226.0	233.45	5.12
d1	60	**62.0**	64.42	**3.33**	64.0	65.97	6.67	64.0	66.0	6.67	64.0	65.81	6.67	63.0	66.06	5.0
d2	66	**67.0**	69.52	**1.52**	69.0	69.97	4.55	68.0	69.97	3.03	69.0	70.26	4.55	68.0	70.03	3.03
d3	72	**75.0**	78.0	**4.17**	76.0	78.86	5.56	**75.0**	78.74	**4.17**	**75.0**	78.58	**4.17**	76.0	78.77	5.56
d4	62	**62.0**	64.0	**0.0**	63.0	64.16	1.61	**62.0**	64.13	**0.0**	**62.0**	64.16	**0.0**	63.0	64.29	1.61
d5	61	**63.0**	65.55	**3.28**	64.0	66.35	4.92	**63.0**	66.13	**3.28**	64.0	66.06	4.92	**63.0**	66.16	**3.28**

6 Conclusion

Learnheuristics improve the balance between exploration and exploitation to obtain high quality solutions. In particular, they perturb the solutions by generating an exploration and exploitation balance but it is not a tenuous and controlled balance. This evidences the great interest of researchers in Learnheuristics or Ambidextrous Algorithms and the great field of research they open due to the two interacting worlds.

Regarding the reward function, it can be concluded from the statistical evaluation that, to solve the Set Covering Problem with the metaheuristic Sine Cosine Algorithm applying Q-Learning as a binarization scheme selector, there is an influence of these but there is not a better one than the others.

In the future, this research can be extended to solve another optimization problem, apply another metaheuristic or apply another reinforcement learning technique to demonstrate the impact of the reward function in other work contexts.

Acknowledgements. Felipe Cisternas-Caneo and Marcelo Becerra-Rozas are supported by Grant DI Investigación Interdisciplinaria del Pregrado/VRIEA/PUCV/039.324/2020. Broderick Crawford and Wenceslao Palma are supported by Grant CONICYT /FONDECYT/REGULAR/1210810. Ricardo Soto is supported by Grant CONICYT/FONDECYT /REGULAR/1190129. Broderick Crawford, Ricardo Soto and Hanns de la Fuente-Mella are supported by Grant Núcleo de Investigación en Data Analytics/VRIEA /PUCV/039.432/2020. José Lemus-Romani is supported by National Agency for Research and Development (ANID)/Scholarship Program/DOCTORADO NACIONAL /2019-21191692.

References

1. Bayliss, C., Juan, A.A., Currie, C.S., Panadero, J.: A learnheuristic approach for the team orienteering problem with aerial drone motion constraints. Appl. Soft Comput. **92**, 106280 (2020)
2. Cisternas-Caneo, F., et al.: A data-driven dynamic discretization framework to solve combinatorial problems using continuous metaheuristics. In: Innovations in Bio-Inspired Computing and Applications, pp. 76–85. Springer International Publishing, Cham (2021)
3. Crawford, B., de la Barra, C.L.: Los algoritmos ambidiestros (2020). https://www.mercuriovalpo.cl/impresa/2020/07/13/full/cuerpo-principal/15/. Accessed 2 December 2021
4. Crawford, B., Soto, R., Astorga, G., García, J., Castro, C., Paredes, F.: Putting continuous metaheuristics to work in binary search spaces. Complexity, 2017 (2017)
5. Dorigo, M., Gambardella, L.M.: A study of some properties of Ant-Q. In: Voigt, H.-M., Ebeling, W., Rechenberg, I., Schwefel, H.-P. (eds.) PPSN 1996. LNCS, vol. 1141, pp. 656–665. Springer, Heidelberg (1996). https://doi.org/10.1007/3-540-61723-X_1029
6. Ecoffet, A., Huizinga, J., Lehman, J., Stanley, K.O., Clune, J.: Go-explore: a new approach for hard-exploration problems. arXiv preprint arXiv:1901.10995 (2019)
7. Fuchida, T., Aung, K.T., Sakuragi, A.: A study of q-learning considering negative rewards. Artif. Life Robot. **15**(3), 351–354 (2010)
8. Michael, R.G., David, S.J.: Computers and intractability: a guide to the theory of np-completeness (1979)
9. Hussain, K., Zhu, W., Salleh, M.N.M.: Long-term memory harris' hawk optimization for high dimensional and optimal power flow problems. IEEE Access **7**, 147596–147616 (2019)
10. Lanza-Gutierrez, J.M., et al.: Exploring further advantages in an alternative formulation for the set covering problem. Mathematical Problems in Engineering (2020)

11. Lazaric, A., Restelli, M., Bonarini, A.: Reinforcement learning in continuous action spaces through sequential monte carlo methods. Adv. Neural. Inf. Process. Syst. **20**, 833–840 (2007)
12. Lemus-Romani, J., et al.: Ambidextrous socio-cultural algorithms. In: Gervasi, O., et al. (eds.) ICCSA 2020. LNCS, vol. 12254, pp. 923–938. Springer, Cham (2020). https://doi.org/10.1007/978-3-030-58817-5_65
13. Mirjalili, S.: SCA: a sine cosine algorithm for solving optimization problems. Knowl. Based Syst. **96**, 120–133 (2016)
14. Morales-Castañeda, B., Zaldivar, D., Cuevas, E., Fausto, F., Rodríguez, A.: A better balance in metaheuristic algorithms: does it exist? Swarm Evolut. Comput. **54**, 100671 (2020)
15. Nareyek, A.: Choosing search heuristics by non-stationary reinforcement learning. In: Metaheuristics: Computer Decision-Making. Applied Optimization, vol. 86. Springer, Boston, MA (2003). https://doi.org/10.1007/978-1-4757-4137-7_25
16. Xinyan, O., Chang, Q., Chakraborty, N.: Simulation study on reward function of reinforcement learning in gantry work cell scheduling. J. Manuf. Syst. **50**, 1–8 (2019)
17. Park, Y., Nielsen, P., Moon, I.: Unmanned aerial vehicle set covering problem considering fixed-radius coverage constraint. Comput. Oper. Res. **119**, 104936 (2020)
18. Smith, B.M.: Impacs-a bus crew scheduling system using integer programming. Math. Program. **42**(1–3), 181–187 (1988)
19. Sutton, R.S., Barto, A.G.: Reinforcement Learning: An Introduction. MIT press, Cambridge (2018)
20. Tapia, D., et al.: A q-learning hyperheuristic binarization framework to balance exploration and exploitation. In: Florez, H., Misra, S. (eds.) ICAI 2020. CCIS, vol. 1277, pp. 14–28. Springer, Cham (2020). https://doi.org/10.1007/978-3-030-61702-8_2
21. Tapia, D., et al.: Embedding q-learning in the selection of metaheuristic operators: the enhanced binary grey wolf optimizar case. In: Proceeding of 2021 IEEE International Conference on Automation/XXIV Congress of the Chilean Association of Automatic Control (ICA-ACCA), IEEE ICA/ACCA 2021, ARTICLE IN PRESS (2021)
22. Vianna, S.S.V.: The set covering problem applied to optimisation of gas detectors in chemical process plants. Comput. Chem. Eng. **121**, 388–395 (2019)
23. Watkins, C.J., Dayan, P.: Q-learning. Mach. Learn. **8**(3–4), 279–292 (1992)
24. Zamli, K.Z., Din, F., Ahmed, B.S., Bures, M.: A hybrid q-learning sine-cosine-based strategy for addressing the combinatorial test suite minimization problem. PloS one **13**(5), e0195675 (2018)
25. Zhang, L., Shaffer, B., Brown, T., Scott Samuelsen, G.: The optimization of dc fast charging deployment in california. Appl. Energy **157**, 111–122 (2015)

A Bayesian Optimisation Approach for Multidimensional Knapsack Problem

Hanyu Gu(✉), Alireza Etminaniesfahani, and Amir Salehipour

School of Mathematical and Physical Sciences, University of Technology Sydney, 15 Broadway, Ultimo, NSW 2007, Australia
{Hanyu.Gu,amir.salehipour}@uts.edu.au,
Alireza.Etminaniesfahani@student.uts.edu.au

Abstract. This paper considers the application of Bayesian optimisation to the well-known multidimensional knapsack problem which is strongly NP-hard. For the multidimensional knapsack problem with a large number of items and knapsack constraints, a two-level formulation is presented to take advantage of the global optimisation capability of the Bayesian optimisation approach, and the efficiency of integer programming solvers on small problems. The first level makes the decisions about the optimal allocation of knapsack capacities to different item groups, while the second level solves a multidimensional knapsack problem of reduced size for each item group. To accelerate the Bayesian optimisation guided search process, various techniques are proposed including variable domain tightening, initialisation by the Genetic Algorithm, and optimisation landscape smoothing by local search. Computational experiments are carried out on the widely used benchmark instances with up to 100 items and 30 knapsack constraints. The preliminary results demonstrate the effectiveness of the proposed solution approach.

Keywords: Bayesian optimisation · Multidimensional knapsack problem · Meta-heuristics

1 Introduction

The Bayesian optimisation (BO) is a powerful machine learning based method for the optimisation of expensive black-box functions, which typically only allow point-wise function evaluation [22,23]. Although BO has been widely used in the experimental design community since the 1990s [13,15], it is not until the last decade that BO has become extremely popular in the machine learning community as an efficient tool for tuning hyper-parameters in various algorithms, e.g., deep learning [5,7], natural language processing [29], and preference learning [10]. The BO is also embraced by new areas such as robotics [16], automatic control [1], and pharmaceutical product development [21].

The Multidimensional Knapsack Problem (MKP) is an extension of the classic Knapsack Problem (KP). It comprises of n items and m knapsacks with

B. Dorronsoro et al. (Eds.): OLA 2021, CCIS 1443, pp. 86–97, 2021.
https://doi.org/10.1007/978-3-030-85672-4_7

limited capacities. Each item contributes a certain amount of profit if selected and consumes "resources" simultaneously in each knapsack. The MKP aims for a subset of items that achieves the highest total profit while abiding by the capacities of all knapsacks. The MKP is a well-known, and strongly NP-hard combinatorial optimisation problem, and has found applications in many practical areas involving resource allocation [11,17]. In spite of the tremendous progress made in exact solution techniques, many instances from the widely used Chu and Beasley MKP test set [4] cannot be solved to optimality [8,12,28], especially when the number of knapsacks is large. The best known solutions on the hard instances are all obtained by specialised meta-heuristics which require exorbitant computation time [3,6,24,25,27]. The simplicity of problem statement and computational hardness makes the MKP an ideal test bed for new solution ideas and techniques [14,18].

The BO encounters insurmountable issues to solve the MKP. Firstly, the BO is designed to solve problems with simple feasible set of continuous variables [9], while the MKP has only binary variables with many knapsack constraints. Whereas a lot of efforts have been committed to consider feasible set with combinatorial structures, all the reported computational studies investigated problems with just a few dozen categorical/integer/binary variables [2,19]. Secondly, the BO is only efficient for low dimensional problems with less than 20 variables, while the MKP can have hundreds of binary variables. Although the BO with random embedding can solve problems with billions of variables, it relies on the "low effective dimensionality" which can be an issue for MKP [26]. Finally, the MKP has a linear function which is "cheap" to calculate, which makes it hard for the BO to compete with other meta-heuristic and artificial intelligence algorithms.

Based on the idea of divide and conquer, a novel two-level model for MKP (TL-MKP) is proposed in this paper to take advantage of the special structure of MKP, i.e., the number of items (variables) is much larger than the number of knapsacks (constraints). In particular, the items are divided into groups, and the knapsack capacities allocated to each group are determined by the first level, or master problem, of the TL-MKP. With assigned knapsack capacities, each group can be solved as a MKP of reduced size in the second-level of TL-MKP, or subproblem. It is shown in Sect. 2 that the master problem has a non-continuous, multi-modal, and expensive to evaluate objective function with simple feasible set, which is suitable for the application of BO. Since the subproblem has a much smaller number of binary variables, it can be efficiently solved to optimality with commercial integer programming solvers.

It is essential to incorporate prior knowledge in the BO, which was designed to be a black-box global optimisation method. Two novel techniques are presented in this paper to make use of the information provided by mathematical programming solver and meta-heuristics. Indeed, when a good solution is known, e.g., by using other meta-heuristics, an efficient heuristic is proposed in this paper to tighten the domain bounds of the master problem in the TL-MKP. Inspired by the simulation approach used in robotics control algorithms to initialise the

BO [20], the Genetic Algorithm (GA) is used in this paper to generate initial trial points for the BO. To take advantage of the linear structure of the objective function of MKP, the GA is run on the MKP instead of the master problem of TL-MKP. These techniques can significantly accelerate the search process of BO as demonstrated by the computational experiments in this paper.

The paper is organised as follows. The novel two-level model for MKP is presented with discussion of the properties of the master problem in Sect. 2. The BO based optimisation approach and some acceleration techniques are described in Sect. 3. The implementation details are discussed in Sect. 4. Computational results are presented in Sect. 5. The conclusion is given in Sect. 6.

2 Two-Level Model for MKP

Given m knapsacks with capacities b_i, $i = 1, \dots, m$, and a set of n items $I = \{1, 2, \dots, n\}$, each item j requires a resource consumption of $a_{i,j}$ units in the i-th knapsack, $i = 1, \dots, m$, and yields c_j units of profit upon inclusion, $j = 1, \dots, n$. The goal is to find a subset of items that yields maximum profit, denoted by z^*, without exceeding the knapsack capacities. The MKP can be defined by the following integer linear programming model:

$$\text{(MKP)} \qquad z^* = \max\{c^T x : Ax \le b, x \in \{0,1\}^n\}, \tag{1}$$

where $c = [c_1, c_2, \dots, c_n]^T$ is an n-dimensional vector of profits, $x = [x_1, x_2, \dots, x_n]^T$ is an n-dimensional vector of 0–1 decision variables indicating whether an item is included or not, $A = [a_{i,j}]$, $i = 1, 2, \dots, m$, $j = 1, 2, \dots, n$ is an $m \times n$ coefficient matrix of resource requirements, and $b = [b_1, b_2, \dots, b_m]^T$ is an m-dimensional vector of resource capacities. It is further assumed that all parameters are non-negative integers.

Assume the items are divided into two groups, i.e., $I = I_1 \cup I_2$, and $I_1 \cap I_2 = \emptyset$. Each group is formulated as a MKP with profit vector $c^i = c_{I_i}$, resource requirement matrix $A^i = A_{I_i}$, and capacity vector $b^i \in \mathbb{R}^m$. The two groups share the capacities of the m knapsacks, i.e.,

$$b^1 + b^2 = b \tag{2}$$

The first level of the TL-MKP (the two-level model for MKP), or the master problem is defined as

$$\text{(L1-MKP)} \qquad f^* = \max\{f(t) : t \in \mathbb{R}^m, 0 \le t \le b\}, \tag{3}$$

where

$$f(t) = z_1^*(t) + z_2^*(b - t) \tag{4}$$

is calculated by solving the second level of the TL-MKP, or subproblems:

$$\text{(L2-MKP)} \qquad z_i^*(u) = \max\{c^i x^i : A^i x^i \le u, x^i \in \{0,1\}^{|I_i|}\}, \quad i = 1, 2 \tag{5}$$

Since each solution of the TL-MKP can be easily converted to a solution to the MKP with the same objective value, and each solution of the MKP can be used to define a value of t for the master problem of TL-MKP (3), the following proposition holds.

Proposition 1. *($t^* = A^1 x^{1*}, x^{1*}, x^{2*}$) is an optimal solution of TL-MKP if and only if x^*, defined as $x^*_{N_1} = x^{1*}$ and $x^*_{N_2} = x^{2*}$, is an optimal solution of MKP. Furthermore, $f^* = z^*$.*

Example 1. Consider an instance of MKP with three items and one knapsack, where $c = [1, 2, 3]$, $A = [1, 2, 3]$, and $b = 4$. The two groups are $I_1 = \{1, 2\}$ and $I_2 = \{3\}$. It is straightforward to show that the first level objective function is

$$f(t) = \begin{cases} 3 & t \in [0, 1) \\ 4 & t = 1 \\ 1 & t \in (1, 2) \\ 2 & t \in [2, 3) \\ 3 & t \in [3, 4] \end{cases}.$$

Example 2. Consider an instance of MKP with 20 items and two knapsacks. The two groups have the same number of items. Figure 1 shows the contour graph of the first level objective function $f(t)$. The optimal value is equal to 75.

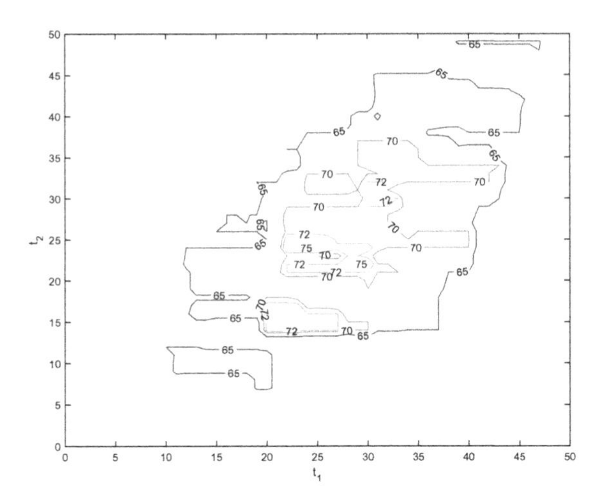

Fig. 1. Contour of the first level objective function $f(t)$; $t_1(t_2)$ is the capacity allocated to group 1 from knapsack 1 (2).

Examples (1) and (2) clearly demonstrate that the objective function of the master problem in TL-MKP is non-continuous, and can have many local optima.

Although the subproblems have much smaller sizes, they are still more expensive to evaluate than the linear function of MKP.

It can be observed that $f(t)$ is not differentiable when at least one knapsack has no slack capacity in one of the subproblems. That leads to the following proposition,

Proposition 2. *$f(t)$ is differentiable almost everywhere in the sense of Lebesque measure with $f'(t) = 0$.*

Although $f(t)$ is differentiable almost everywhere, the derivative is constantly zero and consequently, useless for the design of optmisation algorithms.

3 Bayesian Optimisation and Acceleration

The BO is a promising option to deal with the challenges presented by the master problem of TL-MKP such as no closed form, non-continuity, multiple local optima, absense of useful derivatives, and high cost of function evaluation. In this section, the basic principles of BO are described first [9], then followed by techniques to incorporate prior knowledge to accelerate the search process.

The BO builds a probabilistic model for the unknown $f(t)$ of the master problem of TL-MKP. In particular, $f(t)$ is assumed to be drawn from a Gaussian process (GP), which is determined by a mean function $\mu_0 : \mathbb{R}^m \to \mathbb{R}$, and a positive definite covariance function $k_0 : \mathbb{R}^m \times \mathbb{R}^m \to \mathbb{R}$, also known as the kernel of the GP. The BO sequentially generates points to evaluate within the feasible region of TL-MKP. Assume that n points have been evaluated with observations $\mathcal{D}_n = \{(t^1, f(t^1)), (t^2, f(t^2)), \dots, (t^n, f(t^n))\}$. Using Bayes' rule, the conditional distribution of $f(t)$ is derived as a Normal distribution:

$$P(f(t)|D_n, t) = \mathcal{N}(\mu_n(t), \sigma_n^2(t)) \tag{6}$$

$$\mu_n(t) = \Sigma_0(t, t_{1:n})\Sigma_0(t_{1:n}, t_{1:n})^{-1}(f(t_{1:n}) - \mu_0(t_{1:n})) + \mu_0(t) \tag{7}$$

$$\sigma_n^2(t) = k_0(t, t) - \Sigma_0(t, t_{1:n})\Sigma_0(t_{1:n}, t_{1:n})^{-1})\Sigma_0(t_{1:n}, t) \tag{8}$$

where $f(t_{1:n}) = [f(t^1), \dots, f(t^n)]^T$, $\mu_0(t_{1:n}) = [\mu_0(t^1), \dots, \mu_0(t^n)]^T$, and

$$\Sigma_0(t_{1:n}, t_{1:n}) = \begin{pmatrix} k_0(t^1, t^1) & \cdots & k_0(t^1, t^n) \\ \vdots & \ddots & \vdots \\ k_0(t^n, t^1) & \cdots & k_0(t^n, t^n) \end{pmatrix}.$$

The BO selects the next most promising point to evaluate, i.e., t^{n+1}, by optimising an acquisition function, which balances exploration (uncertainty $\sigma_n(t^{n+1})$ is large) against exploitation (objective expected value $\mu_n(t^{n+1})$ is large). Different types of acquisition function have been proposed in the literature, while the most commonly used is *Expected Improvement* (EI). The EI acquisition function is defined as

$$\mathbf{EI}_n(t) = E_n(max(f(t) - max_{i=1}^n f(t^i), 0)), \tag{9}$$

where $E_n(\cdot)$ is the expectation taken under the posterior distribution (6).

The next point to evaluate is selected as

$$t^{n+1} = \text{argmax}_t \mathbf{EI}_n(t). \tag{10}$$

With new point $(t^{n+1}, f(t^{n+1}))$, the conditional probability of $f(t)$ can be updated according to (6), and the iterative process stops when a sampling budget is reached.

3.1 Variable Domain Tightening

The efficiency of BO depends on the size and dimensionality of the search space of TL-MKP, which is defined in (3) as $[0, b] \subset \mathbb{R}^m$. If a good lower bound of MKP f_L is known, e.g., through a quick meta-heuristic, the search space can be reduced to $F = \{t | f(t) \geq f_L, t \in [0, b] \subset \mathbb{R}^m\}$. However, this will make the EI acquisition function harder to optimise in (10) since F has no simple representation. In this paper, an optimisation based approach is employed to find the smallest hypercube $H = [t^L, t^U]$ that contains F, i.e., $F \subset H$. The upper bound of H along the i-th coordinate, t_i^U, $i = 1, \ldots, m$, can be obtained by solving

$$t_i^U = \max\{A^1 x^1 : c^T x \geq f_L, Ax \leq b, x^1 = x_{I_1}, x \in \{0, 1\}^n\}. \tag{11}$$

The lower bound of H along the i-th coordinate, t_i^L, $i1, \ldots, m$, can be obtained by solving

$$t_i^L = \min\{A^1 x^1 : c^T x \geq f_L, Ax \leq b, x^1 = x_{I_1}, x \in \{0, 1\}^n\}. \tag{12}$$

The exact solution of (11) and (12) is time-consuming. Therefore, t^U (t^L) can be replaced by a upper (lower) bound of (11) ((12)), e.g., using the linear programming relaxation by replacing $x \in \{0, 1\}^n$ with $x \in [0, 1]^n$.

3.2 Initialisation with Genetic Algorithm

The BO randomly generates the initial trial points in the search space which can lead to slow convergence. In this paper, The GA is used to generate initial points that have good solution quality as well as diversity in the search space. The GA is a population based meta-heuristic which evolves by generations through genetic operators such as cross-over and mutation. In the early stage of GA the population has good diversity but low percentage of good solutions; while in the later stage, the population has high percentage of good solutions but with less diversity.

It is computationally infeasible to run GA on the TL-MKP since the objective evaluation involves solving two MIP problems and consequently expensive. Instead, the GA is directly run on the MKP, and the population is mapped to initialise the BO for TL-MKP. In particular, let $\tilde{x}$ be a solution from a population of GA. The mapped solution for TL-MKP becomes

$$\tilde{t} = A^1 \tilde{x}_{N_1}. \tag{13}$$

It is easy to see that

$$f(\tilde{t}) \geq c^T \tilde{x}. \tag{14}$$

3.3 Optimisation Landscape Smoothing

At each sampling point of BO, a feasible solution to the MKP is also generated according to Proposition 1. This solution can be improved by a local search which is efficient to cope with large number of items and constraints. We define the neighbourhood of a solution x as the set of solutions with at most k different items:

$$N_k(x) = \{y \in \{0,1\}^n : Ay \leq b, |||x - y|||_1 \leq k\}. \tag{15}$$

For Example 1, with $k = 1$, the first level objective function becomes

$$f(t) = \begin{cases} 4 & t \in [0,2) \\ 3 & t \in [2,4] \end{cases},$$

which is "smoother" in terms of the optimisation landscape.

4 Implementation

The BO approach for the MKP (BO-MKP) can be described as in Algorithm 1, and a prototype of BO-MKP was implemented in Matlab R2020b. In Step 1 of BO-MKP, the linear relaxation of (11) and (12) are solved to tighten the bounds of the feasible set of TL-MKP using the function *linprog* in Matlab *Optimization Toolbox*. Using the function *ga* in the *Global Optimization Toolbox* of Matlab, an initial set of trial points are generated in Step 2 as input for BO according to (13). In Step 3, the BO is implemented with the function *bayesopt* in the *Global Optimization Toolbox* of Matlab. The acquisition function is set to "expected-improvement", and the maximum number of evaluation, "MaxObjectiveEvaluations", is set to N which is a user specified parameter. The subproblems of TL-MKP (5) are solved by the mixed integer programming solver *intlinprog* in Matlab *Optimization Toolbox*. In Step 4, The best solution of TL-MKP found by BO is converted to a solution of MKP with the same objective function value according to Proposition 1.

The selection of kernel function for GP can have a strong influence on the performance of BO. *bayesopt* uses the ARD Matérn 5/2 kernel

$$k(x_i, x_j | \sigma_f, \sigma_l) = \sigma_f^2 (1 + \frac{\sqrt{5}r}{\sigma_l} + \frac{5r^2}{3\sigma_l^2}) \exp(-\frac{\sqrt{5}r}{\sigma_l})$$

where $r = \sqrt{(x_i - x_j)^T (x_i - x_j)}$, and the parameters are estimated by Gaussian process regression *fitrgp*.

Algorithm 1: The BO approach for the MKP (BO-MKP).

Input: MKP, item groups I_1 and I_2, lower bound of MKP f_L, maximum number of evaluation N for BO.
Output: a feasible solution of MKP.
Step 1: tighten the bounds of feasible set of TL-MKP based on f_L;
Step 2: generate initial trial points using GA;
Step 3: search for the global optimum of TL-MKP using BO within a sampling budget of N evaluations;
Step 4: convert the best solution found by BO to the solution of MKP;
return

5 Computational Experiments

All experiments are carried out on the widely used Chu and Beasley MKP test set in [4]. The Chu and Beasley test set contains classes of randomly generated instances for each combination of $n \in \{100, 250, 500\}$ items, $m \in \{5, 10, 30\}$ constraints, and tightness ratios $\alpha \in \{0.25, 0.5, 0.75\}$ with smaller α representing tighter resource capacities. In the Chu and Beasley MKP test set, the resource consumption values a_{ij} are integers uniformly chosen from (0, 1000), which leads to large values of the knapsack capacities b. Since the search space of BO for TL-MKP is defined by b in (3), the Chu and Beasley MKP test set is an challenging test bed for the proposed BO approach.

To show the effect of tightening bounds in Sect. 3.1, the BO is tested on three selected instances with $n = 100$, and the results are reported in Table 1. The rows correspond to the instances with the number of knapsack constraints $m =$ 5, 10 and 30. The optimal values of these instances are obtained by CPLEX and reported in the column titled "Opt." The columns are divided into two groups for the BO results, one for the cases without bound tightening ("Without tightening") and the other one for the cases with bound tightening ("With tightening"). To have a better understanding of the convergence behavior of BO, two values are applied for the maximum number of evaluations, i.e., $N = 25, 50$. Since the BO is a stochastic algorithm, the average objective function value of 5 runs is reported for each pair of (m, N) in the columns titled "Ave.". The relative gap for the solution found by the BO is calculated as $100 \times (z^* - f)/z^*$ and reported in the columns titled "gap(%)". It can be seen that the performance of BO deteriorates dramatically when m increases. When $m = 30$, the BO reaches a massive relative gap of 63.8% after 50 function evaluations. This observation is consistent with BO's behavior for other optimisation problems. When bound tightening technique is applied, the performance of BO is improved on all (m, N) pairs. The improvement is more dramatic when m becomes large. For $m = 10$ and $N = 50$ the relative gap is reduced from 9.8% to 4%. However, the solution quality for $m = 30$ is still not satisfactory with a large gap of 31.8%.

Table 2 presents the results of BO-MKP which initialises the BO with GA. The initial trial points provided by the GA should be diverse enough while also having good solution quality. Therefore, the maximum number of iterations of

Table 1. Effects of bound tightening for BO on the TL-MKP.

m	Opt.	Without tightening				With tightening			
		N = 25		N = 50		N = 25		N = 50	
		Ave.	gap(%)	Ave.	gap(%)	Ave.	gap(%)	Ave.	gap(%)
5	24381	22897	6.1	23849	2.2	23913	1.9	24017	1.5
10	23064	17145	25.7	20806	9.8	20581	10.8	22149	4.0
30	21946	5710.2	74.0	7955	63.8	14659	33.2	14978	31.8

GA is limited to 55 in BO-MKP. It can be seen that the GA initialisation is not helpful when $m = 5$, which suggests that the BO has strong global search capability when the dimension is low. In sharp contrast, the BO-MKP dramatically reduces the relative gap for larger dimension. Indeed, the relative gap is just 4% for $m = 30$ with 50 function evaluations.

Table 2. Effects of GA initialisation for BO on the TL-MKP.

	Opt.	$N = 25$			$N = 50$		
		w/o GA	Ave.	gap(%)	w/o GA	Ave.	gap(%)
$m = 5$	24381	23913	23928	1.9	24017	24063	1.3
$m = 10$	23064	20581	22471	2.6	22149	22396	2.9
$m = 30$	21946	14659	20727	5.6	14978	21060	4.0

Table 3 shows the impact of employing the local search in solving the BO-MKP. With $k = 5$ for the neighbourhood defined in (15), the three instances with $m = 5$, 10 and 30 are all solved to optimality.

Table 3. Effects of local search for BO on the TL-MKP.

	Opt.	$N = 25$		
		w/o LS	with LS	gap(%)
$m = 5$	24381	23928	24381	0.0
$m = 10$	23064	22471	23064	0.0
$m = 30$	21946	20727	21946	0.0

The overall performance of BO-MKP on all the 90 instances with 100 items, i.e., $n = 100$ is presented in Table 4. For the groups with $m = 5$ and $m = 10$, we set $N = 25$ and $k = 5$. For all instances with $m = 5$ and 26 instances with $m = 10$, the optimal solutions are obtained. The remaining 4 instances in the group with $m = 10$ can also be solved to optimality by increasing N

to 50. We set $k = 10$ and $N = 50$ for all instances with $m = 30$. This group of instances is particularly challenging to BO due to the high dimensions of the search space. However, with a strong local search procedure to smooth the optimisation landscape, high quality solutions are obtained on all instances.

Table 4. Computational results for all instances with 100 items.

		$\alpha = 0.25$	$\alpha = 0.5$	$\alpha = 0.75$
$m = 5$	Average	24197.2	43252.9	60471.0
	Best	24197.2	43252.9	60471.0
	Opt.	24197.2	43252.9	60471.0
	gap %	0.0	0.0	0.0
	Time	146.7 s	128.8 s	83.3 s
$m = 10$	Average	22601.0	42660.2	59555.6
	Best	22601.9	42660.6	59555.6
	Opt.	22601.9	42660.6	59555.6
	gap %	0.0	0.0	0.0
	Time	191.5 s	195.3 s	152.3 s
$m = 30$	Average	21638.2	41420.3	59201.8
	Best	21652.9	41427.2	59201.8
	Opt.	21660.4	41440.4	59201.8
	gap %	0.1	0.0	0.0
	Time	359.0 s	359.3 s	311.3 s

6 Conclusion and Future Work

In this paper, a two-level model is presented for the multidimensional knapsack problem. The master problem has much smaller dimensions, which makes it amenable to Bayesian optimisation. Three techniques are introduced to accelerate the search process of BO. Preliminary test results show the effectiveness of the proposed approach. It strongly demonstrates that incorporating prior knowledge and smoothing the optimisation landscape by the local search are crucial for the success of BO for large MKP.

Future work includes the investigation of the proper kernels in BO for combinatorial optimisation problems, the automatic tuning of hyper-parameters, and comparison with other meta-heuristics. It is also interesting to extend the models to combinatorial optimisation problems with more complex structures.

References

1. Baheri, A., Bin-Karim, S., Bafandeh, A., Vermillion, C.: Real-time control using bayesian optimization: a case study in airborne wind energy systems. Control Eng. Pract. **69**, 131–140 (2017)

2. Baptista, R., Poloczek, M.: Bayesian optimization of combinatorial structures. In: Dy, J., Krause, A. (eds.) Proceedings of the 35th International Conference on Machine Learning. Proceedings of Machine Learning Research, vol. 80, pp. 462–471. PMLR, Stockholmsmässan, Stockholm Sweden, 10–15 July 2018
3. Boussier, S., Vasquez, M., Vimont, Y., Hanafi, S., Michelon, P.: A multi-level search strategy for the 0–1 multidimensional knapsack problem. Discret. Appl. Math. **158**(2), 97–109 (2010)
4. Chu, P.C., Beasley, J.E.: A genetic algorithm for the multidimensional knapsack problem. J. Heuristics **4**(1), 63–86 (1998)
5. Dean, J., et al.: Large scale distributed deep networks. In: Pereira, F., Burges, C.J.C., Bottou, L., Weinberger, K.Q. (eds.) Advances in Neural Information Processing Systems 25, pp. 1223–1231. Curran Associates, Inc. (2012)
6. Della Croce, F., Grosso, A.: Improved core problem based heuristics for the 0/1 multi-dimensional knapsack problem. Comput. Oper. Res. **39**(1), 27–31 (2012)
7. Domhan, T., Springenberg, J.T., Hutter, F.: Speeding up automatic hyperparameter optimization of deep neural networks by extrapolation of learning curves. In: Proceedings of the 24th International Conference on Artificial Intelligence, pp. 3460–3468. IJCAI 2015, AAAI Press (2015)
8. Drake, J.: Or library mkp - best known solutions. http://www.cs.nott.ac.uk/~jqd/mkp/bestresults.html
9. Frazier, P.I.: Bayesian optimization. INFORMS TutORials in Operations Research, pp. 255–278 (2018)
10. Freno, A., Saveski, M., Jenatton, R., Archambeau, C.: One-pass ranking models for low-latency product recommendations. In: Proceedings of the 21th ACM SIGKDD International Conference on Knowledge Discovery and Data Mining , pp. 1789–1798. Association for Computing Machinery, New York, NY, USA (2015)
11. Fréville, A.: The multidimensional 0–1 knapsack problem: An overview. Eur. J. Oper. Res. **155**(1), 1–21 (2004)
12. Fréville, A., Hanafi, S.: The multidimensional 0–1 knapsack problem - bounds and computational aspects. Ann. Oper. Res. **139**(1), 195–227 (2005)
13. Greenhill, S., Rana, S., Gupta, S., Vellanki, P., Venkatesh, S.: Bayesian optimization for adaptive experimental design: A review. IEEE Access **8**, 13937–13948 (2020)
14. Gu, H.: Improving problem reduction for 0–1 multidimensional knapsack problems with valid inequalities. Comput. Oper. Res. **71**(C), 82–89 (2016)
15. Jones, D.R., Schonlau, M., Welch, W.J.: Efficient global optimization of expensive black-box functions. J. Global Optim. **13**(4), 455–492 (1998)
16. Junge, K., Hughes, J., Thuruthel, T.G., Iida, F.: Improving robotic cooking using batch bayesian optimization. IEEE Robot. Autom. Lett. **5**(2), 760–765 (2020)
17. Kellerer, H., Pferschy, U., Pisinger, D.: Knapsack problems. Springer, Heidelberg (2004)
18. Lai, X., Hao, J., Glover, F.W., Lü, Z.: A two-phase tabu-evolutionary algorithm for the 0–1 multidimensional knapsack problem. Inf. Sci. **436–437**, 282–301 (2018)
19. Oh, C., Tomczak, J., Gavves, E., Welling, M.: Combinatorial bayesian optimization using the graph cartesian product. In: Wallach, H., Larochelle, H., Beygelzimer, A., d' Alché-Buc, F., Fox, E., Garnett, R. (eds.) Advances in Neural Information Processing Systems 32, pp. 2914–2924. Curran Associates, Inc. (2019)
20. Rai, A., Antonova, R., Meier, F., Atkeson, C.G.: Using simulation to improve sample-efficiency of bayesian optimization for bipedal robots. J. Mach. Learn. Res. **20**(49), 1–24 (2019)

21. Sano, S., Kadowaki, T., Tsuda, K., Kimura, S.: Application of Bayesian optimization for pharmaceutical product development. J. Pharm. Innov. **15**(3), 333–343 (2019)
22. Shahriari, B., Swersky, K., Wang, Z., Adams, R.P., de Freitas, N.: Taking the human out of the loop: a review of bayesian optimization. Proc. IEEE **104**(1), 148–175 (2016)
23. Snoek, J., Larochelle, H., Adams, R.P.: Practical bayesian optimization of machine learning algorithms. In: Pereira, F., Burges, C.J.C., Bottou, L., Weinberger, K.Q. (eds.) Advances in Neural Information Processing Systems 25, pp. 2951–2959. Curran Associates, Inc. (2012)
24. Vasquez, M., Vimont, Y.: Improved results on the 0–1 multidimensional knapsack problem. Eur. J. Oper. Res. **165**(1), 70–81 (2005)
25. Vimont, Y., Boussier, S., Vasquez, M.: Reduced costs propagation in an efficient implicit enumeration for the 01 multidimensional knapsack problem. J. Comb. Optim. **15**(2), 165–178 (2008)
26. Wang, Z., Hutter, F., Zoghi, M., Matheson, D., De Freitas, N.: Bayesian optimization in a billion dimensions via random embeddings. J. Artif. Int. Res. **55**(1), 361–387 (2016)
27. Wilbaut, C., Hanafi, S.: New convergent heuristics for 0–1 mixed integer programming. Eur. J. Oper. Res. **195**(1), 62–74 (2009)
28. Wilbaut, C., Hanafi, S., Salhi, S.: A survey of effective heuristics and their application to a variety of knapsack problems. IMA J. Manag. Math. **19**(3), 227–244 (2008)
29. Yogatama, D., Kong, L., Smith, N.A.: Bayesian optimization of text representations. In: Proceedings of the 2015 Conference on Empirical Methods in Natural Language Processing, pp. 2100–2105. Association for Computational Linguistics, Lisbon, Portugal

Machine Learning and Deep Learning

Robustness of Adversarial Images Against Filters

Raluca Chitic[1(✉)], Nathan Deridder[1], Franck Leprévost[1], and Nicolas Bernard[2]

[1] University of Luxembourg, House of Numbers, 6, avenue de la Fonte, 4364 Esch-sur-Alzette, Grand Duchy of Luxembourg
{Raluca.Chitic,Franck.Leprevost}@uni.lu
nathan.deridder.001@student.uni.lum

[2] La Fraze, 1288, chemin de la Fraze, 88380 Arches, France
Nicolas.Bernard@lafraze.com

Abstract. This article addresses the robustness issue of adversarial images against filters. Given an image $\mathcal{A}$, that a convolutional neural network and a human both classify as belonging to a category $c_{\mathcal{A}}$, one considers an adversarial image $\mathcal{D}$ that the neural network classifies in a category $c_t \neq c_{\mathcal{A}}$, although a human would not notice any difference between $\mathcal{D}$ and $\mathcal{A}$. Would the application of a filter F (such as the Gaussian blur filter) to $\mathcal{D}$ still lead to an adversarial image $F(\mathcal{D})$ that fools the neural network? To address this issue, we perform a study on VGG-16 trained on CIFAR-10, with adversarial images obtained thanks to an evolutionary algorithm run on a specific image $\mathcal{A}$ taken in one category of CIFAR-10. Exposed to 4 individual filters, we show that the outputted filtered adversarial images essentially do remain adversarial in some sense. We also show that combining filters may render our EA attack less effective. We therefore design a new evolutionary algorithm, whose aim is to create adversarial images that do pass the filter test, do fool VGG-16 and do remain close enough to $\mathcal{A}$ that a human would not notice any difference. We show that this is indeed the case by running this new algorithm on the same image $\mathcal{A}$.

1 Introduction

During the last decade, Neural Networks, and particularly Convolutional Neural Networks (CNNs), have established themselves as the leading way to recognise objects in images. From there, they can be applied to automated image classification, image segmentation, video feed monitoring, etc. However, they are not absolutely foolproof. *Trompe-l'œil* can fool a human into seeing something that is not really there. In the same way, a CNN can be wrong from time to time, misclassifying an object in a picture as something else. *Adversarial* images are specially crafted to this purpose.

Significant work has been performed on adversarial attacks which are designed to fool CNNs trained for object recognition. Among the different types

B. Dorronsoro et al. (Eds.): OLA 2021, CCIS 1443, pp. 101–114, 2021.
https://doi.org/10.1007/978-3-030-85672-4_8

of successful adversarial attacks are those based on Evolutionary Algorithms (EAs). Although EA-based attacks produce adversarial images that are misclassified by CNNs, these images often contain noise-like artefacts. This can pose an issue for the similarity between the original, unmodified and the adversarial images, which is a requirement for adversarial attacks [2]. Moreover, noise-removing filters are a staple of image processing. It raises questions regarding the robustness of noisy adversarial images: Would it be enough to add a filter in front of a CNN to protect it against such existing attacks? Or from the attacker's point of view, given an adversarial image fooling a CNN, is it robust? Does the filtered adversarial image remain adversarial? If it does not, then is it possible to modify an EA-based attack to fool the combination Filter + CNN? This article addresses these questions in a specific context, and in elaborated scenarios.

The considered CNN, briefly described in Sect. 2, is VGG-16 [8,12] trained on the CIFAR-10 [9] dataset to classify images according to 10 categories. The adversarial images are obtained by the evolutionary algorithm $\text{EA}_{L_2}^{\text{target}}$ introduced in [4,5] for the *target scenario*. In a nutshell, this scenario considers two different categories c_t and $c_{\mathcal{A}}$, and an image $\mathcal{A}$ classified by a trained CNN in $c_{\mathcal{A}}$. An EA then aims at evolving $\mathcal{A}$ into an adversarial image $\mathcal{D}$, that the CNN classifies as belonging to c_t, while remaining close to $\mathcal{A}$ for a human eye. Section 3 summarizes the main features of $\text{EA}_{L_2}^{\text{target}}$ in the context of the *target scenario* instantiated on VGG-16 trained on CIFAR-10. Section 4 first explicits the implementation aspects and the parameters of $\text{EA}_{L_2}^{\text{target}}$. Then, one example is detailed. This section displays both the ancestor image in the category *dog*, and the adversarial descendent image in each of the 9 distinct remaining target categories of CIFAR-10, obtained by explicitly running $\text{EA}_{L_2}^{\text{target}}$ on the chosen ancestor with these parameters. To address the questions at the origin of this paper, a series of filters are compared in Sect. 5, and then applied in Sect. 6 to the images of Sect. 4.

A first outcome is that the adversarial images created by $\text{EA}_{L_2}^{\text{target}}$ and the ancestor essentially have the same pattern once exposed to individual filters. Depending on the filter, filtered adversarial images remain adversarial, either for the *target* scenario or for the *untargeted* scenario, for which one only requires the adversarial image to be classified in a different category than the original one. However, using composition of different filters render the $\text{EA}_{L_2}^{\text{target}}$ attack less effective, not only for the *target* but also for the *untargeted* scenario. This outcome leads to the construction in Sect. 7 of the variant $\text{EA}_{L_2}^{\text{target},F}$ of the EA, whose fitness function natively includes the robustness against the filter F for the *target* scenario. Section 7 shows the adversarial images obtained by running $\text{EA}_{L_2}^{\text{target},F}$ for a specific composition of filters on the same *dog* ancestor image, and the behavior of these new adversarial images towards filters. Section 8 summarizes the conclusions of this case study, some characteristics of the new black-box, targeted, non-parametric creation process $\text{EA}_{L_2}^{\text{target},F}$ of adversarial images robust against filters, and provides a series of research directions.

This article, formalizing some aspects of the bachelor student project of the second author, is an additional contribution to the research program announced in [2].

2 VGG-16 Trained on CIFAR-10

Although applicable to any CNN trained at image classification on some dataset, we instantiate our approach on a concrete case: VGG-16 trained on CIFAR-10. On the one hand, the dataset CIFAR-10 [9] encompasses 50,000 training images, and 10,000 test images of size $32 \times 32 \times 3$, meaning that each image has a width and height of 32 pixels, each pixel having a color resulting from the 3 RGB values. The images are sorted according to $\ell = 10$ categories (see Table 1).

Table 1. CIFAR-10.- For $1 \leq i \leq 10$, the 2nd row specifies the category c_i of CIFAR-10. In our experiment, we shall use the picture $n^\circ 16$ in the *dog* category from the test set of CIFAR-10 as ancestor.

i	1	2	3	4	5	6	7	8	9	10
c_i	plane	car	bird	cat	deer	dog	frog	horse	ship	truck

On the other hand, an input image $\mathcal{I}$ given to VGG-16 [12] is processed through 16 layers to produce a classification output vector $\mathbf{o}_\mathcal{I}$ of size $\ell = 10$ in the case considered, namely $\mathbf{o}_\mathcal{I} = (\mathbf{o}_\mathcal{I}[1], \cdots, \mathbf{o}_\mathcal{I}[10])$, where $0 \leq \mathbf{o}_\mathcal{I}[i] \leq 1$, and $\sum_{i=1}^{10} \mathbf{o}_\mathcal{I}[i] = 1$. Each value $\mathbf{o}_\mathcal{I}[i]$ measures the probability that the image $\mathcal{I}$ belongs to the category c_i. As a consequence, an image $\mathcal{I}$ is classified as belonging to the category c_k if $k = \arg\max_{1 \leq i \leq 10}(\mathbf{o}_\mathcal{I}[i])$.

3 Target and Untargeted Scenarios, and Design of $\mathbf{EA}_{L_2}^{\mathbf{target}}$

The *target* scenario consists in first choosing two different categories $c_t \neq c_\mathcal{A}$ among the 10 categories of CIFAR-10. Then one is given an ancestor image $\mathcal{A}$ labelled by VGG-16 as belonging to $c_\mathcal{A}$. Finally one constructs a new image $\mathcal{D}$, classified by VGG-16 as belonging to c_t, although $\mathcal{D}$ remains so close to $\mathcal{A}$ that a human would likely classify $\mathcal{D}$ as belonging to $c_\mathcal{A}$ or even be unable to distinguish $\mathcal{D}$ from $\mathcal{A}$. The classification threshold value is set at 0.95, meaning that such a $\mathcal{D}$ has achieved its purpose if $\mathbf{o}_\mathcal{D}[t] \geq 0.95$. We shall also encounter in Sect. 6 the slightly different *untargeted* scenario. An adversarial image $\mathcal{D}$ is still required to be similar to $\mathcal{A}$ for a human eye, while VGG-16 classifies $\mathcal{D}$ as belonging to a category $c \neq c_\mathcal{A}$, in the sense that the label value of c outputted by VGG-16 for $\mathcal{D}$ is the largest among all label values, and is strictly larger than the label value of $c_\mathcal{A}$. In particular, an image adversarial for the *target* scenario is also adversarial for the *untargeted* scenario, but the inverse may not be true.

Keeping these notations, let us summarize the strategy adopted in [4,5] to construct an evolutionary algorithm $\text{EA}_{L_2}^{\text{target}}$, that creates such adversarial images fooling VGG-16 trained on CIFAR-10 for the target scenario. The main components of our EA are as follows.

Population Initialization. The initial population is set to 160 copies of the ancestor image $\mathcal{A}$.

Evaluation. This operation is performed on each individual image ind of a given generation g_p *via* the fitness function $fit_{L_2}(ind, g_p)$ that takes into account a dual goal made of both the evolution of ind towards the target category c_t, and its proximity with the ancestor $\mathcal{A}$, measured thanks to the L_2-norm:

$$fit_{L_2}(ind, g_p) = A(g_p, ind)\mathbf{o}_{ind}[c_t] - B(g_p, ind)L_2(ind, \mathcal{A}) \geq 0, \tag{1}$$

where the quantities $A(g_p, ind), B(g_p, ind) \geq 0$ weight and balance the dual goal (see Sect. 4 for their values). The L_2-norm is used to calculate the difference between the pixel values of the ancestor and of the considered image ind:

$$L_2(ind, \mathcal{A}) = \sum_{p_j} |ind[p_j] - \mathcal{A}[p_j]|^2, \tag{2}$$

where p_j is the pixel in the jth position, and $0 \leq ind[p_j], \mathcal{A}[p_j] \leq 255$ are the corresponding pixel values of the images ind and $\mathcal{A}$.

Evolution. Once the fitness function of each individual in the population is computed (starting with the first generation made of the initial population), the on-going generation is split into 3 groups. The "elite" consists of the 10 best individuals in the population. The "didn't make it", consisting of the lower scored half of the population, is discarded. The "middle class" consists of the remaining individuals. The "elite" is kept unchanged. Each of the 80 individuals of the "didn't make it" group is replaced by an individual resulting of the mutation of elements from the "elite" and the "middle-class". All "middle-class" individuals are mutated. The performed mutations are those described in [4] (they remain similar to some extent to those of [1]). Cross-overs (see [4]) are applied to all individuals except those of the "elite". Pixel values are modified in a range ± 3 in the version used here of $\text{EA}_{L_2}^{\text{target}}$. These operations lead to the 160 individuals composing the new generation subject to the next round of evaluation.

This loop is performed as many times as necessary to create the adversarial image $\mathcal{D}$ as the result of $\text{EA}_{L_2}^{\text{target}}$ run on $\mathcal{A}$ for the target category c_t. Hence $\mathcal{D} = \text{EA}_{L_2}^{\text{target}}(\mathcal{A}, c_t)$ satisfies $\mathbf{o}_{\mathcal{D}}[t] \geq 0.95$.

4 Obtention of the Adversarial Images: Running $\text{EA}_{L_2}^{\text{target}}$

Concretely, for any generation g_p, one sets $B(g_p, ind) = 10^{-\log_{10}(L_2(ind, \mathcal{A}))}$. The value of $A(g_p, ind)$ depends on $\boldsymbol{o}_{ind}[c_t]$ (note that $\log_{10} \boldsymbol{o}_{ind}[c_t] \leq 0$).

$$A(g_p, ind) = \begin{cases} 10^{-3+\log_{10} \boldsymbol{o}_{ind}[c_t]} & \text{if } \boldsymbol{o}_{ind}[c_t] < 10^{-3} \\ 10^{-2+\log_{10} \boldsymbol{o}_{ind}[c_t]} & \text{if } 10^{-3} \leq \boldsymbol{o}_{ind}[c_t] < 10^{-2} \\ 10^{-1+\log_{10} \boldsymbol{o}_{ind}[c_t]} & \text{if } 10^{-2} \leq \boldsymbol{o}_{ind}[c_t] \end{cases} \quad (3)$$

$\text{EA}_{L_2}^{\text{target}}$ was implemented in Python 3.7 with the NumPy [11] library. Keras [6] was used to load and run the VGG-16 [12] model. Our experiments were run on a computer with an Nvidia RTX 3080 GPU and an Amd Ryzen 7 5800X CPU.

Figure 1 shows the ancestor image $\mathcal{A}$ taken in the category *dog* (image $n°16$ in the category c_6 of the test set of CIFAR-10, used throughout this article). Besides $\mathcal{A}$, Fig. 1 also presents the 9 evolved adversarial images $\mathcal{D}_i = \text{EA}_{L_2}^{\text{target}}(\mathcal{A}, c_i)$, with $i \neq 6$, classified by VGG-16 as belonging to the category c_i with the notations of Table 1. By slightly changing many pixels instead of heavily changing a few pixels, this approach, that enhances the indistinguishability between the adversarial image and the ancestor image, differs substantially from [13], where one single pixel is changed, but this modification is noticeable for a human without difficulty.

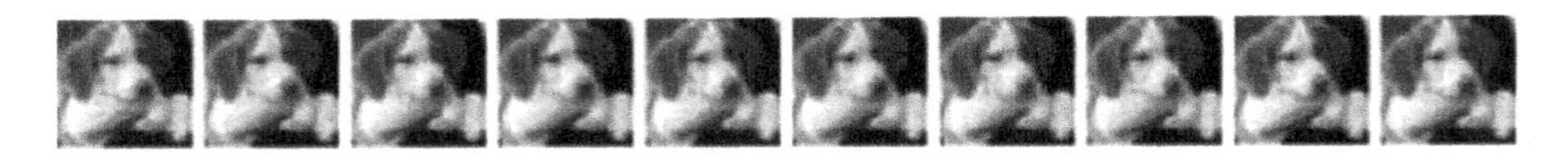

Fig. 1. Comparison of the ancestor $\mathcal{A}$ (chosen as the image $n°16$ in the *dog* category c_6) in the 6th position with the adversarial images $\mathcal{D}_i = \text{EA}_{L_2}^{\text{target}}(\mathcal{A}, c_i)$ in the ith position ($i \neq 6$). VGG-16 classifies $\mathcal{A}$ in the *dog* category with probability 0.9996387, and classifies $\mathcal{D}_i$ in the target category c_i with probability ≥ 0.95.

Table 2 specifies the number of generations and the execution time required by $\text{EA}_{L_2}^{\text{target}}$ to create the adversarial images $\mathcal{D}_i$ of Fig. 1. The images pictured in Fig. 1 are tested in Sect. 6. More precisely, filters performed on these images create new images that are given as input to VGG-16 for classification. The choice of these filters is described in the next section.

5 Selection of Filter

In image processing, a filter or a *Kernel* [14] is essentially given by a square $f \times f$ matrix for an odd integer f. Filtering an image $\mathcal{I}$, say of size $n \times n$, is an operation performed pixel for pixel as follows. For each pixel p of $\mathcal{I}$, one puts in matrix form a $f \times f$ area of the image centered on p. The coefficients of the resulting $f \times f$ matrix $\mathcal{I}_p$ are the RGB values of the corresponding pixels in the considered area. The convolution operation of the kernel matrix and of $\mathcal{I}_p$ leads to a $f \times f$ matrix $F * \mathcal{I}_p$. The values of the pixel p of the filtered image $F(\mathcal{I})$ is the sum of the coefficients of $F * \mathcal{I}_p$. Pixels at a distance $< f$ of an edge of $\mathcal{I}$ require a special treatment to ensure that the size of the filtered image $F(\mathcal{I})$ is also of size $n \times n$ (otherwise, its size would be reduced to $(n-f+1) \times (n-f+1)$).

Table 2. For $1 \leq i \leq 10$, $i \neq 6$, the 2nd row specifies the number of generations required by $\text{EA}_{L_2}^{\text{target}}$ to create the adversarial image $\mathcal{D}_i$ pictured in Fig. 1. The 3rd row shows the total execution time, measured in seconds, while the 4th row represents the average number of generations per second.

i	1	2	3	4	5	7	8	9	10
# of generations	815	960	494	127	1011	376	286	970	526
total time (in seconds)	46.22	69.14	27.02	8.27	53.93	20.67	16.2	52.36	28.21
# of generations/second	17.63	13.88	18.28	15.36	18.74	18.19	17.65	18.53	18.65

Although one could consider a large list of filters, we focus in this article on the following four [10, chapters 7 and 8], that have a significant impact on images. In our computations performed on images of size 32×32, we shall take $f = 1$ for F_1 and $f = 3$ for F_2, F_3, F_4, and used the OpenCV implementation library [3].

The *inverse filter* F_1 replaces all colors by their complementary colors. This operation is performed pixel for pixel by subtracting the RGB value $(255, 255, 255)$ of white by the RGB value of that pixel.

The *Gaussian blur filter* F_2 uses a Gaussian distribution to calculate the Kernel, $G(x, y) = \frac{1}{2\pi\sigma^2} e^{-\frac{x^2+y^2}{2\sigma^2}}$, where x is the distance from the origin on the x-axis, y is the distance from the origin on the y-axis and σ is the standard deviation of the Gaussian distribution. By design, the process gives more priority to the pixels in the center, and blurs around it with a lesser impact as one moves away from the center.

The *median filter* F_3 is used to reduce noice and artefacts in a picture. Though under some conditions it can reduce noise while preserving the edges, this does not really occur for small images like those considered here. In general, one selects a pixel, and one computes the median of all the surrounding pixels.

The *unsharp mask filter* F_4 enhances the sharpness and contrast of images. The unsharp masked image is obtained by blurring a copy of the image using a Gaussian blur, which is then weighted and subtracted from the original image.

Any such filter F, or any combination of filters $F_{i_1}, F_{i_2}, \cdots, F_{i_k}$ operating successively (in that order) on an image $\mathcal{I}$, creates a filtered image $F(\mathcal{I})$ or $F_{i_k} \circ \cdots \circ F_{i_2} \circ F_{i_1}(\mathcal{I})$. In Sect. 6 we make use of the previous four filters taken individually, and of the combination $F_3 \circ F_4$.

6 Filtering the Ancestor and the Adversarial Images

The ancestor image $\mathcal{A}$, and the adversarial images $\mathcal{D}_i = \text{EA}_{L_2}^{\text{target}}(\mathcal{A}, c_i)$ $(i \neq 6)$ represented in Fig. 1 are tested against the filters of Sect. 5. Figure 2 shows the outcome of this process. From left to right, the 10 pictures on the kth row represent $F(\mathcal{A})$ in the 6th position and $F(\mathcal{D}_i)$ in the ith position for $1 \leq i \neq 6 \leq 10$, with $F = F_k$ for $1 \leq k \leq 4$, and $F = F_3 \circ F_4$ for the 5th row. The reason for the choice of $F_3 \circ F_4$ is that F_4 is used to amplify and highlight detail, while

F_3 is used to remove noise from an image without removing detail. Therefore, a combination of these filters could remove the noise created by the EA while maintaining a high level of detail.

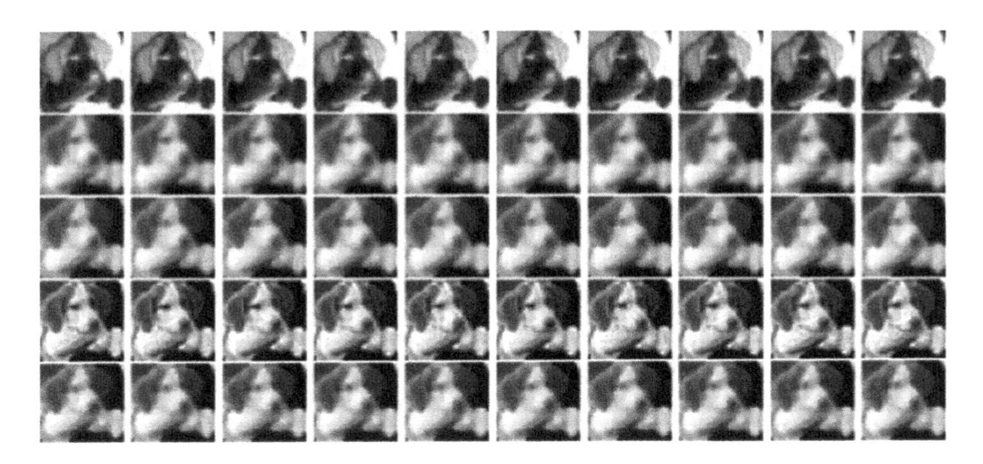

Fig. 2. Comparison of the impact of filters on the ancestor $\mathcal{A}$ and on the adversarial images $\mathcal{D}_i$. The kth row represents $F(\mathcal{A})$ (in 6th position) and $F(\mathcal{D}_i)$ (in i^{th} position, $i \neq 6$), where $F = F_k$ for $1 \leq k \leq 4$, and $F = F_3 \circ F_4$ for $k = 5$.

These filtered images are given to VGG-16 for classification. Table 3 shows the outcome with filters $F = F_k$ for $1 \leq k \leq 3$, while Table 4 is produced with F_4, and Table 5 with $F_3 \circ F_4$. Each Table has (groups of) rows showing the probability of the filtered images for the c_6 category, the target class c_i, the maximum probability and its corresponding class outputted by VGG-16, respectively. In all tables, we set $\mathcal{D}_6 = \mathcal{A}$ to ease the notations.

The inverse, Gaussian blur and median filters (F_1, F_2 and F_3, Table 3) produce images that are adversarial against VGG-16 for the *untargeted* scenario. Indeed, the c_6 probabilities of the $F(\mathcal{D}_i)$ are very low, and there is a category $c \neq c_6$ of probability strictly larger. Only one adversarial image reverts to the *dog* class after being filtered (filter F_3, class 2). The noticeable predominance of the *cat* class as the predicted category of the filtered images is likely due to the similarity in features of the CIFAR-10 *dog* and *cat* images that VGG-16 was trained on. Another reason for the inefficacy of filters F_1, F_2 and F_3 to protect VGG-16 for the *untargeted* scenario is that the c_6 probability of $F(\mathcal{A})$ is drastically reduced from the initial 0.9996387 of $\mathcal{A}$, which is undesired. Although adversarial for the *untargeted* scenario, these filtered images can not be considered adversarial for the *target* scenario. Still, almost all $F(\mathcal{D}_i)$ are classified in the same category *cat* as $F(\mathcal{A})$, hence follow $F(\mathcal{A})$'s pattern.

Filters F_4 and $F_3 \circ F_4$ do not significantly reduce the c_6 probability of $F(\mathcal{A})$ (Tables 4 and 5). The $F_4(\mathcal{D}_i)$s are classified in c_i with high confidence, and are definitively adversarial for the *untargeted* scenario. Moreover, they are all

Table 3. Label values in the category c_6, target class c_i, maximum probability and its corresponding class given by VGG-16 for the filtered ancestor $F(\mathcal{A})$ and the filtered adversarial images $F(\mathcal{D}_i)$ ($i \neq 6$) for $F = F_1$ (1st group of 5 rows), F_2 (2nd group of 5 rows) and F_3 (3rd group of 5 rows).

i	1	2	3	4	5	6	7	8	9	10
$\mathbf{o}_{F_1(\mathcal{D}_i)}[6]$	8e-04	1e-02	8e-03	8e-04	5e-03	1e-03	2e-02	1e-02	3e-03	8e-04
$\mathbf{o}_{F_1(\mathcal{D}_i)}[i]$	5e-04	8e-04	8e-05	0.99	6e-04	1e-03	2e-02	4e-05	1e-02	0.91
$\max(\mathbf{o}_{F_1(\mathcal{D}_i)})$	0.98	0.72	0.98	0.99	0.68	0.99	0.95	0.97	0.82	0.91
$c_{\arg\max(\mathbf{o}_{F_1(\mathcal{D}_i)})}$	truck	frog	cat	cat	frog	cat	cat	cat	truck	truck
i	1	2	3	4	5	6	7	8	9	10
$\mathbf{o}_{F_2(\mathcal{D}_i)}[6]$	1e-04	1e-02	3e-03	1e-04	6e-04	3e-04	1e-03	1e-03	1e-04	2e-04
$\mathbf{o}_{F_2(\mathcal{D}_i)}[i]$	1e-05	2e-06	8e-05	0.99	2e-05	3e-04	1e-05	2e-05	3e-06	2e-06
$\max(\mathbf{o}_{F_2(\mathcal{D}_i)})$	0.99	0.98	0.99	0.99	0.99	0.99	0.99	0.99	0.99	0.99
$c_{\arg\max(\mathbf{o}_{F_2(\mathcal{D}_i)})}$	cat	cat	cat	cat	cat	cat	cat	cat	cat	cat
i	1	2	3	4	5	6	7	8	9	10
$\mathbf{o}_{F_3(\mathcal{D}_i)}[6]$	1e-02	0.84	0.22	6e-03	1e-02	0.26	0.14	0.12	2e-02	0.11
$\mathbf{o}_{F_3(\mathcal{D}_i)}[i]$	1e-05	3e-06	1e-03	0.99	2e-05	0.26	5e-05	6e-05	4e-06	5e-06
$\max(\mathbf{o}_{F_3(\mathcal{D}_i)})$	0.98	0.84	0.77	0.99	0.98	0.73	0.86	0.88	0.97	0.89
$c_{\arg\max(\mathbf{o}_{F_3(\mathcal{D}_i)})}$	cat	dog	cat	cat	cat	cat	cat	cat	cat	cat

Table 4. Label values in the category c_6, target class c_i, maximum probability and its corresponding class given by VGG-16 for the filtered ancestor $F_4(\mathcal{A})$ and the filtered adversarial images $F_4(\mathcal{D}_i)$ ($i \neq 6$).

i	1	2	3	4	5	6	7	8	9	10
$\mathbf{o}_{F_4(\mathcal{D}_i)}[6]$	1e-03	8e-05	1e-02	0.15	3e-04	0.99	3e-04	1e-02	1e-03	5e-05
$\mathbf{o}_{F_4(\mathcal{D}_i)}[i]$	0.94	0.99	0.97	0.84	0.99	0.99	0.99	0.98	0.99	0.99
$\max(\mathbf{o}_{F_4(\mathcal{D}_i)})$	0.94	0.99	0.97	0.84	0.99	0.99	0.99	0.98	0.99	0.99
$c_{\arg\max(\mathbf{o}_{F_4(\mathcal{D}_i)})}$	plane	car	bird	cat	deer	dog	frog	horse	ship	truck

Table 5. Label values in the category c_6, target class c_i, maximum probability and its corresponding class given by VGG-16 for the filtered ancestor $F_3 \circ F_4(\mathcal{A})$ and the filtered adversarial images $F_3 \circ F_4(\mathcal{D}_i)$ ($i \neq 6$).

i	1	2	3	4	5	6	7	8	9	10
$\mathbf{o}_{F_3 \circ F_4(\mathcal{D}_i)}[6]$	0.17	0.99	0.75	0.11	8e-02	0.91	0.51	0.32	0.22	0.58
$\mathbf{o}_{F_3 \circ F_4(\mathcal{D}_i)}[i]$	3e-05	1e-06	3e-03	0.88	3e-05	0.91	1e-04	8e-05	8e-06	7e-06
$\max(\mathbf{o}_{F_3 \circ F_4(\mathcal{D}_i)})$	0.82	0.99	0.75	0.88	0.91	0.91	0.51	0.67	0.77	0.58
$c_{\arg\max(\mathbf{o}_{F_3 \circ F_4(\mathcal{D}_i)})}$	cat	dog	dog	cat	cat	dog	dog	cat	cat	dog

adversarial for the *target* scenario, while $F_4(\mathcal{A})$ simultaneously remains classified as *dog*. In that sense, our $\mathrm{EA}_{L_2}^{\mathrm{target}}$ attack is robust against the unsharp mask filter for the *target* and a fortiori the *untargeted* scenario.

The final filter $F_3 \circ F_4$ has a particular impact. First, $F_3 \circ F_4(\mathcal{A})$ remains classified as *dog*. Second, the c_6 probabilities of the $F_3 \circ F_4(\mathcal{D}_i)$s (Table 5) are either the largest or the second largest. Out of the 9 adversarial images, 4 reverted to the *dog* class after filtering, while the other 5 were classified as *cat*, with *dog* being the second most likely category. Hence the $F_3 \circ F_4$ combination of filters, which brings back a significant proportion of filtered images to the ancestor category, may render our EA-based attack less effective, not only for the *target*, but also for the *untargeted* scenario.

7 The Variant $\mathrm{EA}_{L_2}^{\mathrm{target},F}$

Results of the previous section lead to the conception of $\mathrm{EA}_{L_2}^{\mathrm{target},F}$. This variant of $\mathrm{EA}_{L_2}^{\mathrm{target}}$ natively takes into account the goal to create adversarial images that remain adversarial for the *target* scenario once filtered, in addition to remaining close to the ancestor and being classified as belonging to a target category. The main modification is clearly in the fitness function:

$$fit_{L_2}^{F}(ind, g_p) = A(g_p, ind)(\mathbf{o}_{ind}[c_t] + \mathbf{o}_{F(ind)}[c_t]) - B(g_p, ind)L_2(ind, \mathcal{A}) \quad (4)$$

where the last component measures the probability that the individual filtered with F is classified as the target category. Since $F_3 \circ F_4$ is the only filter which can revert a significant proportion of the adversarial images to c_6, it makes sense to explore $\mathrm{EA}_{L_2}^{\mathrm{target},F_3\circ F_4}$. For the sake of consistency, the range of pixel value modification is set to ± 3 as well for $\mathrm{EA}_{L_2}^{\mathrm{target},F_3\circ F_4}$.

Running $\mathrm{EA}_{L_2}^{\mathrm{target},F_3\circ F_4}$ for the target scenario on the same dog ancestor image as during the previous experiments leads to the $\mathcal{D}_i^{F_3\circ F_4}$ adversarial images pictured in Fig. 3.

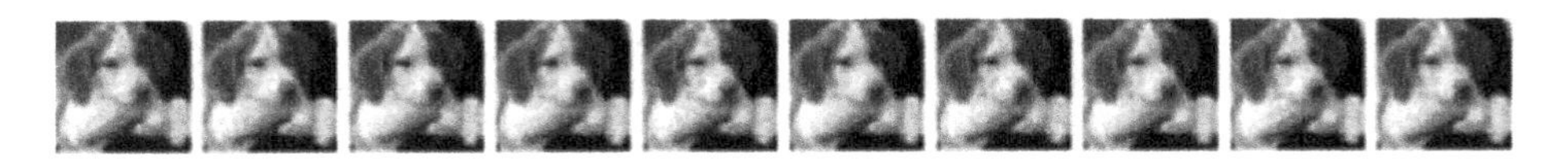

Fig. 3. Comparison of the ancestor $\mathcal{A}$ in the 6th position with the adversarial images $\mathcal{D}_i^{F_3\circ F_4} = EA_{L_2}^{\mathrm{target},F_3\circ F_4}(\mathcal{A}, c_i)$ in the ith position ($i \neq 6$, from left to right). VGG-16 classifies $\mathcal{D}_i^{F_3\circ F_4}$ as belonging to the target category c_i with probability ≥ 0.95.

The ancestor image $\mathcal{A}$, and the adversarial images $\mathcal{D}_i^{F_3\circ F_4} = \mathrm{EA}_{L_2}^{\mathrm{target},F_3\circ F_4}$ $(\mathcal{A})$ ($i \neq 6$) pictured in Fig. 3 are tested against the filters of Sect. 5. Figure 4 shows the outcome of this process. More precisely, from left to right, the picture in the ith position on the kth row represent $F(\mathcal{D}_i^{F_3\circ F_4})$ for $i \neq 6$ and $F(\mathcal{A})$ for $i = 6$, filtered with $F = F_k$ for $1 \leq k \leq 4$, and $F = F_3 \circ F_4$ for $k = 5$.

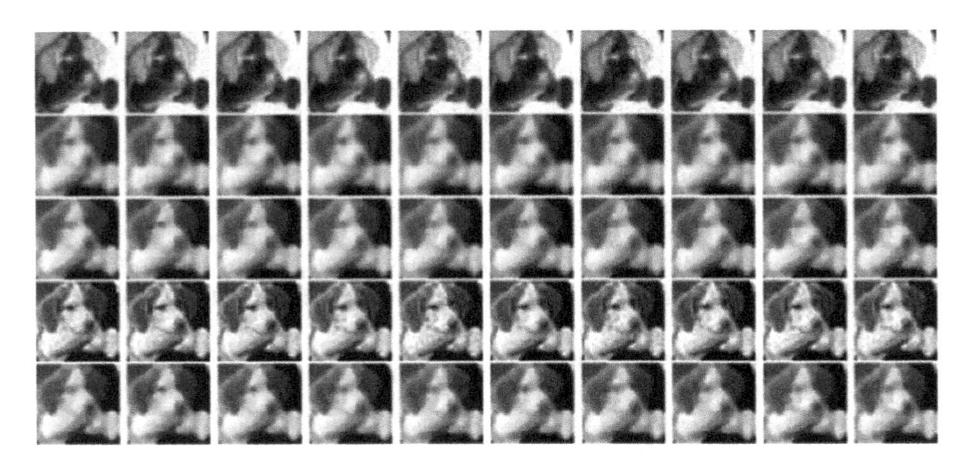

Fig. 4. Comparison of the impact of filters on the ancestor $\mathcal{A}$ and on the adversarial images $\mathcal{D}_i^{F_3 \circ F_4}$. The kth row represents $F(\mathcal{A})$ (in 6th position) and $F(\mathcal{D}_i^{F_3 \circ F_4})$ (in i^{th} position, $i \neq 6$), where $F = F_k$ for $1 \leq k \leq 4$, and $F = F_3 \circ F_4$ for $k = 5$.

These filtered images are given to VGG-16 for classification. Table 6 is produced with the filters F_1, F_2, F_3, while Table 7 is produced with F_4, and Table 8 with $F_3 \circ F_4$. Each Table has (groups of) rows showing the probability of the filtered images for the c_6 category, the target class c_i, the maximum probability and its corresponding class outputted by VGG-16, respectively. In all tables, we set $\mathcal{D}_6^{F_3 \circ F_4} = \mathcal{A}$ to ease the notations.

The first outcome of Tables 6, 7 and 8 is that the $\text{EA}_{L_2}^{\text{target},F_3 \circ F_4}$ attack is robust against all individual filters considered for the *untargeted* scenario. This result is not surprising, since the F_k ($1 \leq k \leq 4$) filters were already not a good defense for the CNN when using adversarial images created with $\text{EA}_{L_2}^{\text{target}}$. Hence, although the $\mathcal{D}_i^{F_3 \circ F_4}$ adversarial images were only designed to circumvent filter $F_3 \circ F_4$, they are also robust against filters F_1 to F_4.

None of the $\text{EA}_{L_2}^{\text{target},F_3 \circ F_4}$ adversarial images filtered with F_1 to F_3 were classified as *dog*. Nonetheless, the predicted categories vary between filters. While the Gaussian filter predicts *cat* for all images, the inverse and median filters also predict the target class for some images. By simply observing the images of Fig. 4, one can see that the Gaussian filter produces the blurriest images, hence reducing not only the object details, but also the noise added by the EA. This might explain why the predictions corresponding to the Gaussian-filtered images do not contain much information related to the target class.

As is the case with $\text{EA}_{L_2}^{\text{target}}$, the adversarial images produced by $\text{EA}_{L_2}^{\text{target},F_3 \circ F_4}$, once filtered with F_4 are all classified as the target class, hence being adversarial for the *target* scenario. This is probably due to the fact that, while the unsharp mask increases the contrast of the *dog* object, it also intensifies the noise added by the EA, which directs the image towards the adversarial class. The noisy aspect of the filtered images can be seen in the 4th row of Fig. 4.

Table 6. Label values in the category c_6, target class c_i, maximum probability $\max \mathbf{o} = \max\left(\mathbf{o}_{F\left(\mathcal{D}_i^{F_3 \circ F_4}\right)}\right)$ and its corresponding class $c_{\max}$, where $c_{\max} = \arg\max\left(\mathbf{o}_{F\left(\mathcal{D}_i^{F_3 \circ F_4}\right)}\right)$ given by VGG-16 for the filtered ancestor $F(\mathcal{A})$ and the filtered adversarial images $F(\mathcal{D}_i^{F_3 \circ F_4})$ $(i \neq 6)$ for $F = F_1$ (1st group of 5 rows), F_2 (2nd group) and F_3 (3rd group).

i	1	2	3	4	5	6	7	8	9	10
$\mathbf{o}_{F_1\left(\mathcal{D}_i^{F_3 \circ F_4}\right)}[6]$	3e-03	2e-03	8e-04	1e-03	1e-03	1e-03	7e-03	0.11	1e-03	1e-04
$\mathbf{o}_{F_1\left(\mathcal{D}_i^{F_3 \circ F_4}\right)}[i]$	2e-03	1e-04	7e-05	0.99	6e-04	1e-03	0.22	5e-05	1e-02	0.99
$\max \mathbf{o}$	0.58	0.66	0.97	0.99	0.85	0.99	0.73	0.86	0.75	0.99
$c_{\max}$	frog	cat	cat	cat	frog	cat	cat	cat	frog	truck
i	1	2	3	4	5	6	7	8	9	10
$\mathbf{o}_{F_2\left(\mathcal{D}_i^{F_3 \circ F_4}\right)}[6]$	2e-04	3e-04	6e-03	1e-04	1e-03	3e-04	6e-03	1e-03	3e-04	1e-03
$\mathbf{o}_{F_2\left(\mathcal{D}_i^{F_3 \circ F_4}\right)}[i]$	3e-05	3e-06	2e-04	0.99	4e-05	3e-04	2e-05	5e-05	4e-06	3e-06
$\max \mathbf{o}$	0.99	0.99	0.99	0.99	0.99	0.99	0.99	0.99	0.99	0.99
$c_{\max}$	cat	cat	cat	cat	cat	cat	cat	cat	cat	cat
i	1	2	3	4	5	6	7	8	9	10
$\mathbf{o}_{F_3\left(\mathcal{D}_i^{F_3 \circ F_4}\right)}[6]$	9e-04	4e-02	1e-03	3e-04	4e-02	0.26	4e-02	4e-03	1e-02	0.10
$\mathbf{o}_{F_3\left(\mathcal{D}_i^{F_3 \circ F_4}\right)}[i]$	0.84	0.17	0.99	0.99	0.88	0.26	0.84	0.98	1e-02	8e-02
$\max \mathbf{o}$	0.84	0.72	0.99	0.99	0.88	0.73	0.84	0.98	0.97	0.80
$c_{\max}$	plane	cat	bird	cat	deer	cat	frog	horse	cat	cat

Table 7. Label values in the category c_6, target class c_i, maximum probability $\max \mathbf{o} = \max\left(\mathbf{o}_{F_4\left(\mathcal{D}_i^{F_3 \circ F_4}\right)}\right)$ and its corresponding class $c_{\max}$, where $c_{\max} = \arg\max\left(\mathbf{o}_{F_4\left(\mathcal{D}_i^{F_3 \circ F_4}\right)}\right)$ given by VGG-16 for the filtered ancestor $F_4(\mathcal{A})$ and the filtered adversarial $F_4(\mathcal{D}_i^{F_3 \circ F_4})$ $(i \neq 6)$.

i	1	2	3	4	5	6	7	8	9	10
$\mathbf{o}_{F_4\left(\mathcal{D}_i^{F_3 \circ F_4}\right)}[6]$	4e-04	6e-05	4e-03	6e-02	4e-04	0.99	3e-04	5e-03	1e-04	1e-04
$\mathbf{o}_{F_4\left(\mathcal{D}_i^{F_3 \circ F_4}\right)}[i]$	0.96	0.99	0.99	0.93	0.97	0.99	0.99	0.99	0.99	0.98
$\max \mathbf{o}$	0.96	0.99	0.99	0.93	0.97	0.99	0.99	0.99	0.99	0.98
$c_{\max}$	plane	car	bird	cat	deer	dog	frog	horse	ship	truck

Table 8. Label values in the category c_6, target class c_i, maximum probability $\max \mathbf{o} = \max\left(\mathbf{o}_{F_3\circ F_4\left(\mathcal{D}_i^{F_3\circ F_4}\right)}\right)$ and its corresponding class $c_{\max}$, where $c_{\max} = \arg\max\left(\mathbf{o}_{F_3\circ F_4\left(\mathcal{D}_i^{F_3\circ F_4}\right)}\right)$ given by VGG-16 for the filtered ancestor $F_3 \circ F_4(\mathcal{A})$ and the filtered adversarial $F_3 \circ F_4(\mathcal{D}_i^{F_3\circ F_4})$ $(i \neq 6)$.

i	1	2	3	4	5	6	7	8	9	10
$\mathbf{o}_{F_3\circ F_4(\mathcal{D}_i^{F_3\circ F_4})}[6]$	6e-05	2e-04	1e-04	2e-04	1e-04	0.91	1e-04	1e-04	5e-04	8e-05
$\mathbf{o}_{F_3\circ F_4(\mathcal{D}_i^{F_3\circ F_4})}[i]$	0.99	0.99	0.99	0.99	0.99	0.91	0.99	0.99	0.99	0.99
$\max \mathbf{o}$	0.99	0.99	0.99	0.99	0.99	0.91	0.99	0.99	0.99	0.99
$c_{\max}$	plane	car	bird	cat	deer	dog	frog	horse	ship	truck

Finally, Table 8, corresponding to $F_3 \circ F_4$, shows a clear improvement compared to Table 5, as $\text{EA}_{L_2}^{\text{target},F_3\circ F_4}$ produces images that are no longer vulnerable to this filter for the *target* scenario.

Although $\text{EA}_{L_2}^{\text{target},F_3\circ F_4}$ provides an increase in robustness compared to $\text{EA}_{L_2}^{\text{target}}$, it is also interesting to compare the time efficiency of the two algorithms. Table 9 gives the number of generations and the amount of time required by $\text{EA}_{L_2}^{\text{target},F_3\circ F_4}$ to create the adversarial images $\mathcal{D}_i^{F_3\circ F_4}$.

Table 9. For $1 \leq i \leq 10$, $i \neq 6$, the 2nd row specifies the number of generations required by $\text{EA}_{L_2}^{\text{target},F_3\circ F_4}$ to create the adversarial images $\mathcal{D}_i^{F_3\circ F_4}$. The 3rd row gives the total execution time, while the 4th row gives the average number of generations per second.

i	1	2	3	4	5	7	8	9	10
# of generations	1495	1164	717	194	1812	820	541	1262	1330
Total time (in seconds)	159.2	136.04	80.23	22.85	199.08	82.98	53.26	133.33	139.01
# of generations/second	9.39	8.56	8.94	8.49	9.1	9.88	10.16	9.47	9.57

Comparing Tables 2 and 9 shows that the higher robustness of $\text{EA}_{L_2}^{\text{target},F_3\circ F_4}$ requires both a longer execution time per generation and more generations. Firstly, $\text{EA}_{L_2}^{\text{target},F_3\circ F_4}$ must satisfy not two, but three conditions. More generations are needed to have not only the plain adversarial images with a target class probability higher than 0.95, but also its filtered version. Secondly, the drop in the average number of generations computed per second is due to the additional filtering step in $\text{EA}_{L_2}^{\text{target},F_3\circ F_4}$.

8 Conclusion

This ongoing work addresses the issue of the robustness against filters of adversarial images fooling CNNs. By considering VGG-16 trained at image classification on CIFAR-10, adversarial images $\mathcal{D}_i$ created by the $\mathrm{EA}_{L_2}^{\text{target}}$ evolutionary algorithm performed on the *dog* instantiation of the "target scenario" with ancestor image $\mathcal{A}$, and specific filters F_1, F_2, F_3, F_4, we first prove that the $F(\mathcal{D}_i)$ images, while no longer adversarial for the *target* scenario for $F = F_1, F_2, F_3$, are not only adversarial for the *untargeted* scenario, but foremost follow $F(\mathcal{A})$'s pattern for these three individual filters. We also prove that the $F_4(\mathcal{D}_i)$ images remain adversarial for the *target* scenario, and a fortiori for the *untargeted* scenario, while $F_4(\mathcal{A})$ is classified in the same category as $\mathcal{A}$. Hence, by essentially following the ancestor's behavior towards these four individual filters, the adversarial images $\mathcal{D}_i$ acquire an additional similarity with $\mathcal{A}$. These results confort their adversarial profile, and enhance the robustness and quality of the $\mathrm{EA}_{L_2}^{\text{target}}$ attack.

We secondly show that the $F_3 \circ F_4$ combination of filters brings back a significant proportion of filtered images in the ancestor category, while $F_3 \circ F_4(\mathcal{A})$ is classified in the same category as $\mathcal{A}$. Since this may render $\mathrm{EA}_{L_2}^{\text{target}}$ less effective, not only for the *target*, but also for the *untargeted* scenario, a third outcome of this work is the construction of the variant $\mathrm{EA}_{L_2}^{\text{target},F}$ of the evolutionary algorithm, that natively takes into account the robustness of adversarial images against a generic filter F. We instantiate this EA on $F = F_3 \circ F_4$. The produced images $\mathcal{D}_i^{F_3 \circ F_4}$ are adversarial against $F_3 \circ F_4$ for the *target* scenario, but also essentially against F_4 and F_3. They are to a large extent adversarial against F_1 and F_2 for the *untargeted* scenario as well. The performance of this new EA is compared to that of $\mathrm{EA}_{L_2}^{\text{target}}$, showing that the provided robustness comes at the cost of more, as well as longer generations.

These preliminary results lead to a series of future work. We intend to extend our methodology to all images of Fig. 10 of [4], beyond the sole *dog* series of the present article, potentially with more than one ancestor and more than one descendant in each given category. We also plan to assess the efficiency of the creation of adversarial images with $\mathrm{EA}_{L_2}^{\text{target},F}$ depending on which ancestor image is provided as input: either the original $\mathcal{A}$ or the processed adversarial image $\mathrm{EA}_{L_2}^{\text{target}}(\mathcal{A}, c_i)$. An important direction would be to consider larger images, such as those of ImageNet [7], since the small 32×32 images of this study are naturally *grainy*. Finally, one could consider to address these issues with a different choice of the measure of proximity between images, for instance with SSIM [15] instead of L_2, and with different scenarios, for instance the *flat scenario* of [4].

References

1. Bernard, N., Leprévost, F.: Evolutionary algorithms for convolutional neural network visualisation. In: High Performance Computing – 5th Latin American Conference, 2018, Bucaramanga, Colombia, 23–28 Sep 2018. Communications in Computer and Information Science, vol. 979, pp. 18–32. Springer, Heidelberg (2018). https://doi.org/10.1007/978-3-030-16205-4_2
2. Bernard, N., Leprévost, F.: How evolutionary algorithms and information hiding deceive machines and humans for image recognition: a research program. In: Proceedings of the OLA 2019 International Conference on Optimization and Learning, Bangkok, Thailand, 29–31 Jan 2019, pp. 12–15. Springer, Heidelberg (2019)

3. Bradski, G.: The OpenCV Library. Dr. Dobb's Journal of Software Tools (2000). https://www.drdobbs.com/open-source/the-opencv-library/184404319
4. Chitic, R., Bernard, N., Leprévost, F.: Evolutionary algorithms deceive humans and machines at image classification: an extended proof of concept on two scenarios. J. Inf. Telecommun. **5**, 121–143 (2020). http://dx.doi.org/10.1080/24751839.2020.1829388
5. Chitic, R., Bernard, N., Leprévost, F.: A proof of concept to deceive humans and machines at image classification with evolutionary algorithms. In: Intelligent Information and Database Systems, 12th Asian Conference, ACIIDS 2020, Phuket, Thailand, 23–26 March 2020, vol. 12034, pp. 467–480. Springer, Heidelberg (2020). https://doi.org/10.1007/978-3-030-42058-1_39
6. Chollet, F.: Keras. GitHub code repository (2015–2020). https://github.com/fchollet/keras
7. Deng, J., Dong, W., Socher, R., Li, L.J., Li, K., Fei-Fei, L.: The imagenet image database (2009). http://image-net.org
8. Geifman, Y.: cifar-vgg (2018). https://github.com/geifmany/cifar-vgg
9. Krizhevsky, A., et al.: The CIFAR datasets (2009). https://www.cs.toronto.edu/~kriz/cifar.html
10. Lim, J.S.: Two-Dimensional Signal and Image Processing. Prentice Hall, Hoboken (1989)
11. Oliphant, T.E.: A guide to NumPy. Trelgol Publishing USA (2006)
12. Simonyan, K., Zisserman, A.: Very deep convolutional networks for large-scale image recognition. CoRR **abs/1409.1556** (2014). http://arxiv.org/abs/1409.1556
13. Su, J., Vargas, D.V., Sakurai, K.: One pixel attack for fooling deep neural networks. CoRR **abs/1710.08864** (2017)
14. Sung Kim, R.C.: Applications of Convolution in Image Processing with MATLAB. University of Washington (2013). https://www.semanticscholar.org/paper/Applications-of-Convolution-in-Image-Processing-Casper/391f4dc0567f671b0718f80834fdc1e83a9fd54b
15. Wang, Z., Bovik, A.C., Sheikh, H.R., Simoncelli, E.P.: Image quality assessment: from error visibility to structural similarity. IEEE Trans. Image Process. **13**(4), 600–612 (2004)

Guiding Representation Learning in Deep Generative Models with Policy Gradients

Luca Lach[1,3](✉), Timo Korthals[2], Francesco Ferro[3], Helge Ritter[1], and Malte Schilling[1]

[1] Neuroinformatics Group, Bielefeld University, Bielefeld, Germany
llach@techfak.uni-bielefeld.de
[2] Cognitronics and Sensor Systems Group, Bielefeld Univeristy, Bielefeld, Germany
tkorthals@techfak.uni-bielefeld.de
[3] PAL Robotics, Barcelona, Spain
luca.lach@pal-robotics.com

Abstract. Variational Auto Encoder (VAE) provide an efficient latent space representation of complex data distributions which is learned in an unsupervised fashion. Using such a representation as input to Reinforcement Learning (RL) approaches may reduce learning time, enable domain transfer or improve interpretability of the model. However, current state-of-the-art approaches that combine VAE with RL fail at learning good performing policies on certain RL domains. Typically, the VAE is pre-trained in isolation and may omit the embedding of task-relevant features due to insufficiencies of its loss. As a result, the RL approach can not successfully maximize the reward on these domains. Therefore, this paper investigates the issues of joint training approaches and explores incorporation of policy gradients from RL into the VAE's latent space to find a task-specific latent space representation. We show that using pre-trained representations can lead to policies being unable to learn any rewarding behaviour in these environments. Subsequently, we introduce two types of models which overcome this deficiency by using policy gradients to learn the representation. Thereby the models are able to embed features into its representation that are crucial for performance on the RL task but would not have been learned with previous methods.

1 Introduction

Reinforcement Learning (RL) gained much popularity in recent years by outperforming humans in games such as *Atari* ([1,2]), *Go* ([2,3]) and *Starcraft 2* [4]. These results were facilitated by combining novel machine learning techniques such as deep neural networks [5] with classical RL methods. The RL framework has shown to be quite flexible and has been applied successfully in many further

This work was supported by the European Union Horizon 2020 Marie Curie Actions under Grant 813713 NeuTouch.

B. Dorronsoro et al. (Eds.): OLA 2021, CCIS 1443, pp. 115–131, 2021.
https://doi.org/10.1007/978-3-030-85672-4_9

domains, for example, robotics [6], resource management [7] or physiologically accurate locomotion [8].

The goal of representation learning is to learn a suitable representation for a given application domain. Such a representation should contain useful information for a particular downstream task and capture the distribution of explanatory factors [9]. Typically, the choice of a downstream task influences the choice of method for representation learning. While Generative Adversarial Network (GAN) are frequently used for tasks that require high-fidelity reconstructions or generation of realistic new data, auto-encoder based methods have been more common in RL. Recently, many such approaches employed the Variational Auto Encoder (VAE) [10] framework which aims to learn a smooth representation of its domain. For a large number of RL environments, the usage of VAEs as a preprocesser improved sample efficiency and performance ([11,12]).

Many of the current methods combining RL with representation learning follow the same pattern, called unsupervised pre-training [13]. First, they build a dataset of states from the RL environment. Second, they train the VAE on this static dataset and lastly train the RL mode using the VAE's representation. While this procedure generates sufficiently good results for certain scenarios, there are some fundamental issues with this method. Such an approach assumes that it is possible to collect enough data and observe all task-relevant states in the environment without knowing how to act in it. As a consequence, when learning to act the agent will only have access to a representation that is optimized for the known and visited states. As soon as the agent becomes more competent, it might experience novel states that have not been visited before and for which there is no good representation (in the sense that the experienced states are out of the original learned distribution and the mapping is not appropriate).

Another issue arises from the manner the representation is learned. Usually, the VAE is trained in isolation, so it decides what features are learned based on its own objective function and not on what is helpful for the downstream task. Mostly, such a model is tuned for good reconstruction. Without the information from the RL model, such a representation does not reflect what is important for the downstream task. As a consequence, the VAE might omit learning features that are crucial for good performance on the task because they appear negligible with respect to reconstruction ([14], Chap. 15, Fig. 15.5). For example, small objects in pixel-space are ignored as they affect a reconstruction based loss only marginally. Thus, any downstream task using such a representation will have no access to information about such objects. A good example for such a task is Atari Breakout, a common RL benchmark. Figures 1a and 1b show an original Breakout frame and its reconstruction. While the original frame contains the ball in the lower right hand corner, this crucial feature is missing completely in the reconstruction.

We approach this issue through simultaneously learning representation and RL task, that is by combining the training of both models. As an advantage, this abolishes the need of collecting data before knowing the environment as it combines VAE and RL objectives. In consequence the VAE has an incentive

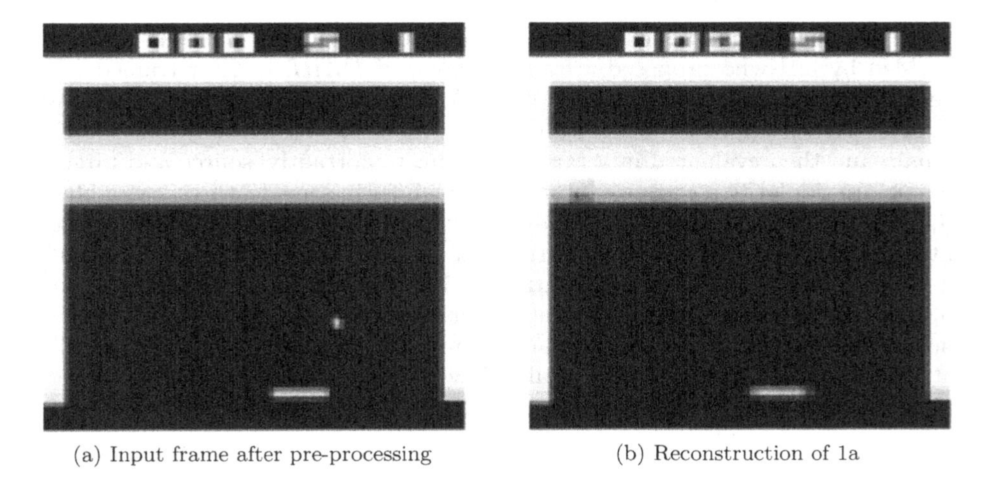

(a) Input frame after pre-processing (b) Reconstruction of 1a

Fig. 1. A frame from Atari Breakout. The original image 1a was passed through a pre-trained VAE yielding the reconstruction 1b. Note the missing ball in the lower right hand corner.

to represent features that are relevant to the RL model. The main contributions of this paper are as follows: First we show, that using unsupervised pre-training on environments that have underrepresented task-relevant features fails to produce good RL policies. Second, we show that by jointly training representation and policy leads to a model that encodes task-relevant information and thus enabling significantly higher performing policies. This will be shown by comparing achieved rewards and by an analysis of the trained model and its representation.

2 Related Work

[15] explored Auto Encoder (AE) ([16–18]) as a possible pre-processor for RL algorithms. The main focus in their work was finding good representations for high dimensional state spaces that enables policy learning. As input, rendered images from the commonly used grid world environment were used. The agent had to manoeuvre through a discretized map using one of four discrete movement actions per timestep. It received a positive reward once reaching the goal tile and negative rewards elsewhere. The AE bottleneck consisted only of two neurons, which corresponds to the dimensionality of the environemnt's state. Fitted Q-Iteration (FQI) [19] was used to estimate the Q-function, which the agent then acted ϵ-greedy upon. Besides RL, they also used the learned representation to classify the agents position given an encoding using a Multi-Layer Perceptron (MLP) [20]. For these experiments, they found that adapting the encoder using MLP gradients lead to an accuracy of 99.46 %. However, they did not apply this approach to their RL task.

A compelling example for separate training of meaningful representation is provided by [21] who proposed a framework called *DARLA*. They trained RL agents on the encoding of a β-VAE ([22,23]) with the goal of zero-shot domain transfer. In their approach, β-VAE and agent were trained separately on a source domain and then evaluated in a target domain. Importantly, source and target domain are similar to a certain extent and only differ in some features, e.g. a blue object in the source domain might be red in the target domain. During training of the β-VAE, the pixel-based reconstruction loss was replaced with a loss calculated in the latent space of a Denoising Auto Encoder (DAE) [24]. Thereby their approach avoids missing task relevant feature encodings at the cost of training another model. For one of their evaluation models, they allowed the RL gradients to adapt the encoder. Their results show that subsequent encoder learning improves performance of Deep Q-Learning (DQN) but decreases performance of Asynchronous Advantage Actor-Critic (A3C) [25].

[26] proposed a combination of VAE, Recurrent Neural Networks (RNN) [27] and a simple policy as a controller. They hypothesized that by learning a good representation of the environment and having the ability to predict future states, learning the policy itself becomes a trivial task. Like in most other models, the VAE was pre-trained on data collected by a random policy. Only the RNN and the controller were trained online. The compressed representation from the VAE was passed into a RNN in order to estimate a probability density for the subsequent state. The controller was deliberately chosen as a single linear layer and could thus be optimized with Covariance Matrix Adaptation - Evolution Strategy (CMA-ES) [28].

This work demonstrated how a VAE can provide a versatile representation that can be utilized in reinforcement learning. In addition, such an approach allows to predict the subsequent encoded state. While these findings encourage the usage of VAE in conjunction with RL, this is only possible in environments where the state space can be explored sufficiently by a random policy. However, if the policy can only discover important features after acquiring a minimal level of skill, sampling the state space using a random policy will not yield high-performing agents. Learning such features would only be possible if the VAE is continuously improved during policy training.

In the work of PlaNet [29], the authors also use a VAE to learn a latent state representation of a pixel input. Based on the learned representation, they use the Cross Entropy Method to learn various robotics control tasks. They refine this method in their subsequent publications Dreamer [30] and DreamerV2 [31] where the agent is trained purely on imagined trajectories from the VAE. Their works are similar to ours to the extent that they also continuously adapt the learned latent state representation. However their environments do not contain task relevant features that are underrepresented, hence their focus does not lie on training them.

Another interesting combination of VAEs and Reinforcement Learning (RL) was recently proposed by [32], with their so called Action-Conditional Variational Auto-Encoder (AC-VAE). Their motivation for creating this model was to train a transparent, interpretable policy network. Usually, the β-VAEs decoder is trained to reconstruct the input based on the representation the encoder produced. In this work though, the decoders objective was to predict the subsequent state s_{t+1}. As input it got the latent space vector $\boldsymbol{z}$ combined with an action-mapping-vector, which is the action vector a_t with a zero-padding to match the latent spaces dimensionality. Inspecting the decoder estimates for s_{t+1} when varying one dimension of the latent space showed, that each dimension encoded a possible subsequent state that is likely to be encountered if the corresponding action from this dimension was taken. Unfortunately, the authors did not report any rewards they achieved on Breakout, hence it was not possible for us to compare model performances.

3 Combination of Reinforcement and Representation Learning Objectives

In this section, we will first revisit the fundamentals of Reinforcement Learning (RL) and VAEs and discuss their different objective functions. Then, we propose a joint objective function that allows for joint training of both models using gradient descent based learning methods.

3.1 Reinforcement Learning with Policy Optimization

RL tries to optimize a Markov Decision Process (MDP) [33] that is given by the tuple $\langle \mathcal{S}, \mathcal{A}, r, p, \gamma \rangle$. $\mathcal{S}$ denotes the state space, $\mathcal{A}$ the action space and $p : \mathcal{S} \times \mathcal{R} \times \mathcal{S} \times \mathcal{A} \rightarrow [0,1]$ the environment's dynamics function that, provided a state-action pair, gives the state distribution for the successor state. $r : \mathcal{S} \times \mathcal{A} \rightarrow \mathcal{R}$ is the reward and $\gamma \in [0,1)$ the scalar discount factor. The policy $\pi_\theta(a|s)$ is a stochastic function that gives a probability distribution over actions for state s. θ denotes the policy's parameter vector which is typically subject to optimization. A trajectory $\tau = (s_0, a_0, ..., s_T, a_T)$ consisting of an alternating sequence of states and actions can be sampled in the environment, where T stands for the final timestep of the trajectory and $a_i \sim \pi_\theta(a_i|s_i)$.

The overarching goal of RL is to find a policy that maximizes the average collected reward over all trajectories. This can be expressed as the optimization problem $\max \mathbb{E}_{\tau \sim p(\tau)} \left[\sum_t r(s,a) \right]$, which can also be written in terms of an optimal policy parameter vector $\theta^* = \arg\max_\theta \mathbb{E}_{\tau \sim p(\tau)} \left[\sum_t r(s,a) \right]$. When trying to optimize the policy directly be searching for θ^*, policy optimization algorithms like Asynchronous Advantage Actor-Critic (A3C), Actor-Critic with Experience Replay (ACER) [34], Trust Region Policy Optimization (TRPO) [35] or Proximal Policy Optimization (PPO) [36] are commonly used. The fundamental idea behind policy optimization techniques is to calculate gradients of the RL objective with respect to the policy parameters:

$$\nabla_\theta J(\theta) = \mathop{\mathbb{E}}_{\tau \sim p(\tau)} \left[\nabla_\theta \log \pi_\theta(\tau)\, r(\tau) \right] \tag{1}$$

where we defined $\sum_{t=0}^{T} r(s, a) = r(\tau)$ for brevity. However, most policy optimization methods introduce heavy modifications to this vanilla gradient in order to achieve more stable policy updates. Throughout our work, we have used PPO as RL algorithm because it is quite sample efficient and usually produces stable policy updates. For an in-depth description of PPO, we refer to our A.1 or the original work [36].

3.2 Learning Representations Using Variational Auto-Encoders

[10] introduced the VAE as a method to perform Variational Inference (VI) [37] using function approximators, e.g. deep neural networks. VI tries to approximate a distribution over the generative factors of a dataset which would otherwise involve calculating an intractable integral. The authors present an algorithm that utilizes the auto encoder framework, an unsupervised learning method which learns data encodings by reconstructing its input. Therefore, the input is first compressed until it reaches a given size and is afterwards decompressed to its original size. When using deep neural networks, these transformations can be achieved by using for example fully connected or convolutional layers. In order for the VAE to approximate a distribution over generative factors, the authors used the so called "reparametrization trick". It allows for gradient based optimization methods to be used in searching for the distribution parameters. For training the VAE, a gradient based optimizer tries to minimize the following loss:

$$\begin{aligned} \mathcal{L}^{VAE}(\boldsymbol{x}, \phi, \psi) = -D_{KL}(q_\phi(\boldsymbol{z}|\boldsymbol{x}) \,||\, p(\boldsymbol{z})) + \mathop{\mathbb{E}}_{q_\phi(\boldsymbol{z}|\boldsymbol{x})} \left[\log p_\psi(\boldsymbol{x}|\boldsymbol{z}) \right] \\ \text{with } \boldsymbol{z} = l(\boldsymbol{\mu}, \boldsymbol{\sigma}, \boldsymbol{\epsilon}) \text{ and } \boldsymbol{\epsilon} \sim p(\boldsymbol{\epsilon}) \end{aligned} \tag{2}$$

where D_{KL} denotes the Kullback-Leibler Divergence (KL) [38] of the approximated distribution over generative factors produced by the encoder $q_\phi(\boldsymbol{z}|\boldsymbol{x})$ and some prior distribution $p(\boldsymbol{z})$. The expectation is often referred to as reconstruction loss that is typically calculated on a per-pixel basis. Lastly, $l(\boldsymbol{\mu}, \boldsymbol{\sigma}, \boldsymbol{\epsilon})$ is a sampling function that is differentiable w.r.t. the distribution parameters, for example $\boldsymbol{z} = \boldsymbol{u} + \boldsymbol{\sigma}\boldsymbol{\epsilon}$.

3.3 Joint Objective Function

Combining both loss functions such that both models can be trained at the same time is rather straight-forward. Adding both individual losses and using an optimizer such as ADAM [39] to minimize them is sufficient to achieve joint training. During backpropagation, gradients from the policy and the VAE are combined in the latent space. Due to different topologies of the networks, gradient magnitudes differ significantly. Therefore, we introduced the hyperparameter κ which can be used to either amplify or dampen the gradients and we arrive at the following loss:

$$\mathcal{L}^{\text{joint}} = \kappa \mathcal{L}^{\text{PG}}(\theta_k, \theta_{k-1}, \phi_k) + \mathcal{L}^{VAE}(\boldsymbol{x}, \phi, \psi, \beta) \tag{3}$$

where $\mathcal{L}^{\text{PG}}$ is some policy gradient algorithm's objective function. As mentioned before, we used PPO's loss $\mathcal{L}^{\text{PPO}}$ (Eq. 4 in the appendix).

4 Experiments

In order to test our model with the combined objective function given by Eq. 3, we have used the well-known benchmark of Atari Breakout. This environment has several properties that make it appealing to use: it is easily understandable by humans, used often as a RL task and the conventional pre-trained methods fail at mastering it. The ball is the most important feature that is required to be encoded in order to perform well, is heavily underrepresented (approximately 0.1% of the observation space). Therefore, the VAE's incentive to encode it is very low whereas our model succeeds in encoding it. In the following, we compare the pre-trained approach to two different continuously trained models that use the loss from Eq. 3.

4.1 Data Collection and Pre-Processing

The raw RGB image data produced by the environment has a dimensionality of $210 \times 160 \times 3$ pixels. We employ a similar pre-precessing as [1], but instead of cropping the grey-scaled frames, we simply resize them to 84×84 pixels. As we will first train models similar to those introduced in previous works with a pre-trained VAE, we needed to construct a dataset containing Breakout states. We used an already trained policy to collect a total of 25,000 frames, the approximate equivalent of 50 episodes.

4.2 Pre-training the Variational Auto-Encoder

Our first model is based on those of the previously introduced works which involve isolated pre-training the VAE on a static dataset. Figure 2 shows the individual parts of the complete training process. For the first model, $\text{PPO}^{\text{fixed}}$, the encoder and decoder (shown in orange and red) are pre-trained before policy training. During this phase, there is no influence from the RL loss. Once the VAE training is finished, the decoder shown in red in Fig. 2 is discarded completely. Later during policy training, we use n instances of the same encoder with shared weights that receive a sequence of the last n frames as input. Stacking allows us to incorporate temporal information and for the policy to predict the ball's trajectory. By sharing the weights, we ensure that the resulting encodings originate from the same function. $\mathbf{U}$ then represents the concatenated encodings of the sequence.

Prior to policy training, we trained the VAE on the dataset we have collected before, with hyperparameters from Table 1. Once pre-training was finished, we discarded the decoder weights and used the stacked encoder as input for the

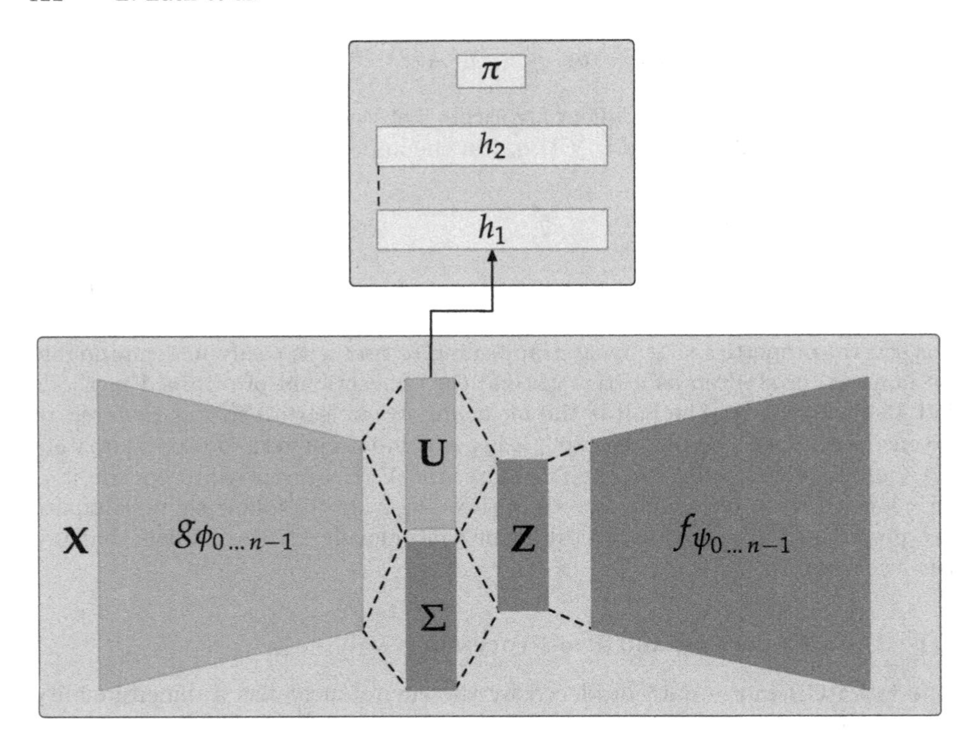

Fig. 2. Model combining PPO and a VAE. Depending on the model configuration, the colored parts are trained differently. $\mathbf{X}$ is the VAE's input and $\hat{\mathbf{X}}$ the reconstructions. PPO receives the mean vectors $\mathbf{U}$ as input and calculates a distribution over actions π. Note that we use capital letters in the VAE to emphasize that we pass n frames at the same time when a policy is trained.

policy Multi-Layer Perceptron (MLP). The MLP was then trained 10M steps with hyperparameters from Table 2. During this training, the encoder weights were not changed by gradient updates anymore but remained fixed.

The second model we introduce is called PPO^{adapt}, which has the same structure and hyperparameters as the first model. For this model, we also train the VAE in isolation first, however the encoder weights are not fixed anymore during policy training. Gradients from the RL objective are back propagated through the encoder, allowing it to learn throughout policy training. We hypothesize that features that are important for policy performance can be incorporated in an already learned representation.

Figure 3 compares the median rewards of three rollouts with different random seeds for all models. PPO^{fixed} was not once able to achieve a reward of 10 or higher, while PPO^{adapt} steadily improved its performance with final rewards well over 50. The learning curve of PPO^{adapt} shows that the model is able to learn how to act in the environment, whereas PPO^{fixed} does not. The non-zero rewards from PPO^{fixed} are similar to those of random agents in Breakout. From these results, we can assume that training the VAE in isolation on a static dataset for Breakout results in a deficient representation for RL. Therefore, using policy gradients to adapt an already learned representation can be beneficial in environments where the VAE fails to encode task-relevant features.

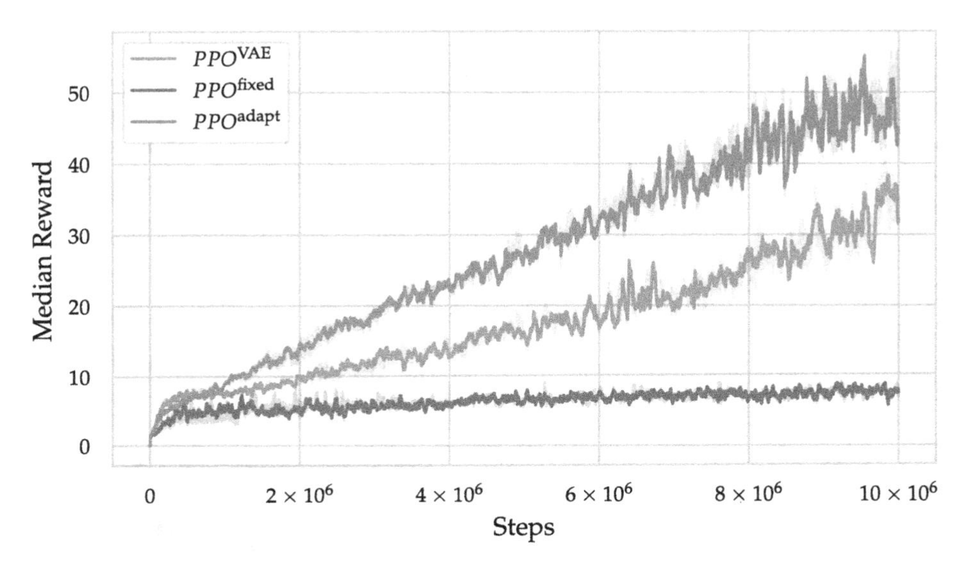

Fig. 3. Reward of the three proposed models across three random seeds each. $\text{PPO}^{\text{fixed}}$ is not able to achieve high rewards while the other two models consistently improve their performance.

4.3 Jointly Learning Representation and Policy

The last model we introduce, PPO^{VAE}, combines a complete VAE with a policy MLP that receives $\mathbf{U}$, the concatenated state encodings, as input. As opposed to the first two models, all weights are initialized randomly before policy training and the VAE is not pre-trained. For this procedure an already trained agent that gathers a dataset for the VAE beforehand is not necessary. The decoder is trained exactly as in the isolated setting, meaning its gradients are also only computed using the VAE's loss function. During backpropagation, the gradients coming from $\mathbf{Z}$ and h_1 are added together and passed through the encoder. This model has the same network configuration and hyperparameters as the first two, with the only difference that we also evaluated different values for κ from the joint loss 3 (see A.3). For the results reported here, we chose $\kappa = 20$. All hyperparameters can be found in Table 3.

By simultaneously training representation and policy, we expect the VAE to learn task-relevant features from the beginning of training. This assumption is supported by the learning curve shown in Fig. 3, which compares PPO^{VAE} to the previous two models. The curve shows a steady increase in reward over the course of training with PPO^{VAE} achieving slightly higher rewards than $\text{PPO}^{\text{adapt}}$ in the beginning. This characteristic changes after less than 1M steps and from that point on $\text{PPO}^{\text{adapt}}$ consistently outperforms PPO^{VAE}. This difference in performance is likely attributed to the fact, that in PPO^{VAE} the decoder is trained throughout the complete training. The gradients of $\text{PPO}^{\text{adapt}}$ can change the latent space without restrictions and they only optimize the RL objective. In PPO^{VAE} however, gradients are also produced by the decoder that presumably do not contain information about the ball. Therefore PPO^{VAE}'s latent space is constantly changed by two different objectives, thus leading to lower rewards for the RL part.

4.4 Analyzing the Value Function Gradients

So far, the results imply that PPO^{VAE} and $\text{PPO}^{\text{adapt}}$ do indeed learn encodings of the ball. One difficulty when analyzing the representation is, that the decoder still has no incentive to reconstruct the ball, even if it is present in the latent space. In a work that enhances Deep Q-Learning (DQN) algorithm [40], the authors visualized the Jacobian of the value function w.r.t. the input images. These visualizations showed which features or regions from the input space are considered as important in terms of future reward. As we also learn a value function, we did the same and visualized what our model considered important and what not.

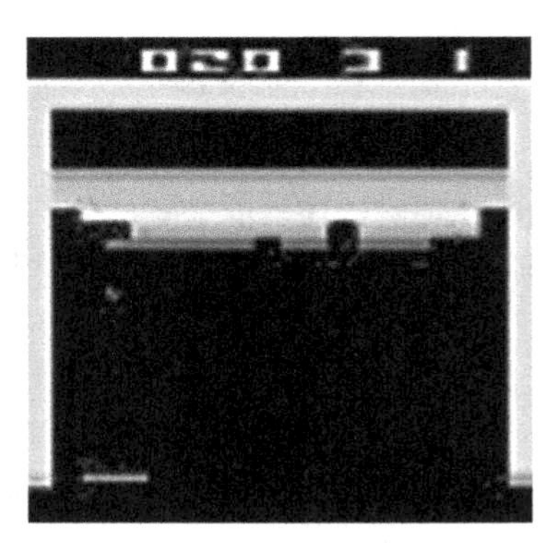

Fig. 4. The Jacobian of PPO's value function. Highlighted areas mean high importance in terms of future rewards. Note the high Jacobian values around the ball and the blocks.

In Fig. 4 we illustrate a pre-processed frame and added the values of the Jacobian to the blue channel if the were greater than the mean value of the Jacobian. Only visualizing above-mean Jacobian values removes some noise in the blue channel makes the images much easier to interpret and only highlights regions of high relevance. We can clearly see, that the Jacobian has high values at missing blocks as well as around the ball, meaning that these regions are considered to have high impact on future rewards. By visualizing the Jacobian we have confirmed that the policy gradients encourage the VAE to embed task-relevant features.

5 Conclusion

This paper focused on the issue of pre-training VAEs with the purpose of learning a policy for a downstream task based on the VAE's representation. In many environments, the VAE has little to no incentive to learn task-relevant features if they are small in observation space. Another issue arises if the observation of these features depends on policy performance and as a result, they are underrepresented in a dataset sampled by a random agent. In both cases, fixing encoder weights during policy training prevents the VAE to learn these important features and policy performance will be underwhelming.

We carried out experiments on the popular RL benchmark Atari Breakout. The goal was to analyze whether policy gradients guide representation learning towards incorporating performance-critic features that a VAE would not learn on a pre-recorded dataset. First experiments confirmed, that the common pre-trained approach did not yield well-performing policies in this environment. Allowing the policy gradients to adapt encoder weights in two different models showed significant improvements in terms of rewards. With policy gradients guiding the learned representation, agents consistently outperformed those that were trained on a fixed representation.

Out work verifies the fundamental issue with pre-trained representations and provides methods that overcome this issue. Nonetheless, future work can still explore a variety of improvements to our models. For once, training not only the encoder but also the decoder with RL gradients can improve interpretability of the VAE and enable it to be used as a generator again that also generates task-relevant features. Another direction is to impose further restrictions on the latent space during joint training of VAE and policy. The goal there would be to maintain the desired latent space characteristics of VAEs while still encoding task-relevant features.

A Appendix

A.1 Stable Policy Learning with Proximal Policy Optimization

Most actor-critic algorithms successfully reduce the variance of the policy gradient, however they show high variance in policy performance during learning and are at the same time very sample inefficient. Natural gradient ([41]) methods such as Trust Region Policy Optimization (TRPO) from [35] greatly increase sample efficiency and learning robustness. Unfortunately, they are relatively complicated to implement and are computationally expensive as the require some second order approximations. PPO ([36]) is a family of policy gradient methods that form pessimistic estimates of the policy performance. By clipping and therefore restricting the policy updates, PPO prohibits too large of a policy change as they have been found to be harmful to policy performance in practice. PPO is often combined with another type of advantage estimation ([42]) that produces high accuracy advantage function estimates.

We define the PPO-Clip objective is defined as

$$J^{\mathrm{PPO}}(\theta_k, \theta_{k-1}) = \mathbb{E}\left[\min\Big(o(\theta) A^{\pi_{\theta_k}}(s,a),\ \mathrm{clip}\Big(o(\theta), 1-\epsilon, 1+\epsilon\Big) A^{\pi_{\theta_k}}(s,a)\Big)\right]$$
$$\text{s.t. } \delta_{\mathrm{MB}} < \delta_{\mathrm{target}} \tag{4}$$

where $o(\theta) = \frac{\pi_{\theta_k}(a|s)}{\pi_{\theta_{k-1}}(a|s)}$ denotes the probability ratio of two policies.

This objective is motivated by the hard KL constraint that TRPO enforces on policy updates. Should a policy update result in a policy that deviates too much from its predecessor, TRPO performs a line search along the policy gradient direction that decreases the gradient magnitude. If the constraint is satisfied during the line search, the policy is updated using that smaller gradient step. Otherwise the update is rejected after a certain number of steps. This method requires to calculate the second order derivative of the KL divergence, which is computationally costly. PPO uses its clipping objective to implicitly constrain the deviation of consecutive policies. In some settings, PPO still suffers from diverging policy updates ([43]), so we included a hard KL constrained on policy updates. The constraint can be checked after each mini-batch update analytically and is therefore not very computationally demanding.

PPO extends the policy gradient objective function from [44]. With the probability ratio $o(\theta)$, we utilize importance sampling in order to use samples collected with any policy to update our current one. Thereby we can use samples more often than in other algorithms, making PPO more sample efficient. Using importance sampling, we still have a correct gradient estimate. Combining the new objective with actor-critic methods yields algorithm 1. K denotes the number of optimization epoch per set of trajectories and B denotes the mini-batch size. In the original paper, a combined objective function is also given with:

$$\begin{aligned} \mathcal{L}^{\text{PPO}}(\theta_k, \theta_{k-1}, \phi_k) = \mathbb{E}\left[c_1 J^{\text{PPO}}(\theta_k, \theta_{k-1}) - c_2 \mathcal{L}^{V^{\pi_\theta}}(\phi_k) + \mathcal{H}(\pi_{\theta_k})\right] \\ \text{s.t.} \;\; \delta_{\text{MB}} < \delta_{\text{target}} \end{aligned} \tag{5}$$

where $\mathcal{H}(\pi_{\theta_k})$ denotes the policy entropy. Encouraging the policy entropy not to decrease too much prohibits the policy from specializing on one action. As discussed in [43], there are two cases for $J^{\text{PPO}}(\theta_k, \theta)$: either the advantage function was positive or negative. In case the advantage is positive, it can be written as:

$$J^{\text{PPO}}(\theta_k, \theta) = \mathbb{E}\left[\min\left(o(\theta), (1+\epsilon)\right) A^{\pi_{\theta_k}}(s, a)\right] \tag{6}$$

$A^{\pi_{\theta_k}}(s, a) > 0$ indicates that the action yields higher reward than other actions in this state, hence we want its probability $\pi_{\theta_k}(a|s)$ to increase. This increase is clipped to $(1+\epsilon)$ once $\pi_{\theta_k}(a|s) > \pi_{\theta_{k-1}}(a|s)(1+\epsilon)$. Note however, that updates that would worsen policy performance are neither clipped nor bound. If the the advantage is negative, it can be expressed as:

$$J^{\text{PPO}}(\theta_k, \theta) = \mathbb{E}\left[\max\left(o(\theta), (1-\epsilon)\right) A^{\pi_{\theta_k}}(s, a)\right] \tag{7}$$

This equation behaves conversely to 6: $A^{\pi_{\theta_k}}(s, a) < 0$ indicates that we chose a suboptimal action, thus we want to decrease its probability. Once $\pi_{\theta_k}(a|s) < \pi_{\theta_{k-1}}(a|s)(1-\epsilon)$, the max bounds the magnitude by which the action's probability can be decreased.

Algorithm 1. Proximal Policy Optimisation with KL constraint

1: Initialize policy parameters θ_0 and value function parameters ϕ_0
2: **for** $k = 0, 1, 2, ...$ **do**
3: Collect set of trajectories $\mathcal{D}_k = \{\tau_i\}$ with π_{θ_k} and compute $\hat{R}_t$
4: $\delta_{\mathrm{MB}} \leftarrow 0$
5: **for** $0, 1, 2, ...K$ **do**
6: **for each** mini-batch of size B in $\{\tau_i\}$ **do**
7: Update the policy by maximizing the PPO-Clip objective 4
8: Minimize $\mathcal{L}^{V^{\pi_\theta}}$ on the mini-batch
9: **end for**
10: **end for**
11: **if** $\delta_{\mathrm{MB}} > \delta_{\mathrm{target}}$ **then**
12: $\theta_{k+1} = \theta_k$
13: **end if**
14: **end for**

A.2 Hyperparameter Tables

Table 1. Hyperparameter table for VAE training on Breakout

Parameter	Value
Epochs	100
Batch size	128
Input size	$(84, 84, 1)$
Optimizer	ADAM
Learning rate	1×10^{-4}
Encoder	Conv2D $32 \times 4 \times 4$ (stride 2) - $64 \times 4 \times 4$ (stride 2) - FC 512 (ReLU)
Latents	20 (linear)
Decoder	FC 512 (ReLU) - $64 \times 4 \times 4$ (stride 2) - $32 \times 4 \times 4$ (stride 2) Conv2D Transpose

Table 2. Policy hyperparameters of PPO^{fixed} and PPO^{adapt}

Parameter	Value
Timesteps	1×10^7
Environments	16
Batch size	32
$t_{\max}$	2048
K	10
c_1	1.0
c_2	0.5
c_3	0.0
γ	0.99
λ	0.95
Network	FC 64 (tanh) - FC 64 (tanh)
Optimizer	ADAM
Learning rate	3×10^{-4}

Table 3. Policy hyperparameter table of PPO^{VAE}

Parameter	Value
Timesteps	1×10^7
Environments	16
Batch size	32
$t_{\max}$	2048
K	10
c_1	1.0
c_2	0.5
c_3	0.0
γ	0.99
λ	0.95
Network	FC 64 (tanh) - FC 64 (tanh)
Optimizer	ADAM
Learning rate	3×10^{-4}
κ	(1, 10, 20)

A.3 Choosing Appropriate Values for κ

In Eq. 3, we introduced the hyperparameter κ to balance VAE and PPO gradients. We found empirically, that tuning κ is straight forward and requires only few trials. In order to simplify the search for κ, one can evaluate gradient magnitudes of the different losses at the point where they are merged at $\mathbf{U}$. Our experiments showed PPO's gradients to be significantly smaller, thus scaling up the loss function was appropriate. This will likely differ if the networks are configured differently. Increasing κ from 1 to 10 led to considerably higher rewards, however the difference in performance was small when increasing κ further to 20. Therefore, we chose $\kappa = 20$ in our reported model performances (Fig. 5).

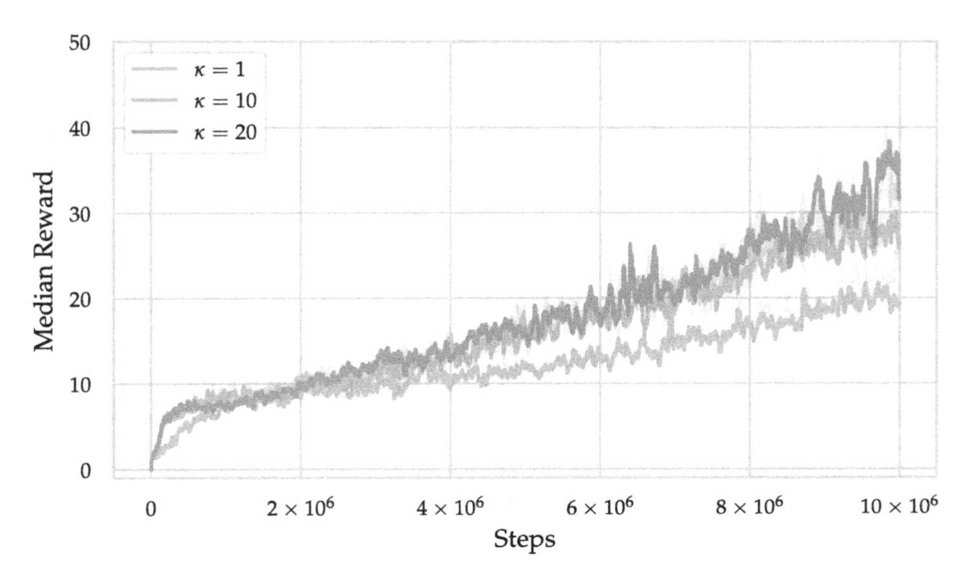

Fig. 5. Performance comparison of PPO^{VAE} with different values for κ

References

1. Mnih, V., et al.: Human-level control through deep reinforcement learning. Nature **518**(7540), 529 (2015)
2. Schrittwieser, J., et al.: Mastering atari, go, chess and shogi by planning with a learned model. arXiv preprint arXiv:1911.08265 (2019)
3. Silver, D., et al.: Mastering the game of go with deep neural networks and tree search. Nature **529**(7587), 484 (2016)
4. Vinyals, O., et al.: Starcraft ii: A new challenge for reinforcement learning. *arXiv preprint*arXiv:1708.04782 (2017)
5. LeCun, Y., Bengio, Y., Hinton, G.: Deep learning. Nature **521**(7553), 436 (2015)
6. Andrychowicz, O.M., et al.: Learning dexterous in-hand manipulation. Int. J. Robot. Res. **39**(1), 3–20 (2020)
7. Mao, H., Alizadeh, M., Menache, I., Kandula, S.: Resource management with deep reinforcement learning. In: Proceedings of the 15th ACM Workshop on Hot Topics in Networks, pp. 50–56. ACM (2016)
8. Kidziński, Ł., et al.: Scott: learning to run challenge solutions: adapting reinforcement learning methods for Neuromusculoskeletal environments. In: Escalera, Sergio, Weimer, Markus (eds.) The NIPS 2017 Competition: Building Intelligent Systems. TSSCML, pp. 121–153. Springer, Cham (2018). https://doi.org/10.1007/978-3-319-94042-7_7
9. Bengio, Y., Courville, A., Vincent, P.: Representation learning: a review and new perspectives. IEEE Trans. Pattern Anal. Mach. Intell. **35**(8), 1798–1828 (2013)
10. Kingma, D.P., Welling, M.: Auto-encoding variational bayes. arXiv preprint arXiv:1312.6114 (2013)
11. Achiam, J., Edwards, H., Amodei, D., Abbeel, P.: Variational option discovery algorithms. arXiv preprint arXiv:1807.10299 (2018)

12. Gregor, K., Rezende, D.J., Besse, F., Wu, Y., Merzic, H., van den Oord, A.: Shaping belief states with generative environment models for Rl. Adv. Neural Inf. Proc. Syst. **32**, 13475–13487 (2019)
13. Erhan, D., Courville, A., Bengio, Y., Vincent, P.: Why does unsupervised pre-training help deep learning? In: Proceedings of the Thirteenth International Conference on Artificial Intelligence and Statistics, pp. 201–208. JMLR Workshop and Conference Proceedings (2010)
14. Goodfellow, I., Bengio, Y., Courville, A.: Deep Learning. MIT press, Cambridge (2016)
15. Lange, S., Riedmiller, M.A.: Deep auto-encoder neural networks in reinforcement learning. In: IJCNN, pp. 1–8. IEEE (2010)
16. Lecun, Y.: Ph.D. Thesis: Modeles connexionnistes de l'apprentissage (Connectionist learning models). Universite P. et M. Curie (Paris 6), June 1987
17. Bourlard, H., Kamp, Y.: Auto-association by multilayer perceptrons and singular value decomposition. Biol. Cybern. **59**, 291–294 (1988). https://doi.org/10.1007/BF00332918
18. Hinton, G.E., Zemel, R.S.: Autoencoders, minimum description length and helmholtz free energy, pp. 3–10 (1994)
19. Ernst, D., Geurts, P., Wehenkel, L.: Tree-based batch mode reinforcement learning. J. Mach. Learn. Res. **6**, 503–556 (2005)
20. Rumelhart, D.E., Hinton, G.E., Williams, R.J.: Learning internal representations by error propagation. Technical report, California Univ San Diego La Jolla Inst for Cognitive Science (1985)
21. Higgins, I., et al.: Darla: improving zero-shot transfer in reinforcement learning. In: Proceedings of the 34th International Conference on Machine Learning-Volume 70, pp. 1480–1490. JMLR. org (2017)
22. Higgins, I., et al.: Early visual concept learning with unsupervised deep learning. arXiv preprint arXiv:1606.05579 (2016)
23. Higgins, I., et al.: Beta-vae: Learning basic visual concepts with a constrained variational framework. **3** (2017)
24. Vincent, P., Larochelle, H., Bengio, Y., Manzagol, P.-A.: Extracting and composing robust features with denoising autoencoders. In: Proceedings of the 25th international conference on Machine learning, pp. 1096–1103 (2008)
25. Mnih, V., et al.: Asynchronous methods for deep reinforcement learning, pp. 1928–1937 (2016)
26. Ha, D., Schmidhuber, J.: Recurrent world models facilitate policy evolution. In: Bengio, S., Wallach, H., Larochelle, H., Grauman, K., Cesa-Bianchi, N., Garnett, R. (eds) Advances in Neural Information Processing Systems, vol. 31, pp. 2450–2462. Curran Associates Inc (2018)
27. Hochreiter, S., Schmidhuber, J.: Long short-term memory. Neural Comput. **9**(8), 1735–1780 (1997)
28. Hansen, N.: The cma evolution strategy: a comparing review. In: Towards a New Evolutionary Computation, pp. 75–102. Springer (2006). https://doi.org/10.1007/3-540-32494-1_4
29. Hafner, D., et al.: Learning latent dynamics for planning from pixels. In: International Conference on Machine Learning, pp. 2555–2565. PMLR (2019)
30. Hafner, D., Lillicrap, T., Ba, J., Norouzi, M.: Dream to control: learning behaviors by latent imagination. arXiv preprint arXiv:1912.01603 (2019)
31. Hafner, D., Lillicrap, T., Norouzi, M., Ba, J.: Mastering atari with discrete world models. arXiv preprint arXiv:2010.02193 (2020)

32. Yang, J., Lee, G., Chang, S., Kwak, N.: Towards governing agent's efficacy: action-conditional β-vae for deep transparent reinforcement learning. volume 101 of Proceedings of Machine Learning Research, pp. 32–47. PMLR, Nagoya, Japan, 17–19 Nov 2019
33. Bellman, R.: A Markovian decision process. J. Math. Mech. 679-684 (1957)
34. Wang, Z., et al.: Sample efficient actor-critic with experience replay. arXiv preprint arXiv:1611.01224 (2016)
35. Schulman, J., Levine, S., Abbeel, P., Jordan, M., Moritz, P.: Trust region policy optimization. pp. 1889–1897 (2015)
36. Schulman, J., Wolski, F., Dhariwal, P., Radford, A., Klimov, O.: Proximal policy optimization algorithms. arXiv preprint arXiv:1707.06347 (2017)
37. Jordan, M.I., Ghahramani, Z., Jaakkola, T.S., Saul, L.K.: An introduction to variational methods for graphical models. Mach. Learn. **37**(2), 183–233 (1999)
38. Kullback, S., Leibler, R.A.: On information and sufficiency. Ann. Math. Stat. **22**(1), 79–86 (1951)
39. Kingma, D.P., Ba, J.: Adam: a method for stochastic optimization. arXiv preprint arXiv:1412.6980 (2014)
40. Wang, Z., Schaul, T., Hessel, M., Hasselt, H., Lanctot, M., Freitas, N.: Dueling network architectures for deep reinforcement learning. In: International conference on Machine Learning, pp. 1995–2003 (2016)
41. Amari, S.-I.: Natural gradient works efficiently in learning. Neural Comput. **10**(2), 251–276 (1998)
42. Schulman, J., Moritz, P., Levine, S., Jordan, M., Abbeel, P.: High-dimensional continuous control using generalized advantage estimation. arXiv preprint arXiv:1506.02438 (2015)
43. OpenAI Spinning Up - PPO. Spinning up explanation of ppo (2018)
44. Kakade, S.M.: A natural policy gradient, pp. 1531–1538 (2002)

Deep Reinforcement Learning for Dynamic Pricing of Perishable Products

Vibhati Burman(✉), Rajesh Kumar Vashishtha(✉), Rajan Kumar, and Sharadha Ramanan(✉)

TCS Research, Chennai, India
{vibhati.b,r.vashishtha,sharadha.ramanan}@tcs.com

Abstract. Dynamic pricing is a strategy for setting flexible prices for products based on existing market demand. In this paper, we address the problem of dynamic pricing of perishable products using DQN value function approximator. A model-free reinforcement learning approach is used to maximize revenue for a perishable item with fixed initial inventory and selling horizon. The demand is influenced by the price and freshness of the product. The conventional tabular Q-learning method involves storing the Q-values for each state-action pair in a lookup table. This approach is not suitable for control problems with large state spaces. Hence, we use function approximation approach to address the limitations of a tabular Q-learning method. Using DQN function approximator we generalize the unseen states from the seen states, which reduces the space requirements for storing value function for each state-action combination. We show that using DQN we can model the problem of pricing perishable products. Our results demonstrate that the DQN based dynamic pricing algorithm generates higher revenue when compared with conventional one-step price optimization and constant pricing strategy.

Keywords: Dynamic pricing · Deep reinforcement learning · Perishable items · Retail · Grocery · Fashion industry · Deep Q-network · Revenue management

1 Introduction

Dynamic pricing, also referred as revenue management, is a strategy to adjust the selling prices of products at the right time for maximizing revenue under changing circumstances. These changing circumstances are the factors which affect the demand and supply. Examples include amount of inventory available, age of the product, weather and customer preferences.

Perishable items are those likely to spoil after a fixed time period. The scope of perishable item spans various industries including grocery, pharmaceuticals, fashion and airlines. The consumer spend on perishables is increasing

Rajan was an employee of TCS when this work was done.

B. Dorronsoro et al. (Eds.): OLA 2021, CCIS 1443, pp. 132–143, 2021.
https://doi.org/10.1007/978-3-030-85672-4_10

and expected to further increase in the next few years. In particular, the demand for perishable food items such as vegetables, fruits, milk and eggs, are continuing to increase as consumers are progressively becoming health conscious. In case of fashion items, the demand is often short-lived due to several factors including the impact of social media. The retailers are under great pressure to price perishables optimally to offload inventory as well as maximize overall revenue. Hence, there is a need to develop a dynamic pricing policy.

Feng et al. [8] developed an economic order quantity (EOQ) inventory model for perishable items and hypothesize that for perishable products the demand is dependent on its price, freshness and stock level. They have considered the following characteristics for perishable food items:

1. Demand for a perishable product is dependent on its price, age and inventory level.
2. The age of the product not only reduces the stocks but also decreases the demand rate.
3. Product can not be sold after its expiration date.

Some work has been done for dynamic pricing through reinforcement learning (RL), but there still exist some challenges. In this paper, we identify and address three important challenges. First, the existing work on dynamic pricing of perishable products using RL employ tabular Q-learning approach. Pricing policy using tabular Q-learning approach cannot be generalized to previously unseen scenarios, i.e., we cannot estimate a pricing policy for a state unless it has been visited several times. Second, the exisiting literature on dynamic pricing through RL makes use of incremental method of learning a policy [22]. Incremental methods are not sample efficient and may lead to slower convergence of policy. Third, most of the existing work on dynamic pricing makes use of myopic approaches that try to optimize the immediate revenue. Harrison et al. [14] showed that myopic policies can lead to incorrect policies. Furthermore, Ravi Ganti et al. [12] have shown that far-sighted policies lead to increased profit in the long term.

In this paper, we have made the following important and original contributions in order to overcome the aforementioned challenges. First, we make use of a model-free DQN function approximator [18,26]. By making use of a function approximator we address the problem of pricing for unseen states. Also, using a function approximator drastically reduces the space requirement for obtaining the pricing policy. To the best of our knowledge, this is the first application of DQN for dynamic pricing of perishables. Second, reinforcement learning provides an alternative approach for optimizing the revenue over the entire selling horizon of the item. This leads to better proftability in the long run.

The remainder of this paper is organized as follows. Section 2 provides a review of related work. Section 3 gives some background on reinforcement learning. Section 4 discusses MDP formulation. We discuss the methodology in Sect. 5. Experimental results and findings are explained in Sect. 6.

2 Related Work

In this section, we review literature relevant to our work. Dynamic pricing has been receiving a lot of attention due to increase in competition in the market and advancement of AI based modeling. Gallego et al. [11] formulated dynamic pricing model for perishables having stochastic demand with an arrival rate as a function of price over finite horizon. Bitran et al. [3] extended this work by considering demand as a Poisson process with an arrival rate as a function of general purchasing patterns. Feng et al. [9,10] considered demand explicitly as a multivariate function of price, freshness and displayed stocks to obtained optimal price. Lu et al. [19] have maximized total profit to obtain the optimal joint dynamic pricing and replenishment policy for perishable items by applying pontryagins maximum principle. Duan et al. [7] proposed a dynamic pricing model for perishable food with quantity and quality deteriorating simultaneously. Their demand depends on the quality, the sales price and the reference price. They formulated an optimal control model to maximize the total profit and solved it by applying pontryagins maximum principle. Diaz et al. [1] studied the relation between dynamic price strategy and relevant factors such as price elasticity of demand, age-sensitivity of demand and age profile of initial inventory for perishables. They proposed a deterministic mathematical model that studied the influence of these factors on revenue and spoilage. Xiong et al. [27] studied the dynamic pricing problem of selling fixed stock of perishable items over a finite horizon, using fuzzy variables to model uncertain demand. They claim the effectiveness and robustness of their algorithm using a real world example. Robust optimization methods have been used to address dynamic pricing for perishable products [2,17,21]. These are one step optimization methods that result in myopic and static solutions, which focus only on maximizing immediate revenue rather than for a long term.

Dynamic pricing problem have also been addressed by using Reinforcement learning algorithms. Gosavi et al. [13] have used reinforcement learning as stochastic optimization for dynamically pricing the airline tickets. They formulated a semi-Markov Decision problem for their single leg problem over an infinite time horizon by involving some of the important factors affecting the pricing of tickets. Raju et al. [13] have developed dynamic pricing model for single seller and two seller market scenario. Customer segmentation is an important aspect for their model. For single seller market, pricing decision is taken using Q-learning algorithm while for 2 seller market, actor-critic algorithm is used to decide the optimal price. Cheng [5] integrated real-time demand learning with look up table based Q-learning algorithm to optimaly price the identical products by a deadline. Rana et al. [22] used Q-learning and Q-leaning with eligibility traces ($Q(\lambda)$) to establish a pricing policy for products having fixed inventory and fixed time horizon. They defined state as remaining inventory, action as set of price points and reward as revenue. Rana et al. [23] established a pricing policy for inter-dependent perishable items or services. Inter-dependent products are the ones whose demand and prices are affected by one another. $Q(\lambda)$ is used to obtain the optimal policy. The Markov Decision Process (MDP) formulation is done as: set of all possible amount of inventory available for each item is considered as state, set of price points are actions and total revenue gathered is considered as reward. Chen et al. [4] used Q-learning algorithm for dynamic pricing of perishables in a competitive multi-agent retailer market. However, they pointed out that their pricing strategy was not always optimal in every market. There are also several papers that uses multi-agent reinforcement learning to learn the optimal pricing strategy [15,16].

3 MDP Formulation for Dynamic Pricing of Perishables

We consider the dynamic pricing problem of a single perishable product, with a given initial inventory. The objective is to price the product dynamically so as to maximize the total expected revenue over a finite selling horizon.

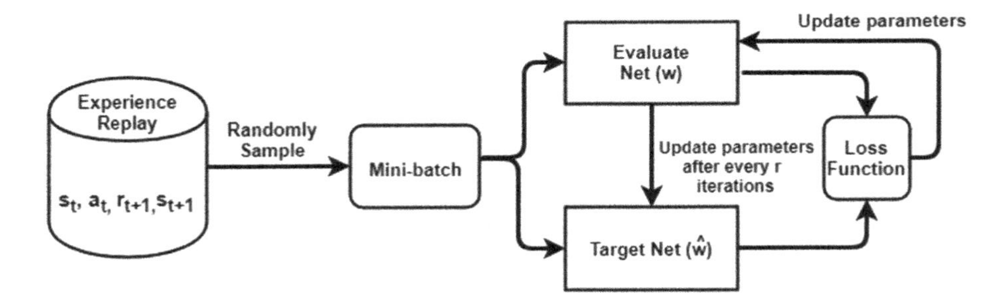

Fig. 1. DQN methodology

We formulate the dynamic pricing problem as a Markov Decision Process. Since we do not know the environment dynamics, we use a model-free reinforcement learning approach to solve the MDP. Specifically, we use DQN value function approximator to solve the MDP and develop an efficient pricing method.

A reinforcement learning task that satisfies the Markov property is called a Markov Decision Process, or MDP [24]. The task is a finite MDP if the state and action spaces are finite. A finite MDP is defined by set of states, set of actions and the one-step environment dynamics. In our MDP, the agent (decision maker) will choose a pricing action from the set A (action set) at each time step t. Since the environments response at time $t+1$ depends only on the state-action representation at time step t, the sequence of states s_t, with t = 1, 2, 3,..., m satisfies the Markov property.

We are dealing with a finite horizon problem that is treated as an episodic task in reinforcement learning. Each episode ends in a terminal state, which occurs at the end of the product lifetime. After termination of an episode, the state values are reset to the default initial value.

Selling horizon is the period of time during which the product is sold. The selling horizon is separated into m discrete decision times. Let t = 1, 2, 3,..., m

denote the index of decision time. A pricing decision is made at the beginning of every decision time.

The elements of the MDP are described below:

1) **State:** The state is formulated as a vector of length two. The state at any time step t, for a product, is given by s_t = (inventory left, time since product launch). Here, the time since the product launch is considered as the age of the product. The pricing action is selected based on the current state. The current state is determined by the amount of inventory remaining and the age of the product. This state formulation contains all information about the past agent-environment interaction and thus, follows the Markov property.
2) **Action:** The agent transitions between different states by performing actions. Actions represents the choices the agent makes based on the current state. Here, the actions space includes all the discrete prices the product can be assigned. The action space A is given by $A = \{a_1, a_2, .., a_n\}$ for a product, where n is the number of discrete actions the agent can execute. Selecting a pricing action affects the future state of the environment and thus requires foresightedness.
3) **Reward:** The reward obtained at time $t + 1$ is determined by the state s_t and action a_t at time step t. In our case, the reward is given as the revenue generated by taking an action a_t at time step t. The reward, $r_{t+1} = a_t * demand_t$, is the product of price and demand. Demand is the units sold. We aim to maximize the expected return i.e., cummulative sum of rewards.

4 Methodology

Although classical reinforcement learning algorithms such as Q-learning and SARSA have been applied successfully for various applications [20,25] in the past, these algorithms fail to scale up for real life problems involving large state spaces. A classical RL algorithm makes use of lookup table for storing the Q-values for every state-action combination it encounters. The size of this lookup table grows in proportion to the size of state space. Also, the information obtained from one state-action pair can not be propagated to other state-action pairs. So, rather than remembering the solutions, we are finding a general function to estimate the Q-value of a state-action pair. Neural networks, being a good function approximator, have been used here. Recently, a neural network based Deep Q-network or DQN technique proposed by Mnih et al. [26] was shown to successfully learn control policies from large state spaces. They applied DQN method to Atari games and demonstarted that it outperforms all previous approaches. We have used a DQN with similar structure as presented in [26] for dynamic pricing of perishables.

When reinforcement learning control algorithms are used with a non-linear function approximator, like a neural network, they are liable to instability. To improve the stability of the DQN, two neural networks of the same structure are used [26], evaluate network and target network. Target network is used for generating the targets y_j in the Q-learning updates. The target network's parameters,

$\hat{w}$, remain fixed for r steps after which they are updated to the latest values of the evaluate network parameters w. The evaluate network parameters, w, are updated at every time step.

Experience replay is another key feature of the DQN, introduced by Mnih et al. [26], that enhances its stability. The last N transitions of the form $(s_t, a_t, r_{t+1}, s_{t+1})$ are stored in memory. Here, s_t, a_t are the state, action at time step t, and r_{t+1}, s_{t+1} is the reward and state at time step $t+1$. From these N transitions, we randomly sample a mini-batch of D transitions. By sampling uniformly from a large memory, we can avoid the temporal correlations and make the data nearly i.i.d.

The methodology used for training our DQN based dynamic pricing algorithm is shown in Fig. 1. We use an architecture where the input to the neural network is the state representation. The output layer consist of a separate neuron for each discrete action. Each output neuron represents the predicted state action value function ($Q(s,a)$). The Q values for all the actions for a given state are calculated in one forward pass of the neural network. A linear activation is used for the output layer and rectified linear unit, ($ReLU$), is used as activation function for the hidden layers.

The loss function used for Q-learning update is shown in Eq. 1. Adam optimizer is used for updating the network weights.

$$L = \frac{1}{2n}\sum_{i=1}^{n}[\hat{Q}(s_t, a_t^i; \hat{w}) - Q(s_t, a_t^i; w)]^2 \tag{1}$$

where n is the total number of available control actions, Q is the action-value function of the evaluate network, $\hat{Q}$ is the action-value function of the target network.

Learning: The learning of DQN is outlined in Algorithm 1. The training of the DQN takes place offline. The algorithm starts by randomly initiaizing the parameters w of the evaluate network and the target network's parameters $\hat{w}$. We then initialize a zero filled replay memory D of size (N, (length of state vector * 2) + 2). The outer for loop controls the number of simulation episodes M. We reset the state to its initial value after reaching the end of an episode. Each episode lasts for a maximum of T time steps. Within each episode, we follow the following procedure. At each time step, an action a_t is selected at random with a probability ϵ. Otherwise, action is selected greedily. Next, the sale at a particular pricing action is given by the demand function. The reward r_{t+1} is the revenue generated by taking action a_t at state s_t. The inventory is now updated to obtain the next state s_{t+1}.The state transition tuple $(s_t, a_t, r_{t+1}, s_{t+1})$ is stored in the replay memory D. Then a mini batch is randomly sampled from this replay memory. The target vector is calculated as given in line 15 of Algorithm 1. The parameters w of evaluate neural network Q are updated by using the Adam optimizer. The parameters $\hat{w}$ of the target network remain fixed for r time steps after which they are updated to most recent value of evaluate network's parameters w.

Algorithm 1. DQN based pricing algorithm

1: Initialize replay memory D to size N
2: Initialize action-value function Q (evaluate network) with random weights w
3: Initialize target action-value function $\hat{Q}$ (target network) with random weights $\hat{w} = w$
4: **for** $i = 1$ to M **do**
5: Reset state to initial value ($s_{t=1}$ = (initial inventory, 1))
6: **for** $t = 1$ to T **do**
7: With probability ε select a random action a_t
8: Otherwise, select $a_t = argmax_a\, Q(s_t, a; w)$
9: Execute action a_t
10: sale = min (inventory, demand(a_t,t))
11: reward: r_{t+1} = sale*a_t
12: inventory=inventory-sale
13: store transition $(s_t, a_t, r_{t+1}, s_{t+1})$ in D.
14: randomly sample minibatch of transitions $(s_j, a_j, r_{j+1}, s_{j+1})$ from D
15:

$$Set\ y_j = \begin{cases} r_{j+1}, \text{if } \textit{episode terminates at step } j+1 \\ r_{j+1} + \gamma \max_{a'} \hat{Q}(s_{j+1}, a'; \hat{w}), \quad \text{otherwise} \end{cases}$$

16: Perform a gradient descent step on $[y_j - Q(s_j, a_j; w)]^2$ with respect to the evaluate net parameter w.
17: every r steps, reset $\hat{w} = w$
18: **end for**
19: **end for**

5 Experimental Results

In this section, numerical results are provided to evaluate the performance of our dynamic pricing method using DQN. We study the results of the DQN based pricing model by simulating the demand function.

Scenario 1: In practice, the demand for perishables is a function of various factors such as age of the product, its price, competitor price etc. Here, the demand function is considered to depend upon the price and freshness of the product. The demand of a perishable food item is, in general, observed to decrease as its age increases. Also, the price elasticity of perishable food items is mostly negative, i.e., as price increases the demand decreases. So, we have modelled demand as an exponentially decreasing function of time and price. Figure 2a shows the simulated demand function. We have introduced randomness in the demand function to account for dynamic market behaviour.

We consider a real-life instance of a grocery retailer who needs to sell a grocery product with a fixed initial inventory. The retailer has 580 units of a product with no replenishment and with shelf life of 40 days. According to [6] the prices of perishable food items are normally discounted by 20–50% during the last few days of the product's shelf-life. Therefore, we have selected the following discrete price points - 20, 25, 30, 35, 40. Also, here the prices can be changed daily.

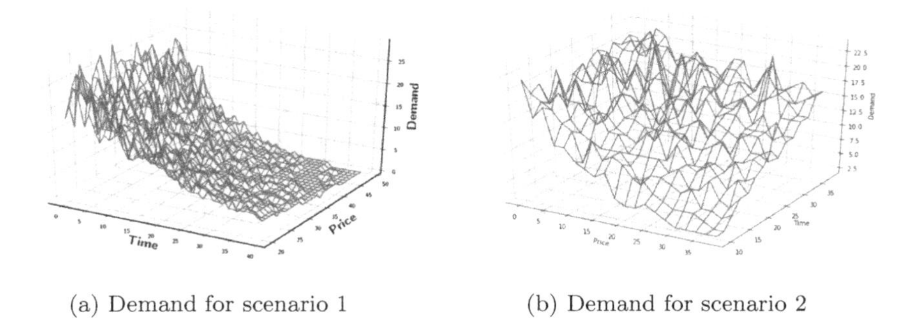

(a) Demand for scenario 1 (b) Demand for scenario 2

Fig. 2. Demand function.

We use DQN for learning the best pricing strategy for the simulated demand function. The time-steps are given by $t = 1, 2, 3, ..., 40$. The action set is the set of discrete prices, $A = \{ 20, 25, 30, 35, 40 \}$. Since the state is defined by (inventory left, age of product), we have 40×580 states in the state-space.

The DQN involves two similar networks-target network and evaluate network. Both these networks have the same architecture. The layout of the network and the hyper-parameters are listed in Table 1. We have trained the DQN for 200000 episodes with the simulated demand function. The DQN algorithm has been implemented using Python and Tensorflow on a MACOS Catalina system with 64-bit i5 processor @1.60 GHz and 8 GB DDR3 RAM. The average execution time for training the DQN on 50000 episodes is 60 min and for training the DQN on 200000 episodes is 3 h 50 min. GPU can be employed to significantly reduce this training time. After training the DQN, we evaluate the pricing policy given by it.

We compare the performance of the dynamic pricing algorithm using DQN with the performance of myopic (single-step) optimization algorithm. The single step price optimization model is formulated here as a non-linear function of sales and price elasticity that satisfies the price constraints and maximizes the revenue. We run this single-step optimization on the simulated demand function. The overall revenue is calculated as the sum of daily revenue. This is compared to the DQN results. Figure 3a shows the comparative plot of the cummulative revenue for RL algorithm and myopic optimization algorithm at different time steps. We can observe that the net revenue generated at the end of the product shelf-life is significantly higher for the RL based algorithm. We also observe that the cummulative revenue gap increases as the time increases. This shows the long term effectiveness of DQN in revenue maximization.

We also compare the performance of the DQN based pricing strategy with a fixed pricing strategy, which is a typical scenario in real world. In a fixed pricing strategy, the product price remains constant irrespective of the inventory left and the freshness of the product. Results for the net revenue generated at different conatant prices at the end of the shelf life, are tabulated in Table 2a. We observe

Table 1. Hyper-parameters of the DQN network

S. No.	Parameter name	Value
1	Number of input layer neurons	2
2	Number of output layer neurons	5
3	Number of hidden layer	3
4	Number of hidden layer neurons	35, 20, 10
5	Optimizer	Adam
6	Learning rate (α)	0.00001
7	Discount factor (γ)	.9
8	ϵ	0.5
9	Mini-batch size	64
10	Memory size	3000
11	r	200

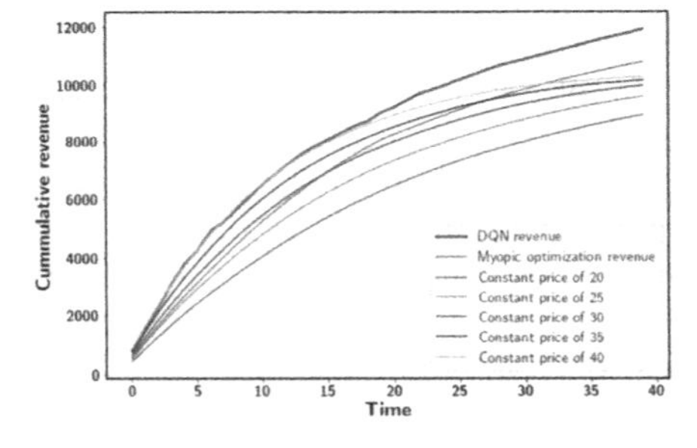

(a) Cummulative revenue comparision for different pricing strategies, for scenario 1

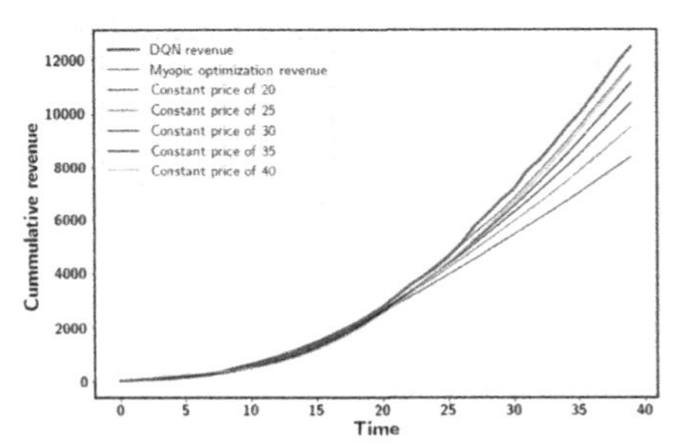

(b) Cummulative revenue comparision for different pricing strategies, for scenario 2

Fig. 3. Cummulative revenue comparision.

that the RL based pricing policy generates the maximum revenue at the end of the product life cycle. A 7.8% increase in revenue is observed by using DQN based pricing policy, when compared with one-step price optimization.

Figure 4a exbhits the pricing strategy suggested by the trained DQN model. We observe that the DQN pricing strategy suggests different prices for different time periods. The pricing policy gradually decreases with time as the demand decreases. This seems rational as a perishable product near its expiry cannot be sold at the maximum selling price. Moreover, according to Chung and Li [6], 88% of consumers will check expiry dates before buying a perishable product. This also implies that the products cannot be sold at maximum price near their expiry.

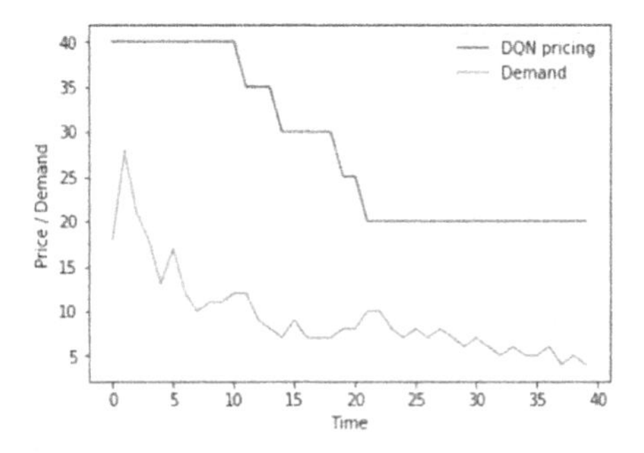

(a) Pricing policy suggested by DQN based dynamic pricing model, for scenario 1

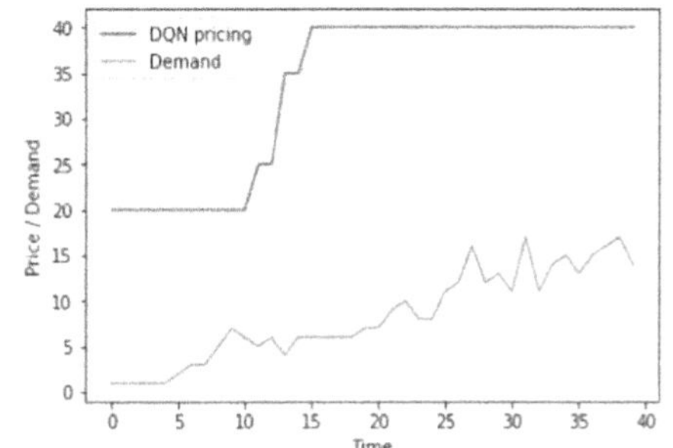

(b) Pricing policy suggested by DQN based dynamic pricing model, for scenario 2

Fig. 4. Pricing policy suggested by DQN.

Table 2. Revenue comparison

(a) Revenue comparison, for scenario 1

Pricing strategy	Total Revenue
DQN	11635
1 step optimization	10792
constant price of 40	10283
constant price of 35	10137
constant price of 30	9940
constant price of 25	9552
constant price of 20	8901

(b) Revenue comparison, for scenario 2

Pricing strategy	Total Revenue
DQN	12445
1 step optimization	11771
constant price of 40	11700
constant price of 35	11112
constant price of 30	10384
constant price of 25	9449
constant price of 20	8400

Scenario 2: We have also experimented with the DQN based dynamic pricing algorithm for the case where the demand increases with time and decreases with price. The simulated demand function is shown in Fig. 2b. This type of demand is frequently encountered in airline ticketing and hotel room booking. The number of seats available in a flight and the number of rooms available in a hotel can be considered analogous to inventory available for a perishable product.

Here, we consider a scenario where the initial inventory level is 650 and the distinct prices available are 20, 25, 30, 35, 40. Forty distinct time steps are considered and the pricing decision can be taken at each time step. The DQN architecture and hyper-parameter values are same as in scenario 1. The results for total revenue generated at the end of the selling horizon, for different pricing strategies are in Table 2b.

In this scenario also, we observe that the DQN based pricing strategy generates the highest revenue when compared with single-step and constant pricing strategy. This can also be observed in Fig. 3b. We observe a 5.7% increase in revenue by using DQN based pricing policy, when compared with one-step price optimization. The pricing policy given by DQN is shown in Fig. 4b. We see that as the demand increases with time, the prices also increase. This is typically observed in airline ticketing also, where the ticket prices increases as the date of journey approaches.

6 Conclusion

In the near future, consumer spend is only expected to increase exponentially. Social media impacts demand in continuous time and retailers need to respond with near real-time optimal pricing to persuade consumers to spend. Hence, dynamic pricing for perishables is a critical problem for retailers. In conclusion, this paper presents a deep reinforcement learning based approach to implement dynamic pricing for perishables. This approach is chiefly suitable for control problems with large state space. We have formulated the dynamic pricing problem as a Markov decision process and our results demonstrate that the DQN based dynamic pricing algorithm generates higher revenue when compared with constant pricing strategy and one-step price optimization. We are working to scale this approach to price multiple products for a real world use-case.

References

1. Adenso-Díaz, B., Lozano, S., Palacio, A.: Effects of dynamic pricing of perishable products on revenue and waste. Appl. Math. Model. **45**, 148–164 (2017)
2. Adida, E., Perakis, G.: A robust optimization approach to dynamic pricing and inventory control with no backorders. Math. Program. **107**(1–2), 97–129 (2006)
3. Bitran, G.R., Mondschein, S.V.: Periodic pricing of seasonal products in retailing. Manage. Sci. **43**(1), 64–79 (1997)
4. Chen, W., Liu, H., Xu, D.: Dynamic pricing strategies for perishable product in a competitive multi-agent retailers market. J. Artif. Soc. Soc. Simula. **21**(2) (2018)
5. Cheng, Y.: Real time demand learning-based q-learning approach for dynamic pricing in e-retailing setting. In: 2009 International Symposium on Information Engineering and Electronic Commerce, pp. 594–598. IEEE (2009)
6. Chung, J., Li, D.: A simulation of the impacts of dynamic price management for perishable foods on retailer performance in the presence of need-driven purchasing consumers. J. Oper. Res. Soc. **65**(8), 1177–1188 (2014)
7. Duan, Y., Liu, J.: Optimal dynamic pricing for perishable foods with quality and quantity deteriorating simultaneously under reference price effects. Int. J. Syst. Sci.: Oper. Logist. **6**(4), 346–355 (2019)
8. Feng, L., Chan, Y.-L., Cárdenas-Barrón, L.E.: Pricing and lot-sizing polices for perishable goods when the demand depends on selling price, displayed stocks, and expiration date. Int. J. Prod. Econ. **185**, 11–20 (2017)
9. Feng, Y., Xiao, B.: A continuous-time yield management model with multiple prices and reversible price changes. Manage. Sci. **46**(5), 644–657 (2000)
10. Feng, Y., Xiao, B.: Optimal policies of yield management with multiple predetermined prices. Oper. Res. **48**(2), 332–343 (2000)
11. Gallego, G., Van Ryzin, G.: Optimal dynamic pricing of inventories with stochastic demand over finite horizons. Manage. Sci. **40**(8), 999–1020 (1994)

12. Ganti, R., Sustik, M., Tran, Q., Seaman, B.: Thompson sampling for dynamic pricing. arXiv preprint arXiv:1802.03050 (2018)
13. Gosavii, A., Bandla, N., Das, T.K.: A reinforcement learning approach to a single leg airline revenue management problem with multiple fare classes and overbooking. IIE Trans. **34**(9), 729–742 (2002)
14. Michael Harrison, J., Bora Keskin, N., Zeevi, A.: Bayesian dynamic pricing policies: learning and earning under a binary prior distribution. Manage. Sci. **58**(3), 570–586 (2012)
15. Könönen, V.: Dynamic pricing based on asymmetric multiagent reinforcement learning. Int. J. Intell. Syst. **21**(1), 73–98 (2006)
16. Kutschinski, E., Uthmann, T., Polani, D.: Learning competitive pricing strategies by multi-agent reinforcement learning. J. Econ. Dyn. Control **27**(11–12), 2207–2218 (2003)
17. Lim, A.E.B., Shanthikumar, J.G.: Relative entropy, exponential utility, and robust dynamic pricing. Oper. Res. **55**(2), 198–214 (2007)
18. Lin, L.-J.: Self-improving reactive agents based on reinforcement learning, planning and teaching. Mach. Learn. **8**(3–4), 293–321 (1992)
19. Lihao, L., Zhang, J., Tang, W.: Optimal dynamic pricing and replenishment policy for perishable items with inventory-level-dependent demand. Int. J. Syst. Sci. **47**(6), 1480–1494 (2016)
20. Ng, A.Y., et al.: Autonomous inverted helicopter flight via reinforcement learning. In: Ang, M.H., Khatib, O. (eds.) Experimental Robotics IX. STAR, vol. 21, pp. 363–372. Springer, Heidelberg (2006). https://doi.org/10.1007/11552246_35
21. Perakis, G., Sood, A.: Competitive multi-period pricing for perishable products: a robust optimization approach. Math. Program. **107**(1–2), 295–335 (2006)
22. Rana, R., Oliveira, F.S.: Real-time dynamic pricing in a non-stationary environment using model-free reinforcement learning. Omega **47**, 116–126 (2014)
23. Rana, R., Oliveira, F.S.: Dynamic pricing policies for interdependent perishable products or services using reinforcement learning. Expert Syst. Appl. **42**(1), 426–436 (2015)
24. Sutton, R.S., Barto, A.G.: Reinforcement Learning: An Introduction (2018)
25. Tesauro, G.: Temporal difference learning and td-gammon. Commun. ACM **38**(3), 58–68 (1995)
26. Mnih, V., Kavukcuoglu, K., Silver, D., Rusu, A.A., Veness, J.: Human-level control through deep reinforcement learning. Nature **518**(7540), 529–533 (2015)
27. Xiong, Y., Li, G., Fernandes, K.J.: Dynamic pricing model and algorithm for perishable products with fuzzy demand. Appl. Stoch. Model. Bus. Ind. **26**(6), 758–774 (2010)

An Exploratory Analysis on a Disinformation Dataset

Matheus Marinho[1], Carmelo J. A. Bastos-Filho[1(✉)], and Anthony Lins[2]

[1] Universidade de Pernambuco, Recife, PE, Brazil
{mblm,carmelofilho}@ecomp.poli.br
[2] Universidade Catolica de Pernambuco, Recife, PE, Brazil
anthony.lins@unicap.br

Abstract. Understanding the phenomenon of disinformation and its spread through the internet has been an increasingly challenging task, but it is necessary since the effects of this type of content have their impacts in the most diverse areas and generate more and more impacts within society. Automated fact-checking systems have been proposed by applying supervised machine learning techniques to assist in filtering fake news. However, two challenges are still present, the first related to understanding disinformation in its subgroups. The second challenge is related to the availability of datasets containing news classified between true and false. This article proposes an exploratory analysis through unsupervised algorithms and the t-SNE technique to visualize data with high dimensionality, identify the subgroups present in the disinformation, and the identification of possible outsiders between the classes. We also propose a new Corpus in Portuguese containing 19446 news, classified as true and false, and 15 linguistic features extracted from this dataset. Finally, we propose to use two classification models using the Random Forest techniques, with and without intruders. In the end, the model without intruder achieved superior performance, reaching an accuracy of 97.33%.

Keywords: Disinformation · Unsupervised learning · Clustering

1 Introduction

Disinformation can change political opinions, influencing the results of elections, contributing to the spread of diseases, causing problems in public health, and even causing deaths with hate campaigns and generating extremist groups [13]. The growth of "Fake News" was stimulated by the change in society, which started to live more and more in a network [3], organized around interconnected universes of digital communication driven by the internet. The network society has generated a reduction in complexity and agility in the dispersion of content, thus allowing news, whether true or false, written by any individual, to reach an audience of hundreds of millions of readers [1].

One of the solutions that emerged in this fight against the disinformation ecosystem is fact-checking, a service proposed by journalists that seek to identify

B. Dorronsoro et al. (Eds.): OLA 2021, CCIS 1443, pp. 144–155, 2021.
https://doi.org/10.1007/978-3-030-85672-4_11

evidence, understand the context of the information and what can be inferred from that evidence through a fact verification process [14]. However, due to the complexity linked to this solution, the conclusion of the analyzed fact's veracity can take days [7]. This complexity creates a challenge, as human fact-checkers are unable to keep up with the amount of disinformation and the speed with which they spread, and in this way, the opportunity arises for the creation of automated fact-checking systems (AFC).

Typically, AFCs are developed to assist in the classification activity, where the news is labeled as true or false. In this type of activity, supervised machine learning techniques are used in most cases, with a model previously trained from a set of data containing the types of news evaluated. However, for the successful construction of these models used in the AFC, two things need to be taken into account: (i) identify the main characteristics that need to be taken into account when evaluating the news and the strengths of these features in the process [12] and identify the relationships between the variables input and output, as well as the relationships between features and; (ii) the use of a diverse dataset since the precision of the model generated by supervised machine learning is directly related to the quality of the data contained in the classifier training dataset.

This paper aims to make an exploratory analysis of Fake.br Corpus [9], a dataset related to disinformation, through the generation of clusters, to understand: how the features are distributed in each cluster; the main characteristics of these groups and; the relations of the features with the type of news (true/fake). In this paper, we also present a new set of disinformation data with news in Portuguese, increase the existing corpus of Fake.br, and perform analysis of possible noisy data, which may hinder the process of generating the classification model. In the end, we present the characteristics of these noisy data and compare the results with and without noisy data.

Although the data already have a classification, the present paper seeks to identify possible intrusive data existing in the data sets and understand the characteristics existing in the subgroups related to disinformation and true news.

The rest of the paper is organized as follows. Section 2 reviews the related work. Section 3 offers the background theory about Hierarchical Clustering and t-SNE. Section 4 describes the methodology. Section 5 shows the results. Finally, Sect. 6 presents conclusions and future work.

2 Related Work

In [11], the authors analyzed news features: n-grams, punctuation, psycholinguistics, legibility, and syntax. For each class, they proposed different classifiers using a linear support vector machine (SVM). The authors observed that depending on the dataset analyzed, a different category has the best performance. In this case, the legibility and punctuation classes had a better performance. The authors also considered only semantic characteristics and can conclude that actual news tends to have more function words, negations, and express relativity. On the other hand, the language used to report fake content uses more social and positive words and focuses on the present and future actions. Also, fake news authors use more adverbs, verbs, and punctuation characters than legitimate news authors.

Reis et al. [12] listed 11 categories for the evaluation of news and proposed 294,292 models, each with 20 features randomly selected from the existing classes. The work concluded that features related to the social media data class, such as the number of shares or reactions related to the news, the credibility of the domain, and characteristics that indicate political bias, are the most present in the models. In these cases, the models obtained better performance in the separation of trustworthy news and disinformation.

For the application of machine learning techniques, a fundamental requirement is a dataset capable of representing the problem in question. However, on the theme of Disinformation, it was only in 2018 that the authors [9] were the first to build a Corpus containing true and false news in Portuguese. Fake.br Corpus, contains 7200 news items, divided equally between true news and disinformation, collected between the years 2016 and 2018. Among the themes present in the data set, are politics and economics. The authors also made available a set containing 25 characteristics, 21 of which related to linguistic issues.

3 Background Theory

In this section, we present the relevant concepts to allow the reader to understand the proposal properly.

3.1 Hierarchical Clustering

The process of unsupervised learning in the generation of clusters consists of dividing the data into groups so that the data present in the same cluster are as similar as possible, and the difference between groups is as significant as possible [6]. Clustering techniques can be divided into two categories: hierarchical and non-hierarchical. Unsupervised techniques are distinguished, as they do not constitute a specific number of groupings; however, they generate groups through an increasing sequence of divisions or continuous group connections.

Hierarchical methods are composed of two classes of algorithms for generating clusters: agglomerative clustering and divisive clustering [5].

Agglomerative clustering is a conventional clustering method that can produce an informative hierarchical structure of clusters. The algorithm starts with a large number of small initial clusters. The agglomerative cluster iteratively merges a pair of clusters with the highest affinity under a given criterion until some stop condition is reached [4]. There are many conventional methods for calculating the affinity between a pair of clusters, such as single linkage, full linkage, medium linkage, and ward [2]. In this work, we use the ward link to reduce the variance between the merged clusters. We used agglomerative clustering in this work because we aim to analyze the purity of the formed clusters in different aggregation levels.

3.2 t-SNE

T-Distributed Stochastic Neighbor Embedding is a method for exploring high-dimensional data, proposed by Maaten and Hinton [8]. Visual exploration is an essential component of data analysis, as it allows the development of intuitions and hypotheses for the processes that generated the data. Stochastic neighbor embedding techniques calculate an N $\times$ N similarity matrix both in the original data space and in the small dimension embedding space so that similarities form a probability distribution over pairs of objects, these probabilities are usually provided by a kernel Normalized Gaussian or Student-t calculated from the input data [15].

T-SNE minimizes the divergence of the distribution that measures similarities between pairs of the input objects and a distribution that measures similarities between pairs of the corresponding low-dimension points in the embedding, through a function that calculates the distance between a pair of objects, typically uses the Euclidean distance. Another critical parameter is perplexity, which is used to compare probability models, the performance of the SNE is quite robust to changes in perplexity, and the typical values are between 5 and 50.

4 Methodology

In this section, we present the Fake.br Corpus, how the data preprocessing methods and the use of agglomerative clustering. After that, we discuss the dataset proposed in this paper, how it was built, its final configuration, and the existing features. Finally, the process of applying the t-SNE technique to identify regions of conflict between true and false news will be discussed, as well as how we apply the supervised technique to verify if there was a better convergence of the classifier after removing this news and applying the statistic tests in the validation process between the existing classifiers.

4.1 Fake.br Corpus

This paper aims to understand the construction of disinformation in Brazil. In this context, we found that there is a wide variety of datasets in the English language. However, it is not easy to obtain this type of information in the Portuguese language. The Fake.Br dataset [9] is the first with this purpose. The language of the data set is Portuguese and contains 7200 news collected from websites. The dataset is divided equally between fake and true news and grouped into six categories: politics; tv and celebrities; society and daily news; science and technology; economy and; religion.

The authors also made available a set with 25 features regarding each news. Among them, we can cite the number of words, the number of verbs, and the average sentence size. In the data set, the features' types are distributed as follows: 21 features are numeric with linguistic information, and four are categorical, which are related to the date of publication, author, news link, and category.

4.2 Pre-processing

After analyzing the distribution of features, we discarded the categorical features. Since we observed that the information in them is irrelevant and does not influence the model's response, for example, taking the author as a parameter could bias the proposed model. Thus, instead of having clusters that group true and false news, groups divided the authors or the categories of each news item. Because of this, we used the 21 numerical features related to linguistic characteristics.

Finally, we normalized the data using a Min-Max algorithm that outputs a value in the range between 0 and 1. It allows an easy way to compare measured values using different scales or measures.

4.3 Agglomerative Clustering

The clustering algorithm was applied using the scikit-learn toolkit [10]. We used the following configurations: the amount of cluster equal to 16; the affinity is the ward, and; the metric is the Euclidean distance. Then, we calculate each cluster's purity according to the majority of news belonging to a specific class, which, in turn, labels that cluster. For example, suppose that we have a cluster containing 100 news, 80 of which are true, and 20 are fake. Then, we label the cluster as true with a purity of 80%.

We propose an evaluation of the purity of each cluster to understand: (i) what are the characteristics of the purest, that is, that it contains only news of the same classification; and (ii) identify which subgroups exist within each class of classification, in order to understand the different behaviors that may exist within disinformation and true news.

4.4 Brazilian Disinformation Corpus

To increase diversity and update Fake.br Corpus, this article proposed an update with 12246 more news, collected in the following ways: (i) for a part of the true news, three web crawlers were proposed, to automate the extraction of content from the Globo, Sistema Jornal do Commercio and Diário de Pernambuco. Among the types of news, those related to politics and health were extracted, as they are the most relevant topics at the moment due to the state elections in November 2020 and the amount of disinformation generated in the health area related mainly to treatments and ways of propagation of COVID-19. These three communication vehicles were chosen because they are companies that do not publish disinformation. (ii) For collecting a part of the disinformation news, a web crawler was developed to extract content from Boatos.org, a service managed by journalists to compile the fake content that circulates on the internet. Finally, (iii) texts were extracted through Monitor do Whatsapp, a project that provides texts extracted from public groups within the communication platform, with themes related to politics. The classification of these texts was carried out by a team of journalists with experience in the fact verification process, and only news related to disinformation was used in this work, as it was the class with the least amount of examples. At the end of this collection process, the new data set configuration can be seen in Table 1, containing 12127 true news and 7319 disinformation.

Table 1. Brazilian disinformation corpus configuration

Source	Quantity
Fake.br Corpus - Fake	3600
Fake.br Corpus - True	3600
Globo	935
Sistema Jornal do Commercio	5661
Diário de Pernambuco	1931
Monitor do Whatsapp	887
Boatos.org	2832

Regarding the extracted characteristics, we initially attempted to replicate those existing in Fake.br Corpus, however of the 21 numerical characteristics, 6 were not capable of replication because they did not beat their value with those existing in the original data set, leaving the new set of data with 15 characteristics, they were: number of tokens; words without punctuation; the number of types; the number of uppercase; the number of verbs; the number of nouns; the number of adjectives; the number of adverbs; the number of pronouns; pausality; the number of characters; average sentence length; average word length; emotiveness; and diversity.

4.5 Outsiders Analysis

We proposed two analyzes using the technique for data exploration and visualization of high-dimensional data, t-SNE, at the Brazilian Disinformation Corpus. In the first evaluation, we used the Corpus classifications, making it possible to identify conflicting regions, with intruders both from the real news inside the fake news region, and the opposite. For the second evaluation, a new classification was proposed using an unsupervised technique, called k-means. This analysis aims to visualize the subgroups existing within each class and understand which groups are more reliable to their class: purer, and which groups contain that intrusive data. For this, different values were tested in the cluster quantity, seeking to find the quantity that best represented the problem, which in this case, is a high purity value in all groups, and this value in the Brazilian Disinformation Corpus was nine groups.

After identifying the intruders and an analysis of their characteristics, two supervised classifiers were proposed: one with all Corpus data and; another without the intruding elements. We aim to validate whether the removal of such news would increase the classification accuracy, as the models would not be trained with data that may be noisy. For this last stage, classifiers were generated using the Random Forest technique, as we have presented the latest work using Fake.br Corpus, this technique performed well in the disinformation classification process. For this evaluation, we used precision, recall, and F1. Besides, we used the t-student test to verify if there was a statistically significant improvement among the models.

5 Results

We carried out the initial analysis using the unsupervised technique at Fake.br Corpus, to identify and understand the subgroups present in the truthful news and the disinformation. We emphasize that it is not our intention to carry out a classification process, as the data are already previously classified. Analyzing the dendrogram (Fig. 1) generated by the hierarchical clustering algorithm, it is possible to observe that this technique was able to separate the two groups contained in the dataset (fake and true news). In the first level (see Table 2), we obtained a cluster with 93.82% purity with its data mostly belonging to the class of true news and a second cluster with 93.70% purity, this being composed mostly by false news. Thus, in general analysis, only 449 news items were grouped erroneously, and this value corresponds to only 6.24% of the total data. In contrast, 93.76% of the news items were correctly separated between the generated clusters.

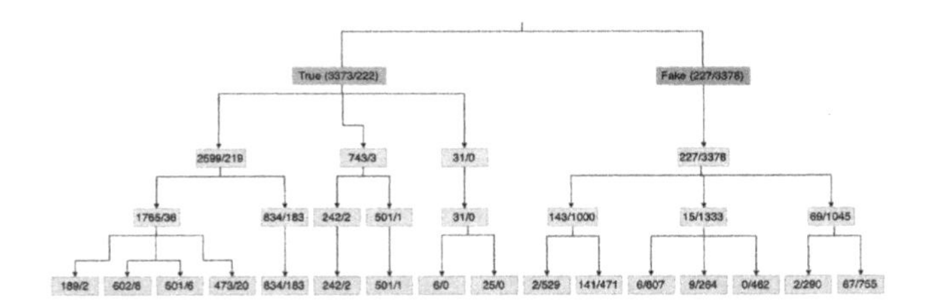

Fig. 1. Dendrogram

Thirty clusters were formed, of which 18 clusters have the most accurate news, and 12 clusters contain mostly fake news. This information reveals more considerable variability in the characteristics belonging to the data of this class within the true news. However, the true clusters ended up being repeated in 4 cases; that is, the lower level cluster is the same as the upper-level cluster. This repetition may be evidence that these clusters contain strong characteristics and therefore did not go through changes in the merges performed by the agglomerative technique.

In general, of the 30 clusters proposed (Fig. 1, 26 are more than 90% pure, of these 13 clusters are more than 99%, and five are 100% pure. Analyzing the perspective of classes: from the total of 26 clusters, 15 have the class of the true ones, and 11 are fake clusters; of the 13 clusters, nine are true, and only four are fake and; of the five clusters with 100% purity, four are true, and only one is fake. It reinforces the initial conclusion that the true clusters have a greater variety in the characteristics' values, so it has a larger number of clusters. However, this information is different, consequently allowing a good separation between the true news.

Analyzing clusters with 100% purity, these are the characteristics present in each cluster: (i) in the only fake cluster, the feature that stands out from the others is the number of links, this value is higher than in all others groups, that is, the 462 news items in this group have a large number of links in the text; (ii) the first of the four true clusters, consists of only six news items and these have a high value in 13 of the 21 characteristics analyzed; (iii) the second cluster has 25 true news items, these have the values of 14 characteristics above the average, with 13 characteristics with lower values than the previous cluster and with the characteristic of the quantity of first and second personal pronouns with the highest value between clusters and; (iv) the other two true clusters with 100% purity is generated from the (ii) and (iii) cluster, that is, they have the same characteristics previously presented.

Table 2. Purity from k = 2

Cluster	Class	True news	Fake news	Purity (%)
0	True	3373	222	93.82
1	Fake	227	3378	93.70

For the analysis of subgroups, we chose to analyze the dendrogram sheet containing 16 groups, as it contains the largest number of divisions, thus generating a more detailed (specific) level among the news within each macro class (fake and true). In the disinformation group, it is possible to observe that there are not many differences between the groups. The group with the most significant difference is the one that has a high value in the number of links, and the rest has a small variation in the characteristics of average word size, average sentence size, and diversity. This was already expected, as stated earlier, were more true than false groups were generated.

After the analysis the subgroups existing in the set of true news, we observed that they are initially divided by the number of tokens, the number of words without punctuation and the number of types. Another thing that distinguishes them is the number of pronouns and the number of characters. Finally, the higher the values in these characteristics mentioned above, the greater their purity related to the true groups.

The second analysis was related to applying the t-SNE technique in the dataset generated by this paper, the Brazilian Disinformation Corpus. In Fig. 2, in blue, we have the points related to true news and, in red, disinformation.

Fig. 2. t-SNE Brazilian disinformation corpus

Table 3. Outsiders characteristics

1	2	3	4	5	6	7	8	9	10	11	12	13	14	15	Class
0,46	0,49	0,49	0,25	0,46	0,42	0,38	0,26	0,33	0,27	0,46	0,35	0,28	0,20	0,60	True
0,57	0,60	0,60	0,33	0,58	0,53	0,51	0,43	0,51	0,29	0,57	0,38	0,29	0,24	0,51	Fake

Note that there is a separation zone between the groups; however, intruders are in both classes. To understand, the characteristics of these news items that do not belong to their zone, a cut was proposed in the reduced plan, where the region of each class of news was determined, and in this way, the news that did not belong to their zone was classified as outsiders. The characteristics of these news items can be seen in Table 3, where the class indicates which one it belongs to and therefore, if the class is true, it means that it is the set of real news that was in the region of fake news and vice versa and columns 1 until 15 represent the features.

Analyzing the table of intruders, it is possible to understand the reason why these news invaded regions that did not belong, because in the case of true intruders, they are news that has a high value of diversity, and as seen earlier, this is a characteristic of related news disinformation. On the other hand, the disinformation classified as an intruder presented high values in most of the features, which was mapped as a characteristic of the real news, for example, the features of number of tokens, number of types, and numbers of nouns. In total, 670 intruders were identified, grouped into 429 false and 241 true.

The third analysis was carried out by applying an unsupervised k-means technique to find the subgroups existing in the Brazilian Disinformation Dataset. In this case, the number of groups that best represented the set was 9, and from that identification, we applied the technique of dimensionality reduction t-SNE again. The result can be seen in Fig. 3. Groups 3, 5, and 8, respectively, dark green, light blue, and pink, are related to disinformation, and the rest related to true news. Again, out of 9 proposed groups, the class with the largest number of groups is true.

Fig. 3. t-SNE Brazilian disinformation corpus with k-means and k = 9

From the image, it is possible to notice that three groups are well separated from the others, they are: 2; 3 and; 7, and 3 groups have their data in a conflict zone, the groups: 1; 5 and; 6. Making a more detailed analysis in groups 3 and 7, which are the groups of the right and left extremities, and they have the following characteristics: group 3, which is true news, has a high value in the first three features (number tokens, word number without punctuation and number types); and group 7, in turn, related to disinformation has a high value in the diversity characteristic and a low value in the others.

Table 4. Supervised results Brazilian disinformation dataset

Dataset	Class	Precision	Recall	F1	Accuracy
Raw	Fake	93.72	92.72	93.07	94.85
	True	95.76	96.14	95.90	
Outsiders	Fake	96.42	96.37	96.35	97.33
	True	97.93	97.89	97.89	

The last analysis was done by applying the supervised machine learning technique to assess the impact of generating a classifier using the intrusive data. For this purpose, the Random Forest algorithm was used in the Brazilian Disinformation Dataset, and the results can be seen in Table 4. In all metrics, the removal of intruders increased the accuracy from 94.85% to 97.33%. In a particular way, in each class, intruders' absence contributed to an increase in precision and recall metrics. Analyzing the F1 measure of the true class, the best result reached 97.89%, and the disinformation class obtained 96.35%.

6 Conclusion

Society is increasingly connected through the internet, and news has been spreading in a massive and fast way. The absence of filters has allowed an increase in the circulation of fake news. Automated fact-checking systems have gained global attention to combat disinformation. It can be implemented by combining computational intelligence techniques and journalistic concepts. Thus, it is possible to create models capable of identifying probable false news and alerting the network readers.

This article had the following objectives. The first is constructing a new Corpus, entitled Brazilian Disinformation Corpus, containing 19446 news items and 15 language features. This new data set allowed for mapping more news related to disinformation. The second objective is to apply unsupervised techniques to identify existing subgroups in the categories of true and false news, reinforcing once again that we did not intend to carry out a classification activity but rather to have a deeper understanding of the Corpus. From the obtained results, we observed that there are more subgroups related to true news than related to disinformation, and it can also be concluded that within the pure subgroups, the features that stood out the most were a high value in the number of links in fake class and in the true news a high value in 13 out of 21 features proposed in [9].

The third objective is the identification of the existing subgroups in the Brazilian Disinformation Dataset. For this, we applied the k-means technique 9 clusters. Despite the difference in the datasets' size, the number of true clusters was greater than that of false ones and what differentiated the subgroups was also the different values in the features of the number of tokens, word number without punctuation, and number types. The true news had a high value, and the fake news groups had a low value, and in the diversity feature, this behavior was reversed. The fourth assessment was related to identifying intrusive news through a technique of visualization of data of high dimensionality. We identified the regions that characterized the real news and the fake news, and finally, we extracted the intruder news. In the end, 670 news items were classified as intruders, and their characteristics show behavior that diverged from the pattern found in each macro group (true and false). The true intruders had a high diversity value and fake content a high value of token quantity features and number of types.

Lastly, two classification models were proposed, generated from the Random Forest algorithm, for the disinformation and true news classification process. The first model contained all the news from the Brazilian Disinformation Corpus and the second model was generated without the outsiders elements identified through the analysis of the regions generated by the t-SNE technique. In the end, the model without the existence of outsiders obtained statistically better performance. Thus, we can conclude that removing the outsiders improved the classifier efficiency, reaching an accuracy of 97.33% and F1 measure of 96.35% for fake news and 97.89% for true news.

We plan to analyze entropy, purity, and statistical tests to discard the features for this work highlighted as irrelevant in the clustering process. We believe we can obtain even more concise results in the formation of supervised models. We will also perform a more detailed analysis of the intruder data to understand whether or not there is a possible miss classification of this data. Finally, all of this will contribute to constructing an automated fact verification system using computational intelligence techniques.

References

1. Allcott, H., Gentzkow, M.: Social media and fake news in the 2016 election. J. Econ. Perspect. **31**(2), 211–36 (2017)
2. Barreto, S., Ferreira, C., Paixao, J., Santos, B.S.: Using clustering analysis in a capacitated location-routing problem. Eur. J. Oper. Res. **179**(3), 968–977 (2007)
3. Castells, M.: A Galáxia Internet: reflexões sobre a Internet, negócios e a sociedade. Zahar (2003)
4. Cimiano, P., Hotho, A., Staab, S.: Comparing conceptual, divise and agglomerative clustering for learning taxonomies from text. In: Proceedings of the 16th European Conference on Artificial Intelligence, ECAI'2004, Including Prestigious Applicants of Intelligent Systems, PAIS 2004 (2004)
5. Fahad, A.: A survey of clustering algorithms for big data: taxonomy and empirical analysis. IEEE Trans. Emerg. Top. Comput. **2**(3), 267–279 (2014)
6. Frigui, H., Krishnapuram, R.: Clustering by competitive agglomeration. Pattern Recogn. **30**(7), 1109–1119 (1997)
7. Hassan, N., et al.: The quest to automate fact-checking. In: Proceedings of the 2015 Computation+ Journalism Symposium (2015)
8. Maaten, L.V.D., Hinton, G.: Visualizing data using t-SNE. J. Mach. Learn. Res. **9**(Nov), 2579–2605 (2008)
9. Monteiro, R.A., Santos, R.L.S., Pardo, T.A.S., de Almeida, T.A., Ruiz, E.E.S., Vale, O.A.: Contributions to the study of fake news in Portuguese: new corpus and automatic detection results. In: Villavicencio, A., et al. (eds.) PROPOR 2018. LNCS (LNAI), vol. 11122, pp. 324–334. Springer, Cham (2018). https://doi.org/10.1007/978-3-319-99722-3_33
10. Pedregosa, F., et al.: Scikit-learn: machine learning in Python. J. Mach. Learn. Res. **12**, 2825–2830 (2011)
11. Pérez-Rosas, V., Kleinberg, B., Lefevre, A., Mihalcea, R.: Automatic detection of fake news. arXiv preprint arXiv:1708.07104 (2017)
12. Reis, J.C., Correia, A., Murai, F., Veloso, A., Benevenuto, F., Cambria, E.: Supervised learning for fake news detection. IEEE Intell. Syst. **34**(2), 76–81 (2019)
13. Silva, R.M., Santos, R.L., Almeida, T.A., Pardo, T.A.: Towards automatically filtering fake news in Portuguese. Expert Syst. Appl. **146**, 113199 (2020)
14. Thorne, J., Vlachos, A.: Automated fact checking: task formulations, methods and future directions. arXiv preprint arXiv:1806.07687 (2018)
15. Van Der Maaten, L.: Accelerating t-SNE using tree-based algorithms. J. Mach. Learn. Res. **15**(1), 3221–3245 (2014)

Automatic Synthesis of Boolean Networks from Biological Knowledge and Data

Athénaïs Vaginay[1,2(✉)], Taha Boukhobza[1], and Malika Smaïl-Tabbone[2]

[1] Université de Lorraine, CNRS, CRAN, 54000 Nancy, France
[2] Université de Lorraine, CNRS, Inria, LORIA, 54000 Nancy, France
athenais.vaginay@loria.fr

Abstract. Boolean Networks (BNs) are a simple formalism used to study complex biological systems when the prediction of exact reaction times is not of interest. They play a key role to understand the dynamics of the studied systems and to predict their disruption in case of complex human diseases. BNs are generally built from experimental data and knowledge from the literature, either manually or with the aid of programs. The automatic synthesis of BNs is still a challenge for which several approaches have been proposed. In this paper, we propose `ASKeD-BN`, a new approach based on Answer-Set Programming to synthesise BNs constrained in their structure and dynamics. By applying our method on several well-known biological systems, we provide empirical evidence that our approach can construct BNs in line with the provided constraints. We compare our approach with three existing methods (`REVEAL`, `Best-Fit` and `caspo-TS`) and show that our approach synthesises a small number of BNs which are covering a good proportion of the dynamical constraints, and that the variance of this coverage is low.

Keywords: Boolean network synthesis · Answer-Set Programming

1 Introduction

Models of biological systems are important to understand the underlying processes in living organisms [10]. Once built, the model is an artefact that can be used to study a system through simulation. Several formalisms have been proposed to model biological systems [11], and they all have their own strengths and weaknesses. The choice of a formalism is guided by the question at hand: the best formalism is the most abstract formalism which can answer the question [3]. For example, differential equations are a formalism suited to run detailed dynamic simulations because they contain information on kinetic parameters. However, they do not scale to large systems.

Boolean Networks (BNs) are a formalism used to study complex biological systems where prediction of exact reaction times is not of interest [1]. They play a key role to understand the dynamics of biological systems and predict their

B. Dorronsoro et al. (Eds.): OLA 2021, CCIS 1443, pp. 156–170, 2021.
https://doi.org/10.1007/978-3-030-85672-4_12

disruption in case of complex human diseases [2]. The key notions of BNs are presented in Sect. 2.2. BNs are built from available knowledge about the structure of the system and data about the behaviour of its components (Sect. 2.3). The knowledge and data are used as constraints for the BN synthesis. The automatic synthesis of BNs from biological data and knowledge is still a challenge for which several methods have been developed. In Sect. 3, we review three state-of-the-art approaches: `REVEAL`, `Best-Fit` and `caspo-TS`.

In Sect. 4, we present `ASKeD-BN`, a new automatic approach for the synthesis of BNs constrained in their structure and dynamics. We rely on the Answer-Set Programming framework to generate non-redundant BNs fulfilling the given constraints. We compare the performances of our approach with `REVEAL`, `Best-Fit` and `caspo-TS` on several biological systems with experimental and synthetic data (Sect. 5). Finally, we discuss the results and conclude.

2 Boolean Networks and Their Synthesis

2.1 Prior Knowledge Network (PKN)

Part of the knowledge one has about a biological system is the list of components (genes, proteins...) constituting the system and how these components influence each other. Influences have a **polarity**: activation (polarity "+") or inhibition (polarity "−"). The **parents** of a component X are the components which influence X. A **Prior Knowledge Network** (PKN) encodes this knowledge. The nodes of the network are the components of the system. The edges are directed from parent components to child components and labelled "+" or "−" according to the polarity of the influences. Figure 1 shows an example PKN for a system of three components. In this PKN, C and A are the parents of C.

- "A activates C"
- "B interacts with itself"
- "C activates A"
- "C interacts with B"
- "C inhibits itself"

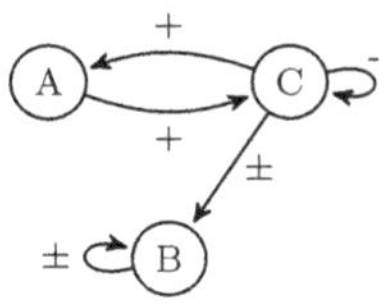

Fig. 1. PKN example of a three-components system.

2.2 Boolean Networks (BNs)

BNs were introduced by Kauffman [7] to model genetic regulatory networks. Concepts used in BNs are described in a recent review [17]. Two examples of BNs are given in Fig. 2.

The components of a BN are the components of the considered biological system. For example, a BN modelling a system of three proteins called A, B and C has three components named A, B and C. A **configuration** of a BN is a vector

$$\mathcal{B}_1 = \begin{cases} f_\mathrm{A} := \mathrm{C} \\ f_\mathrm{B} := \mathrm{B} \wedge \neg \mathrm{C} \\ f_\mathrm{C} := \neg \mathrm{C} \end{cases}$$

(a) Transition functions of $\mathcal{B}_1$

$$\mathcal{B}_2 = \begin{cases} f_\mathrm{A} := 0 \\ f_\mathrm{B} := (\mathrm{B} \wedge \neg \mathrm{C}) \vee (\neg \mathrm{B} \wedge \mathrm{C}) \\ f_\mathrm{C} := \mathrm{A} \end{cases}$$

(b) Transition functions of $\mathcal{B}_2$

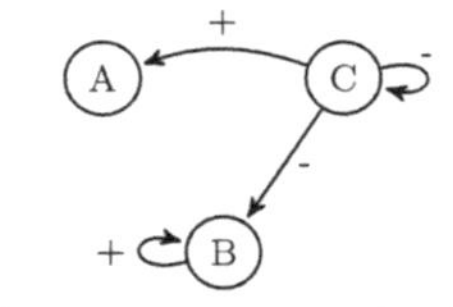

(c) Interaction graph of $\mathcal{B}_1$

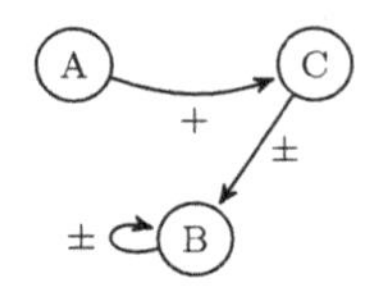

(d) Interaction graph of $\mathcal{B}_2$

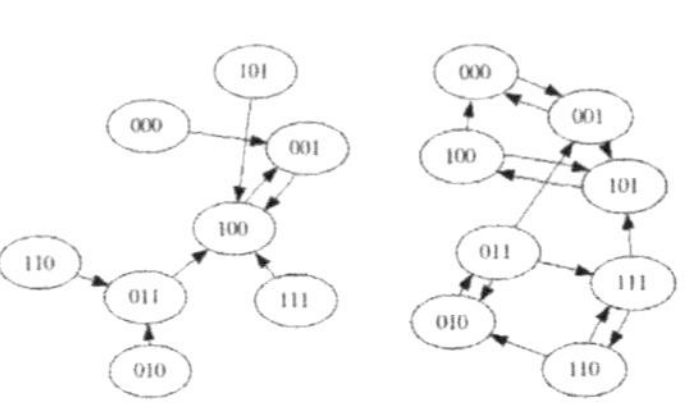

(e) Synchronous (left) and asynchronous (right) state transition graphs of $\mathcal{B}_1$

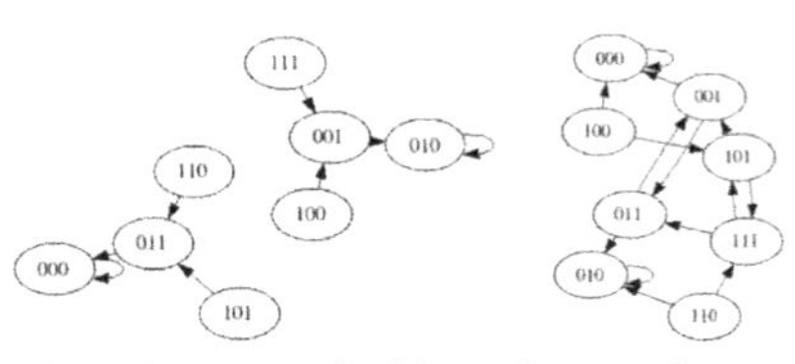

(f) Synchronous (left) and asynchronous (right) state transition graphs of $\mathcal{B}_2$

Fig. 2. The transition functions, derived interaction graph, and state transition graphs according to synchronous and asynchronous update schemes of two BNs.

which associates a Boolean value (1/active or 0/inactive) to each component of the BN. A BN with n components has 2^n possible configurations. For example, the $2^3 = 8$ possible configurations of a BN with 3 components are: 000, 001, 010, 011, 100, 101, 110 and 111.

Each component has an associated **transition function** ($\mathbb{B}^n \rightarrow \mathbb{B}$) which maps the configurations of the BN to the next value of the component. The transition functions are usually written as Boolean expressions. In this paper, these expressions are in Disjunctive Normal Form (DNF), *i.e.*, disjunctions of conjunctions. The conjunctions are *satisfiable*, which means they do not contain a literal and its contrary. The operators $\neg$, $\wedge$, $\vee$ represent respectively negation, conjunction and disjunction. Figures 2a and b show examples of transition functions. The transition function associated with B in $\mathcal{B}_2$ states that the value of B will be 1 if either the value of B or of C was 1 in the previous configuration.

Like for the PKN, the structure of a BN is defined in terms of parent-child relationships between the components. A component P which appears in the transition function of a component X is called a **parent** of X. If the parent is negated in the DNF, we say that the *polarity* of the influence of P on X is negative. Conversely, if the parent is not negated, the polarity is positive. The **Interaction Graph (IG)** summarises these relationships as a directed graph.

The directed edge P $\rightarrow$ X is labelled with "+" or "−" depending on the polarity of the influence P has on X. For example, the IG of $\mathcal{B}_1$ contains B $\xrightarrow{+}$ B and C $\xrightarrow{-}$ B because B appears positively and C appears negatively in the transition function associated with B. As we will see in Sect. 2.3, the PKN will act as a hard constraint on the IG of the BNs we want to synthesise.

The BN **dynamics** is obtained by applying iteratively the transition functions starting from each possible configuration. The order of application of the transition functions is defined by the **update scheme**. The **synchronous**, **asynchronous** and **mixed** update schemes are the most commonly used. In the synchronous update scheme, the transition functions are applied all at once, while in the asynchronous scheme, they are applied one by one. In the mixed update scheme, any number of components can be updated at each step. Thus, the update possibilities from both the synchronous and asynchronous update schemes are included in this third update scheme.

The **State Transition Graph (STG)** is a directed graph whose nodes are the 2^n possible configurations of the BN. In this graph, there is a directed edge from c to c' if c' is the result of applying to c the transition function(s) according to the chosen update scheme. Figure 2 shows examples of synchronous and asynchronous STGs. Later, we discuss how dynamical constraints are enforced in the STGs, and how we use the mixed STG to quantify how well the synthesised BNs match the dynamical constraints.

2.3 Synthesis of BNs from PKN and Multivariate TS

In general, BNs that model biological systems have to satisfy two categories of *constraints*. On one hand, the BNs have to comply with a PKN. The PKN constrains the structure of the synthesised BNs by defining which components can appear as variables in each transition function and the polarity of those variables. Hence, a component P is allowed to appear in the transition function of a component X with a polarity s if the PKN contains an edge P $\xrightarrow{s}$ X. Formally, a BN is compatible with a PKN if its IG is a *spanning subgraph* of the PKN. In other words, the IG of a BN compatible with a given PKN is formed from the vertices and a subset of the edges of the PKN. For example, the two BNs presented in Figs. 2a and b are compatible with the PKN given in Fig. 1. On the contrary, a BN containing the transition function $f_{\mathrm{A}} := \mathrm{B}$ is not, since the IG of this BN contains the edge B $\xrightarrow{+}$ A, which is not in the PKN. A BN having $f_{\mathrm{A}} := \neg\mathrm{C}$ is also incompatible: despite C being a possible parent of A, the negative polarity is not allowed, since the PKN does not contain the edge C $\xrightarrow{-}$ A.

On the other hand, the synthesised BNs are expected to reproduce as well as possible the sequence of configurations extracted from an observed continuous multivariate Time Series (TS) of the concentration of the components over time. An example of a multivariate TS is given in Table 1. Various strategies for extracting the sequence of configurations and fitting the transition functions to the observations have been proposed in the literature, but they all roughly result in enforcing the STG of the synthesised BNs to contain specific edges, corresponding to specific transitions of configuration.

We focus here on the *automatic synthesis* of BNs that respect the structure of a given PKN and are designed to reproduce as well as possible the observations from one given multivariate TS. For each synthesised BN, this ability of reproducing the observations is measured in terms of *coverage* proportion, *i.e.*, the proportion of transitions observed in the multivariate TS that are retrieved by the BN when computing its STG according to the mixed update scheme. Ideally, an identification method would only return BNs with a perfect coverage proportion (*i.e.*, 1).

Table 1. Multivariate TS of the three-components system given as example. The continuous concentrations of the components have been sampled for 20 time steps. Here, all the observations range from 0 to 100. The value resulting from the binarisation with a threshold of 50 is indicated by the colour of the cells: green if the result of the binarisation is 1 and red if 0. The resulting binary vectors are the configurations. Here there are four configurations (010, 011, 100 and 001) lasting respectively 4, 3, 3 and 10 time steps. Vertical bars indicate a change of configuration.

Configurations sequence:

	010				→ 011			→ 100			→ 001									
Time	1	2	3	4	5	6	7	8	9	10	11	12	13	14	15	16	17	18	19	20
A	0	3	7	13	20	30	49	61	100	63	36	25	2	3	1	1	3	0	0	0
B	100	86	64	57	54	53	51	49	45	37	33	28	22	19	14	12	9	5	2	0
C	0	27	36	42	60	75	54	44	38	48	60	72	88	90	100	100	100	100	100	100

3 State-of-the-Art Methods of BN Synthesis from PKN and TS

Several studies have been dedicated to the automatic synthesis of BNs from PKNs and observed multivariate TS. Here, we review three main state-of-the-art approaches: `REVEAL` [12], `Best-Fit` [9] and `caspo-TS` [16].

For each component of the system, `REVEAL` tests all the possible combinations of its parent nodes, and attempts to find the functions that explain *all* the observations of the binarised TS. For example with the multivariate TS from Table 1: `REVEAL` tries to explain 010 → 010 → 010 → 010 → 011 → 011 → 011 → 100 → ... Hence, it cannot handle *inconsistencies*—such as a configuration being associated to distinct successor configurations. Such inconsistencies are frequent when sampling concentrations along time, because the processes involved can have different speeds. In the example (Table 1), observing both 010 → 010 and 010 → 011 is an inconsistency which causes the failure of `REVEAL`. Furthermore, `REVEAL` cannot use the influence signs from the PKN, and since it uses an already binarised TS, it is possibly biased by the chosen binarisation.

Like REVEAL, Best-Fit tests every possible combination of the parent nodes of each component. It cannot use the influence signs and works on the binarised TS as well. Unlike REVEAL it can manage inconsistencies from the TS since it returns the functions that explain the maximal number of time steps. In Table 1, since 010 $\rightarrow$ 010 is observed three times and 010 $\rightarrow$ 011 only once, Best-Fit will focus on explaining the former instead of the latter.

caspo-TS was designed to manage several multivariate TS, corresponding to several experiments where the system is perturbed (forced activation or inhibition of some components), and where some measurements are potentially missing. Unlike REVEAL and Best-Fit, caspo-TS takes the influence signs into account, but it can only generate locally monotonous BNs, *i.e.*, BNs for which a parent of a component cannot be both its activator and its inhibitor. $\mathcal{B}_2$ is an example of a BN caspo-TS cannot generate because it is not locally monotonic. Indeed, in f_B, the components B and C act both as activator and inhibitor of B. caspo-TS works as the following: first, it derives the set of BNs that are compatible with the given PKN and an over-approximation of the dynamics of the TS, using the so-called most-permissive semantics [4]. Because of this over-approximation, the result can contain many false positive BNs, *i.e.*, BNs optimising the cost function used under the hood of caspo-TS, while their asynchronous dynamics is not able to reproduce the configurations sequence of the multivariate TS. These false positive BNs are subsequently ruled out using exact model checking. This filtering is PSPACE-hard, but thanks to the first step, a large set of BNs has already been excluded.

4 Our Approach: ASKeD-BN

4.1 Details of the Approach

We propose an approach for the Automatic Synthesis of Boolean Networks from Knowledge and Data (ASKeD-BN). It computes a non-redundant set of BNs complying with a given PKN and one observed multivariate TS. Unlike REVEAL and Best-Fit, ASKeD-BN is capable of using the influence signs provided in the given Prior Knowledge Network (PKN) and the raw values of the input multivariate Time-Series (TS). Unlike caspo-TS, ASKeD-BN directly fits the behaviour of each component with the TS. Also, it is not limited to the synthesis of locally-monotonous BNs.

For each component of the studied system, our approach searches among all possible transition functions. All the transition functions that do not respect the given PKN are ruled out. Then, every remaining candidate is evaluated on the basis of both their simplicity and their ability to reproduce the given observations. The candidate transition functions for the component X might not be able to explain all the binary state transitions happening at time $t \rightarrow t'$. The set of unexplained t' is denoted $\mathcal{U}$. Every time step t' in $\mathcal{U}$ is associated with a measure stating "how far" the continuous value X'_{t} is from the binarisation threshold θ: $|\theta - \mathrm{X}'_{\mathrm{t}}|$. These spotted errors are then averaged on the T time steps of the TS through the Mean Absolute Error (MAE):

$$\text{MAE}_X = \frac{\sum_{t' \in \mathcal{U}} |\theta - X_{t'}|}{T}$$

Among the candidates having the smallest MAE, we select the ones that has the smallest number of influences. Finally, we create all the possible BNs by generating all the combinations of the selected functions.

We implemented our approach using Python and the Answer-Set Programming framework (ASP) with the system *clingo* [6]. ASP is a declarative programming language oriented towards difficult (NP-hard) search problems. The possible solutions of a problem are described with the constraints they must fulfill. These constraints are written as a logic program. The ASP solver is tasked with finding the solutions of the program. To do so, it uses a Conflict-Driven Clause Learning (CDCL) algorithm inspired by SAT solvers. In our case, the CDCL algorithm avoids the evaluation of all the possible transition functions by learning from conflicts: whenever it finds that a candidate is in conflict with the constraints, it creates a new constraint that explains the conflict. These learned constraints subsequently eliminate other conflicting candidates, pruning the search space. Thanks to these pruning heuristics, our approach is efficient. ASP and in particular *clingo*, have already been used in similar contexts including `caspo-TS`.

4.2 Illustration on the Toy Example

Let us illustrate our approach on the toy example consisting of the PKN in Fig. 1 and the multivariate TS in Table 1.

When no PKN is available, the default PKN is a complete graph assuming that each component can inhibit/activate all the others (including itself). In this setup, a component with n parents have 2^{2^n} possible transition functions. In the toy example, each component can be explained by $2^{2^3} = 256$ distinct functions, which correspond to 16 777 216 potential BNs (formed by all the possible combinations of all the candidates of each component). Thanks to the available PKN, the number of candidate functions for each components A, B and C falls respectively to 3, 16 and 6. Besides the CDCL pruning, `ASKeD-BN` virtually evaluates all the candidates, but for illustration purpose we will focus on the two that are present in $\mathcal{B}_1$ and $\mathcal{B}_2$ (Figs. 2a and b).

For the component A, the candidate $f_\text{A} := 0$ does not contain any literal and it cannot explain the transition of configuration for A at $t_7 \rightarrow t_8$. Hence, the set $\mathcal{U}$ of unexplained time steps is $\{t_8\}$. The concentration of A at time t_8 is 61, and the candidate's MAE is thus $|50-61|/20 = 0.55$. The candidate $f_\text{A} := \text{C}$ involves one literal (which is C). This candidate can explain all transitions. Hence, $\mathcal{U} = \emptyset$ and the MAE associated with this candidate is 0. Despite requiring more literals, $f_\text{A} := \text{C}$ is a better candidate than $f_\text{A} := 0$ because its MAE is smaller. The comparisons of the candidates proposed for the components B and C in $\mathcal{B}_1$ and $\mathcal{B}_2$ are summarised in Table 2.

For the toy example, our approach returns $\mathcal{B}_1$ as the only solution. It retrieves the 3 configuration transitions extracted from the binarised TS, thus its coverage proportion is 1. REVEAL does not find any BN, and the BN returned by Best-Fit does not comply with the PKN. caspo-TS finds 5 BNs with coverage proportions ranging from 0.33 to 1 (standard deviation of 0.25).

Table 2. Number of influences and MAE for the candidate functions in $\mathcal{B}_1$ (Fig. 2a) and $\mathcal{B}_2$ (Fig. 2b). A checkmark indicates the candidate selected by our approach, and the best for each criterion: (1) minimal MAE and (2) minimal number of influences.

candidate	$f_B := B \wedge \neg C$ ✓		$f_B := (B \wedge \neg C) \vee (\neg B \wedge C)$		$f_C := \neg C$ ✓		$f_C := A$	
MAE ($\mathcal{U}$)	0 ($\emptyset$)	✓	0 ($\emptyset$)	✓	0 ($\emptyset$)	✓	0.5 ($\{t_5\}$)	
# influences	2	✓	4		1	✓	1	✓

5 Datasets and Procedure for the Comparative Evaluation

5.1 Datasets

In order to compare our approach with REVEAL, Best-Fit and caspo-TS, we used eight biological systems. For two of these systems (*yeast*'s cell cycle and *A. thaliana*'s circadian clock), their PKN and experimental multivariate TS are taken from [13] and [18] respectively. These two systems are summarised in Table 3. They respectively involve 4 and 5 components.

Table 3. Summary of two biological systems and their corresponding datasets

System	Genes	PKN	TS	Source
yeast (cell cycle)	Fkh2, Swi5, Sic1 & Clb1	Sic1 does not influence itself nor Fkh2	14 time steps 6 transitions	[18]
A. thaliana (circadian clock)	LHY, PRR7, TOC1, X & Y	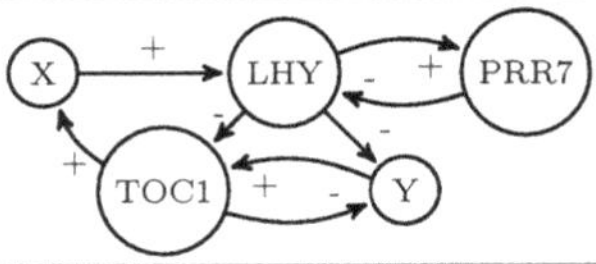 X, LHY, PRR7, TOC1, Y (+, -)	50 time steps 11 transitions	[13]

For the six other systems[1], we conducted our experiments on multivariate TS that we simulated from existing BNs taken from the repository of example BNs of the package PyBoolNet [8]. For these systems, the number of components ranges from 3 to 10. For each system, the used PKN is the IG of the associated BN. As for the generation of the multivariate TS, three parameters are

[1] raf, randomnet_n7k3, xiao_wnt5a, arellano_rootstem, davidich_yeast and faure_cellcycle.

taken into consideration: the update scheme (in {synchronous, asynchronous}), the maximum number of introduced repetitions of each configuration (in $\{1, 4\}$) and the standard deviation of the added noise (in $\{0, 0.1\}$). For each setting of these parameters, we follow a procedure similar to what is implemented in the `generateTimeSeries` function of the R package `BoolNet` [15]:

1. choose randomly a configuration of the considered BN,
2. on this configuration, apply the update function(s) 20 times w.r.t the chosen update scheme,
3. duplicate randomly each configuration in the obtained sequence (added in contrast to `generateTimeSeries`),
4. add a Gaussian noise with a standard deviation of N.

For a given setting of the 3 parameters and a given system, we run the procedure 7 times (with different random seeds). In the following, we denote *ARN* the setting with the Asynchronous update scheme, Repetitions (of 4) and Noise (of 0.1). We believe that this setting allows us to obtain multivariate TS which are quite close to real TS.

We illustrate here how to generate a synthetic multivariate TS in the ARN setting for $\mathcal{B}_1$ (Fig. 2a). We would start from a random configuration. Let it be 010. Then we apply 20 times the transition functions of $\mathcal{B}_1$ with the asynchronous update scheme. This process is not deterministic as any path from Fig. 2e (right) starting from 010 and of length 20 is valid. Let's say we obtain a path starting with 010 $\rightarrow$ 011 $\rightarrow$ 010 $\rightarrow$ 011 $\rightarrow$ 111 $\rightarrow$ 101 $\rightarrow$ ... Then we add a random number of duplications (in bold). The beginning of the sequence could for example look like 010 $\rightarrow$ 011 **$\rightarrow$ 011$\rightarrow$** 010 $\rightarrow$ 011 **$\rightarrow$ 011$\rightarrow$ 011$\rightarrow$** 111 $\rightarrow$ 101 **$\rightarrow$ 101$\rightarrow$ 101$\rightarrow$ 101$\rightarrow$** ... Finally, we add a random Gaussian noise with a standard deviation of 0.1. The synthetic multivariate TS could now start with $(0.02; 0.92; -0.16) \rightarrow (0.04; 0.8; 0.7) \rightarrow (-0.05; 1.06; 0.7) \rightarrow \dots$

5.2 Details on the Evaluation Procedure

For `REVEAL` and `Best-Fit` we use the implementation from the R package `BoolNet` [15]. `caspo-TS` is ran with the option `mincard`, that asks for BNs with functions minimising the number of influences. Note that this is also what our method optimises.

In the following, we define an *experiment* as a BN identification method applied on a system with *one* multivariate TS. The unicity of the multivariate TS makes the problem under-specified and allows us to evaluate the performances of the different approaches in this context.

`REVEAL`, `Best-Fit` and our approach need the binarised multivariate TS in their inputs. We use a simple form of binarisation: the binarisation threshold is defined as $\min + (\max - \min)/2$. All values from the multivariate TS greater or equal to the threshold are binarised to 1, and to 0 otherwise. For the two systems with real TS, the theoretic range of the values is not know in advance, so the binarisation threshold is determined component-wise: the components are binarised taking into account their observed minimum and maximum. For the six systems with the synthetic multivariate TS, we know *a priori* that the values of all the components are between 0 and 1 ($\pm$ the noise). In case of noisy data, the fluctuations of a constant component are interpreted as state changes when using a threshold computed component-wise. However, the identification methods are not capable to detect these spurious transitions in the binarised TS. Hence, we compute the binarisation threshold globally, on all the observations of all the components.

In order to have a fair comparison of the methods, and since `caspo-TS` is making the binarisation itself and is not aware that the theoretical minimum and maximum of the components are 0 and 1 ($\pm$ the noise), we correct *a posteriori* the transition functions it returned. The value of the constant is set to the binarised value that is the most present in the binarised TS of the component concerned. Also, since `caspo-TS` does not return a function for the components without parents in the PKN nor for the components that it founds constant for all the TS (in the case where no noise is involved), we use the same technique to set the transition functions to their correct values. We also added a step to filter out BNs returned by `REVEAL` and `Best-Fit` which do not respect the polarities given in the PKN.

For all the BNs returned by the four methods (and after the PKN-based filtering for `REVEAL` and `Best-Fit`), we use `PyBoolNet` [8] to compute the STG of each retrieved BN according to the mixed update scheme. Finally, we evaluate the results of each experience according to three criteria:

- the number of BNs returned;
- the median of the coverage ratios: the proportion of configuration transitions extracted from the input TS that are present in the mixed STG;
- the standard deviation of the coverage ratios.

All data and programs needed to reproduce the presented results are accessible at https://gitlab.inria.fr/avaginay/OLA2021.

6 Results

6.1 Results on Systems with Real PKN and Experimental Multivariate TS

Yeast (Fig. 3 left). For this system `caspo-TS` find 61 BNs while `Best-Fit` and `ASKeD-BN` both find 16 BNs. As for `REVEAL`, due to inconsistencies in the TS, it does not return any BN. Concerning the coverage, on the 7 transitions observed in the TS, the BNs synthesised by `Best-Fit` recover 4 and the BNs synthesised by `ASKeD-BN` recover five. The best coverage ratio (6 retrieved transitions over 7) is obtained for 8 BNs synthesised by `caspo-TS` (among the total of 61). Nevertheless, as the box plot shows, the BNs synthesised by `caspo-TS` present a large variance in their coverage.

A. thaliana (Fig. 3 right). For this system, REVEAL returns no BN. The only BN returned by Best-Fit has all the components set to 1 and recovers 5 transitions over the 10 observed. ASKeD-BN also returns a single BN with a perfect coverage since the BN recovers all the 10 transitions. As for the 5 BNs synthesised by caspo-TS, we can make the same observation as before: they present a variability in their coverage. The best coverage obtained by caspo-TS are from 2 different BNs including the one synthesised by ASKeD-BN.

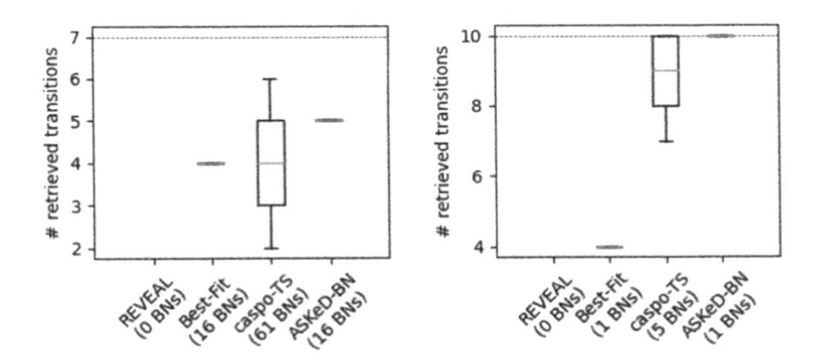

Fig. 3. Number of transitions retrieved by the BNs synthesised using the different methods on the systems *yeast* (left) and *A. thaliana* (right). The blue dashed line indicates the number of transitions that were observed in the multivariate TS. (Color figure online)

To sum up, the results on these two real examples show that:

- REVEAL constantly fails to return any BN. At the opposite, caspo-TS returns more BNs than the other methods;
- the coverage of the BNs returned by both our approach and caspo-TS are better than for Best-Fit;
- caspo-TS presents worse variability in the coverage ratio of its BNs compared to our approach.

6.2 Results on Systems with Generated Multivariate TS

Number of Synthesised BNs: The total number of BNs returned on the synthetic datasets and the number of times the identification methods failed returning any BNs are reported in Table 4. The table shows that a large proportion of BNs generated by REVEAL and Best-Fit were not complying with the influence signs from the input PKN. The following reported results do not take into account these non-compliant BNs. REVEAL is the method which returns the smallest number of BNs, in particular in the ARN setting. This is due to the inconsistencies in the TS, which are frequent in the ARN setting (as in real TS). On the opposite, caspo-TS is the method that returned the largest number of BNs. Moreover, when considering all experiments, there are 18 experiments for which caspo-TS generated more than 100 BNs. In these cases, we stopped the enumeration and analysed the 100 first BNs `caspo-TS` returned. Despite this limit, `caspo-TS` returned between 5 and 7 times more BNs than our method.

Table 4. Number of experiments for which each method failed to return any BN, number of BNs returned over all 336 experiments with synthetic TS and number of BNs returned over the 42 experiments with the ARN setting. The labels "before" and "after" refers to the filtering step which rules out the BNs not respecting the signs of the given PKN (see Sect. 5.2).

measure (setting)	REVEAL before	REVEAL after	Best-Fit before	Best-Fit after	caspo-TS	ASKeD-BN
# failing experiments (all)	230	240	0	64	20	0
# BNs returned (all)	100 677 500	406	100 678 198	724	8481	1210
# BNs returned (ARN)	3	3	51	35	720	85

From here on, we focus on the results of the experiments corresponding to the ARN setting (Asynchronous update scheme, random Repetition of configurations, and Noise addition) after having remove the BNs from `REVEAL` and `Best-Fit` which does not respect the given PKN.

Coverage Ratio: To assess the coverage ratio criterion, instead of plotting the boxplots for the 42 experiments of this setting (6 systems times 7 replicates), we summarised them in Fig. 4. In the scatter plot, each experiment is represented by a point whose coordinates are the coverage ratio median of the synthesised BNs and the associated standard deviation (std). The more top-right a point is, the better the corresponding identification method is (*i.e.*, it produces BNs with high coverage ratio and low std). We can see that for the few experiments for which `REVEAL` was able to return BNs, the median coverage is actually excellent. The median coverage of the BNs returned by `Best-Fit` is almost uniform: `Best-Fit` lacks regularity in finding BNs with good coverage. But the high pick around 0 on the plot of std distribution shows that for a given experiment, the BNs returned by `Best-Fit` have similar coverage rates. `caspo-TS` and our approach have a very similar distribution of median coverage. They are both good at finding BNs with very good coverage. But here again, for a given experiment, the BNs synthesised by `caspo-TS` present a bigger variation of their coverage proportions than the ones synthesised by our approach.

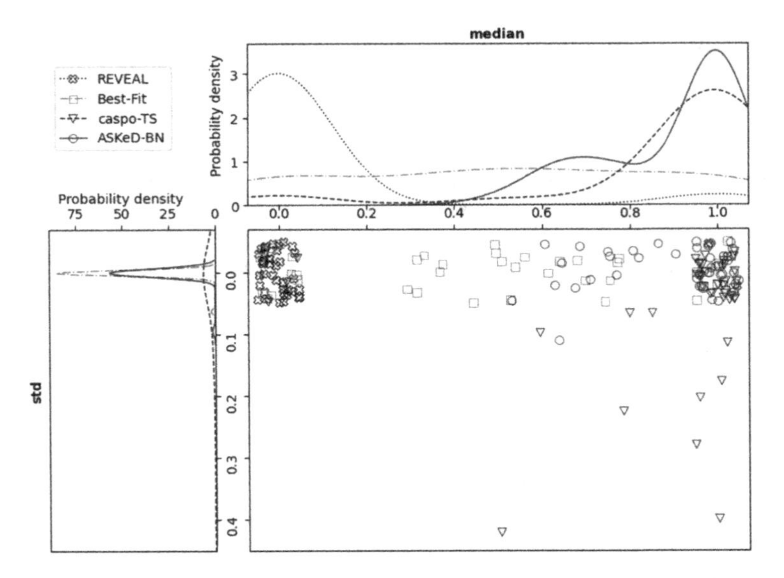

Fig. 4. On the scatter plot, each point represents an experiment in the ARN condition for which a given method potentially returned several BNs with different coverage ratios. The horizontal coordinate of the point is the median of these ratios. The vertical coordinate is their standard deviation (std). For a better visualisation, the coordinates have been jittered with a variance of 0.1 on both axes. The curves on the top (resp. on the left) of the scatter plot are the probability densities of the median (resp. the std) of the points in the scatter plots. The densities have been estimated from the non-jittered coordinates of the points with the Gaussian kernel density estimation method. The smoothing parameter of the estimator was determined automatically (with the Scott method). The areas under all these curves are 1, and the picks show where the points are the most concentrated.

7 Conclusion and Perspectives

We presented `ASKeD-BN`, a novel method to create BNs from a PKN and a multivariate TS. The results on 8 biological systems showed that our approach has the best trade-off on the evaluation criteria: it returns a small set of BNs with a high coverage median and low variance. Our results actually confirm that although `caspo-TS` finds good BNs, too many sub-optimal BNs are also retrieved. Indeed a new version of `caspo-TS` was recently proposed to tackle this problem [5].

We now present two perspectives to improve our approach and the study. First of all, real datasets may contain outlier measurements which could mislead the computation procedure of the binarisation thresholds we used in this paper. It would be interesting to see how such cases impact the performances of the identification methods and to propose a better binarisation procedure with prior outliers detection for instance. Second, contrarily to `REVEAL`, `Best-Fit` and `caspo-TS`, our approach does not handle multiple multivariate TS. However, biologists often have several multivariate TS generated with perturbations forcing some components to stay either active or inactive. However, exploiting such supplementary data gives more information about the behaviour of the studied system in specific conditions (*e.g.*, pathological states). This knowledge allows to constrain even more the space of solutions.

Finally, we are currently working on an automatic pipeline for BN synthesis from a curated mathematical model repository, namely BioModels [14]. This requires (i) automatic extraction of the PKN from the model structure encoded in the SBML[2] file format and (ii) generation of a multivariate TS by simulation of these models.

Acknowledgements. We thank Julie Lao and Hans-Jörg Schurr for their valuable comments and suggestions.

References

1. Albert, R., Thakar, J.: Boolean modeling: a logic-based dynamic approach for understanding signaling and regulatory networks and for making useful predictions. Wiley Interdisc. Rev. Syst. Biol. Med. **6**(5), 353–369 (2014). https://doi.org/10.1002/wsbm.1273
2. Biane, C., Delaplace, F., Melliti, T.: Abductive network action inference for targeted therapy discovery. Electron. Notes Theoret. Comput. Sci. **335**, 3–25 (2018). https://doi.org/10.1016/j.entcs.2018.03.006
3. Bornholdt, S.: Less is more in modeling large genetic networks. Science **310**(5747), 449–451 (2005). https://doi.org/10.1126/science.1119959
4. Chatain, T., Haar, S., Kolčák, J., Paulevé, L.: Most permissive semantics of Boolean networks. Research Report (2020). https://hal.archives-ouvertes.fr/hal-01864693
5. Chevalier, S., Noël, V., Calzone, L., Zinovyev, A., Paulevé, L.: Synthesis and simulation of ensembles of Boolean networks for cell fate decision. In: Abate, A., Petrov, T., Wolf, V. (eds.) CMSB 2020. LNCS, vol. 12314, pp. 193–209. Springer, Cham (2020). https://doi.org/10.1007/978-3-030-60327-4_11
6. Gebser, M., Kaminski, R., Kaufmann, B., Schaub, T.: Answer Set Solving in Practice. Morgan & Claypool Publishers, Williston (2012). ISBN 978-1-60845-971-1
7. Kauffman, S.A.: Metabolic stability and epigenesis in randomly constructed genetic nets. J. Theor. Biol. **22**(3), 437–467 (1969). https://doi.org/10.1016/0022-5193(69)90015-0
8. Klarner, H., Streck, A., Siebert, H.: PyBoolNet: a python package for the generation, analysis and visualization of Boolean networks. Bioinformatics (2016). https://doi.org/10.1093/bioinformatics/btw682
9. Lähdesmäki, H., Shmulevich, I., Yli-Harja, O.: On learning gene regulatory networks under the Boolean network model. Mach. Learn. **52**(1), 147–167 (2003). https://doi.org/10.1023/A:1023905711304
10. Lazebnik, Y.: Can a biologist fix a radio?—Or, what i learned while studying apoptosis. Cancer Cell **2**(3), 179–182 (2002). https://doi.org/10.1016/S1535-6108(02)00133-2

[2] Systems Biology Markup Language.

11. Le Novère, N.: Quantitative and logic modelling of molecular and gene networks. Nat. Rev. Genet. **16**(3), 146–158 (2015). https://doi.org/10.1038/nrg3885
12. Liang, S., Fuhrman, S., Somogyi, R.: REVEAL, a general reverse engineering algorithm for inference of genetic network architectures. In: Pacific Symposium on Biocomputing, pp. 18–29 (1998). ISSN 2335–6928
13. Locke, J.C.W., et al.: Experimental validation of a predicted feedback loop in the multi-oscillator clock of arabidopsis thaliana. Mol. Syst. Biol. **2**(1), 59 (2006). https://doi.org/10.1038/msb4100102
14. Malik-Sheriff, R.S., et al.: BioModels—15 years of sharing computational models in life science. Nucleic Acids Res. **48**(D1), D407–D415 (2020). https://doi.org/10.1093/nar/gkz1055
15. Müssel, C., Hopfensitz, M., Kestler, H.A.: BoolNet—an R package for generation, reconstruction and analysis of Boolean networks. Bioinformatics **26**(10), 1378–1380 (2010). https://doi.org/10.1093/bioinformatics/btq124
16. Ostrowski, M., Paulevé, L., Schaub, T., Siegel, A., Guziolowski, C.: Boolean network identification from perturbation time series data combining dynamics abstraction and logic programming. Biosystems **149**, 139–153 (2016). https://doi.org/10.1016/j.biosystems.2016.07.009
17. Schwab, J.D., Kühlwein, S.D., Ikonomi, N., Kühl, M., Kestler, H.A.: Concepts in Boolean network modeling: what do they all mean? Comput. Struct. Biotechnol. J. **18**, 571–582 (2020). https://doi.org/10.1016/j.csbj.2020.03.001
18. Spellman, P.T., et al.: Comprehensive identification of cell cycle-regulated genes of the yeast Saccharomyces cerevisiae by microarray hybridization. Mol. Biol. Cell **9**(12), 3273–3297 (1998). https://doi.org/10.1091/mbc.9.12.3273

Transportation and Logistics

Solving Inventory Routing Problems with the Gurobi Branch-and-Cut Algorithm

Danny Meier, Benjamin Keller, Markus Kolb, and Thomas Hanne(✉)

University of Applied Sciences and Arts Northwestern Switzerland, Olten, Switzerland
thomas.hanne@fhnw.ch

Abstract. This study is about the implementation and test of a branch-and-cut algorithm for an Inventory-Routing Problem (IRP). The considered mathematical model and the dedicated problem instances have been published by C. Archetti and others. The performance of the implementation on present day computer hardware has been compared to results published in literature several years ago. Also, the influence of the configuration of the used optimizing software package as well as experiences with the publication of the mentioned researchers are documented in this paper.

Keywords: Inventory Routing Problem · Logistics · Numerical optimization · Branch-and-cut · Gurobi

1 Introduction

The supply chain facilitates the procurement of materials and transforms them into a finished product when it can be distributed to the customers. A supply chain interconnects the supplier, the distributor, the manufacturer, the logistic, the retailer and the end customer. This relation between suppliers till end customers involves a lot of processes and the involvement of the electronic supply chain management (e-scm). Supply chain management is nowadays a crucial factor for success. The role of IT is becoming more and more important and to stay competitive it is inevitable to continuously improve the information system. Larger international companies are facing a lot of challenges like the growing complexity of electronic supply chain integration, the increasing customer demands and global competition (Jitpaiboon 2005).

One part of the supply chain is the distribution of goods. In this study we will focus on the outbound logistic. Products have to be distributed between manufacturer, retailers and customers. In the oil industry gas has to be distributed to the gas stations. In the retailer market products have to be shipped to the retailer stores. What these market segments have in common is that customers will eventually buy from a competitor if products are not available. Depending on market segments products should never run out of stock. But costs incur by storing a product. Depending on the storing condition the product storage can be cheaper in the main warehouse than in the retailer's warehouse. Larger companies, especially global operating corporations, have recognized that optimizing their logistic can yield tremendous cost reductions. In the industry today new ways

B. Dorronsoro et al. (Eds.): OLA 2021, CCIS 1443, pp. 173–189, 2021.
https://doi.org/10.1007/978-3-030-85672-4_13

of relationships have been established in the supply chain. The regular scenario for inventory management is that each retailer will place orders to the supplier to prevent stock out (Retailer Managed Inventory). One of the new scenarios is the so-called Vendor Managed Inventory (VMI). This system moves the inventory planning responsibility to the supplier. The supplier has to monitor the inventory of each retailer and will replenish according to a defined replenishment policy to prevent stock out as well as holding inventory costs as low as possible. The applied model is explained in more detail in Sect. 3.

Due to its significance to the global industry the research sector has started to address these issues more than half a century ago. One of the first problems that came up in the 1930s was the mathematical optimization problem of visiting all cities on a map with respect to find the shortest tour where each city must be visited only once. This problem is called Traveling Salesman Problem (TSP) and a solution represents a Hamiltonian Circle. The TSP is one of the most basic problems and has its limitations. For traveling to the destinations only one vehicle is available. The vehicle has unlimited capacity and there is no time limitation when a destination should be delivered (Bertazzi and Speranza 2013). Before the algorithm can start calculating all destinations must be known. Solving this simple model is already NP-hard and is classified in most literature as NP-complete shown by Papadimitriou (Papadimitriou 1977).

On top of the TSP another subproblem is in between to the Inventory Routing Problem (IRP). The Vehicle Routing Problem (VRP) extends the raw routing problem by introducing at least one or more vehicles, demanding customers, and other constraints such as multiple suppliers, capacity constraints or time windows, which increases the complexity dramatically. Every vehicle starts and returns after a defined route to the same depot. Between the depot and every customer as well as between all customers is one link that has a specified cost. The VRP algorithm minimizes the cost to fulfill the customers demand while satisfying the defined constraints (Bertazzi and Speranza 2013). Constraints can be vehicle capacity limitation or the retailer opening hours.

In addition to the VRP the Inventory Routing Problem (IRP) considers time horizon. The algorithm can run over a period of time, for example three days. The calculation can be extended by the decision, if it is better to deliver a customer on day one, two or three. This increases the problem significantly because the quantity to be delivered can vary day by day. To simplify the problem and depending on the products to be delivered a constraint can be defined for example when a customer is visited the stock has to be filled up to the maximum level. The IRP is deciding day by day, which customer has to be served with how much of the specified product (Bertazzi and Speranza 2013). The results of the algorithm are optimized routes and costs. Daily consumption is most of the time an assumption and has to be predicted. This implies in some scenarios a calculation cannot be made in advance. In some environments time can be essential. There are many ways to calculate the companies IRP. The simplest way is the brute-force method which tries out every permutation. This solution needs a lot of computing power and is not solvable for larger instances even not with distributed computing.

The branch-and-cut method is a popular technique to solve many variations of integer linear problems (ILPs) and can be tailored to a specific integer programming problem (Mitchell 1999). There are different libraries or solvers available to find solutions in linear programming. The paper of Archetti et al. (Archetti et al. 2007) has used the CPLEX solver from IBM. This paper will give new benchmark results - produced with up-to-date consumer hardware - for the Vendor-Managed Inventory Routing Problem with a deterministic order-up-to level policy. In this implementation we have used the Gurobi solver package. The implemented exact branch-and-cut approach is explained in the paper of Archetti et al. (Archetti et al. 2007). These benchmarks are necessary to have a sound baseline for other solution approaches. They can be compared against that.

This paper is structured as follows. Section 2 gives a literature survey of research papers which has been consulted and read to become acquainted with the topic of inventory routing problems and the understanding of optimization algorithms. Section 3 introduces the reader into the chosen model for what an efficient algorithm is researched. Explanations and finding to certain algorithms have been given in Sect. 4 whereas Sect. 5 will list the received results and compare them to existing approaches.

2 Literature Survey

This is not meant to be a comprehensive review of literature in the field of inventory routing but rather give an overview of activity in the field and especially put in context some papers which are relevant for the specific model dealt with in this paper. The first studies on inventory routing problems were conducted in the eighties. Mentionable pioneering work was done by Bell et al. (1983), Federgruen and Zipkin (1984) or Blumenfeld et al. (1985). Since then numerous studies about different variants of inventory routing problems have been discussed in literature. A standard version of the problem does not exist. The discussed problems differ in several criteria. There are several literature surveys which make a classification of the problems presented in published research papers. Among them is Coelho et al. (2013) which applies these criteria to about 30 published inventory routing problems: time horizon (finite, infinite), structure (one-to-one, one-to-many, many-to-many), routing (direct, multiple, continuous), inventory policy (maximum level, order-up-to level), inventory decisions (lost sales, back-order, non-negative), fleet composition (homogeneous, heterogeneous), fleet size (single, multiple, unconstrained). Another survey by Andersson et al. (2010) classifies about 70 publications in similar categories and elaborates on industrial aspects of combined inventory management and routing.

The problem considered in this study was developed and solved with an exact algorithm by Archetti et al. (Archetti et al. 2007). This paper describes the first attempt to solve an inventory routing problem exactly. The authors developed a branch-and-cut algorithm and a set of problem instances. The instances have been published and used by other researchers in the field for benchmarking. The algorithm of Archetti et al. has later been analyzed and improved by Solyalı and Süral (2011). They use an a priori tour heuristic to simplify the routing decision and found their algorithm to perform better than the one initially proposed by Archetti et al. Also Coelho and Laporte (2013) reference the same model. They extended the model to a multi vehicle problem and developed an algorithm for solving multi vehicle instances to optimality. Archetti et al. (2012) also developed a hybrid heuristic based on the problem defined by Archetti et al. (2007).

3 Model

The treated model - Vendor Managed Inventory Routing Problem with a deterministic order-up-to level policy - was introduced first by Bertazzi et al. (2002). This kind of problem is still very hard to solve exactly for bigger instances even though the computation power of a single consumer computer has increased massively. The defined model behind this paper is reduced to its main characteristics for simplification and comparability to other benchmarks.

A product is shipped from one common supplier to several retailers, where each retailer has a deterministic consumption in each time frame over a given finite time horizon. The goal is to find tours starting from the supplier to only these retailers which have to be served at the respective time frame. The total costs, consisting of inventory and transportation costs, have to be minimized.

3.1 Problem Description and Formulation

A single product is shipped from a supplier s_O to a set of retailers $M = \{1, 2, \ldots, n\}$ over a time horizon H. At each discrete time $t \in T = \{1, \ldots, H\}$ a single product of quantity r_{Ot} is produced at the supplier and a quantity of r_{st} is consumed at each retailer $s \in M$. A starting inventory level B_O at the supplier and I_{sO} at the retailers is given. The maximum inventory level at the supplier is not restricted whereas a retailer has a given maximum inventory level $U_s \geq I_{sO}$. The current inventory level at time t for the supplier s_O is $B_t \geq 0$ and for each retailer s the current inventory level at time t is $I_{st} \geq 0$. The inventory cost h_O for the supplier is charged in $T' = T \cup \{H + 1\}$. The extended time horizon $H + 1$ considers that for the time frame before start, the inventory cost must be considered. Inventory costs are always calculated at the end of a time frame that means after replenishment and consumption took place. The retailer s, $s \in M$, has inventory cost h_s of and the total inventory cost over time horizon is then $\sum_{t \in T'} h_s I_{st}$. The inventory level after the time horizon is not relevant, which means that after the last time frame an unfeasible status can occur. Shipments can be performed at any time frame $t \in T$ by one vehicle with capacity C. The vehicle's capacity must not be exceeded at any time. The transportation cost c_{ij} where $c_{ij} = c_{ji}$, $i, j \in M' = M \cup \{0\}$ is calculated by the Euclidean distance $NINT = \sqrt{(X_i - X_j)^2 + (Y_i - Y_j)^2}$. The inventory policy at the retailers is denoted by order-up-to level (VMIR-OU) that means if a retailer is visited $y_t^{ij} \in \{0, 1\}$ at time t the inventory must be replenished up to maximum inventory level U_s. In each time frame $t \in T'$ the supplier replenishes its inventory by r_{Ot}. The objective function is to minimize the total transportation and inventory costs without violating any constraint.

$$minimize \sum_{t \in T'} h_o B_t + \sum_{s \in M} \sum_{t \in T'} h_s I_{st} + \sum_{t \in M'} \sum_{j \in M', j \neq i} \sum_{t \in T'} c_{ij} y_{ij}^t \tag{1}$$

3.2 Mathematical Formulation

Variable	**Description**
M	Set of retailers
$M^{\mathrm{I}} = M \cup \{0\}$	Set of supplier and retailers
H	Time horizon
T	Set of discrete time frames
$T^{\mathrm{I}} = T \cup \{H+1\}$	Set of discrete time frames including one time frame before delivery takes place
C	Vehicle capacity
$r_{0\mathrm{t}}$	Production at supplier at time t
r_{st}	Consumption at retailer s at time t
h_0	Unit inventory cost at supplier
h_s	Unit inventory cost at retailer s
B_0	Starting inventory at supplier
B_t	Inventory at supplier at time t
I_{s0}	Starting inventory at retailer s
I_{st}	Inventory at retailer s at time t
U_s	Maximum inventory at retailer s
c_{ij}	Transportation cost from retailer i to retailer j
x_{st}	*S*hipped quantity to the visited retailer s at time t
z_{st}	1 if retailer s at time t is visited. Otherwise, 0
y_{ij}^{t}	Arc from retailer i to retailer j traveled at time t

MIP Formulation. An exact definition of the model with all constraints and inequalities has been given by (Archetti et al. 2007). Next to the VMIR-OU policy they also consider other inventory policies such as VMIR and VMIR-ML.

Constraints. The following constraints have been assessed as useful and efficient.

Inventory and stock-out constraints

$$B_t = B_{t-1} + r_{0t-1} - \sum\nolimits_{s \in M} x_{st-1}, t \in T' \tag{2}$$

The supplier's inventory level is calculated at time t by the level at time $t-1$ plus the produced quantity of products r_{0t-1} at time $t-1$, minus the sum of quantity shipped x_{st-1} to the retailers at time $t-1$.

$$B_t \geq \sum\nolimits_{s \in M} x_{st} \tag{3}$$

Supplier stock-out constraint: This constraint ensures that the inventory level at the supplier has sufficient quantity in stock for the upcoming deliveries at time t.

$$I_{st} = I_{st-1} + x_{st-1} - r_{st-1}, s \in M, t \in T' \text{ where } x_{s0} = r_{s0} = 0, s \in M \tag{4}$$

Retailer inventory definition: The retailer's inventory level is given a time t by the level at time $t-1$, plus the quantity x_{st-1} delivered at time $t-1$, minus the used quantity r_{st-1} at time $t-1$.

$$I_{st} \geq 0, s \in M, t \in T' \tag{5}$$

Replenishment constraints

$$x_{st} \geq U_s z_{st} - I_{st}, s \in M, t \in T \tag{6}$$

$$x_{st} \leq U_s - I_{st}, s \in M, t \in T \tag{7}$$

$$x_{st} \leq U_s z_{st}, s \in M, t \in T \tag{8}$$

Order-up-to level: These constraints define that if a retailer s is served the quantity x_{st} shipped to the retailers will fill up the inventory to the maximum U_s. If a retailer s is served at time t the variable z_{st} is 1 and otherwise equals 0.

Capacity constraint

$$\sum_{s \in M} x_{st} \leq C, t \in T \tag{9}$$

Capacity constraint: This constraint ensures that the delivered product quantity does not exceed the load capacity C of the transportation vehicle at any time $t \in T'$.

Routing constraints

$$\sum_{j \in M', j<i} y_{ij}^t + \sum_{j \in M', j>i} y_{ij}^t = 2z_{it} \tag{10}$$

Routing constraint: When a retailer s is served at time t the vehicle must use a route leaving from retailer s at time t.

(a) Subtour elimination constraint according to Fischetti et al. (1998) and Gendreau et al. (1998)

$$\sum_{i \in S} \sum_{j \in S, j<i} y_{ij}^t \leq \sum_{i \in S} z_{it} - z_{kt}, S \subseteq M, t \in T \text{ for some } k \in S \tag{11}$$

(b) Traditional subtour constraint according to (Dantzig et al. 1954)

$$\sum_{i \in S} \sum_{j \in S, j<i} y_{ij}^t \leq |\mathrm{S}| - 1, S \subseteq M, t \in T \tag{12}$$

(c) Improved subtour constraint (11)

$$\sum_{t \in \mathrm{T}'} \left(\sum_{i \in S} \sum_{j \in S, j<i} y_{ij}^{t'} \leq \sum_{i \in S} z_{it'} - z_{kt'} \right), S \subseteq M, t \in T, k \in S \text{ where } S \text{ is subtour of } t \tag{13}$$

$$x_{st} \geq 0, s \in M, t \in T \tag{14}$$

Non-negativity constraint: When a retailer s is visited at time t the delivered good x_{st} cannot be below zero.

$$y_{ij}^t \in \{0, 1\}, i, j \in M, \ \mathrm{j} < \mathrm{i}, \ t \in T \tag{15}$$

Path constraint: When the path between retailer i to retailer j is traveled at time t the value is equal 1 otherwise 0.

$$y_{i0}^t \in \{0, 1, 2\}, i \in M, t \in T \tag{16}$$

Path constraint: When the path between supplier and retailer i is traveled at time t and continued to another retailer the value y_{i0}^t is equal 1 and when same path is taken back to the supplier the value y_{i0}^t is equal 2. If the path has not been taken yet the value y_{i0}^t is zero.

$$z_{it} \in \{0, 1\}, \quad i \in M', \quad t \in T \tag{17}$$

Retailer visit constraint: When the retailer z_{it} is visited at time t the value is equal 1 otherwise zero.

Inequalities

The following inequalities have been assessed as useful and efficient.

$$I_{st} \geq (1 - z_{st})r_{st}, s \in M, t \in T \tag{18}$$

When the retailer s is not visited (z_{st} is equal zero) at time t, then the inventory level I_{st} must be at least equal to the product quantity r_{st} consumed at time t.

$$I_{st-k} \geq \left(\sum\nolimits_{j=0}^{k} r_{st-j}\right)\left(1 - \sum\nolimits_{j=0}^{k} z_{st-j}\right), s \in M, t \in T, k = 0, \ldots, t-1 \tag{19}$$

This inequality extends the inequality (18) by the value k. The period of time retailers are not served is calculated by $t - k$.

$$I_{st} \geq U_s z_{st-k} - \sum\nolimits_{j=t-k}^{t-1} r_{st}, s \in M, t \in T, k = 1, \ldots, t-1 \tag{20}$$

The inventory at a retailer I_{st} at time t must be equal or greater than the inventory at the last delivery minus the sum of consumed goods since then.

$$z_{st} \leq z_{0t}, s \in M, t \in T \tag{21}$$

When any retailer s is visited at time t, z_{st} is equal 1. Additionally, the supplier has to be included in the route at time t, z_{0t} equals 1. This inequality is to prevent that a retailer can be visited without ever having visited the supplier.

$$y_{i0}^t \leq 2z_{it}, s \in M, t \in T \tag{22}$$

If the arc between supplier and retailer i has been traveled at time t then retailer i must be visited.

$$y_{ij}^t \leq z_{it}, i, j \in M, t \in T \tag{23}$$

If the arc between retailer i and j has been traveled, then both retailers must be visited.

3.3 Test Instances

Archetti et al. (2007) produced a large set of 160 test instances classified into three groups and five test sets. These instances are specific to the Vendor-Managed Inventory Routing Problem and are the most common and most used instances to benchmark. The three parameters which identify the test instance and its respective properties are the time horizon $h = \{3, 6\}$ that describes how many discrete time frames should be taken into account. The inventory costs $c = \{\text{low, high}\}$ are grouped into two ranges of values. Each retailer and the supplier are in this inventory cost range. The low group has inventory costs between 0.01and 0.05 and the high group has inventory costs multiplied by 10 in the range of 0.1 to 0.5. The third parameter describes how many retailers s are involved. A test set $i = \{1, \ldots, 5\}$ contains every combination of the parameters mentioned before. To identify the file the following naming convention has been applied: h[h][c][i]n[s]. For example, a test instance from test set 2 with 6 time frames, 30 retailers and low inventory costs is named as h6low2n30. A complete test set with all combinations is listed at Table 2 on page 10. The whole collection could be obtained from (Coelho n.d.). Therefore, one test set consists of 32 test instances. Table 1 summarizes the properties within one test set.

Table 1. Content of a test set provided by Archetti et al.

Inventory cost c	Time frames h	Retailers s
Low $[0.01, 0.05]$	3	$5 \le x \le 50$, $x \bmod 5 = 0$
Low $[0.01, 0.05]$	6	$5 \le x \le 30$, $x \bmod 5 = 0$
High $[0.1, 0.5]$	3	$5 \le x \le 50$, $x \bmod 5 = 0$
High $[0.1, 0.5]$	6	$5 \le x \le 30$, $x \bmod 5 = 0$

In this paper only the first two test sets have been fully benchmarked and compared. The results of the other three test sets are quite similar. Test set three and four have the same characteristics as test set two. And test set five has the same characteristics as test set one.

4 Algorithm

To solve the inventory routing problem the commercial solver package from Gurobi has been examined. The development has been done in C# and.NET 4.0. The developed implementation fully concentrates on the branch-and-cut algorithm and its limited options to fine tune the behavior of the solving process. The evaluated parameters are specific to the used solver package and do not allow further in-depth controlling.

- CPU cores: The default setting is set to the amount of available physical cores. Intel based CPUs with Hyper-Threading support offer the double number of cores to use.

- Heuristic ratio: The percentage of heuristic influence as a ratio of the total computation time to increase finding feasible solutions.
- Presolve capabilities: How much effort should be invested to try to tighten the model.
- Solving strategy: Sets the focus with which strategy a feasible solution should be found. Strategy can be set to balanced, focus on finding fast feasible solutions, focus on optimality or focus on best bounds.
- Branching priority: Without configuration the branching priority is set automatically and can change during the execution. Setting the branching priority for certain decision variables overrides the automatic behavior and considers the set value with priority.

Next to the configurable properties, optimization has been tried to apply through finding stronger constraints for the subtour elimination (see Sect. 4.1). Adding each subtour elimination constraint at the beginning is not feasible because of the number of available permutations to form the constraint. Next to time issues this will blow up computer memory at the time. According to best practice this kind of constraints will be added as soon as a violated intermediate solution will occur. Adding the additional constraint just after the solving process and resolve the model again has been deemed as insufficient. This has been proven by some simple runs of more complex test instances. Thus, the approach using a callback method is even more efficient. If the callback method recognizes an optimal solution with a violated subtour, a lazy constraint according to (11) will be added. In Applegate et al. (2007) different approaches for subtour elimination are examined. For the test series in this paper the two most promising algorithms have been implemented. The proposed subtour elimination algorithm by (Archetti et al. 2007) is the separation algorithm of (Padberg and Rinaldi 1991) with the constraint definition of (11) that was introduced by (Fischetti et al. 1998) and (Gendreau et al. 1998). The side constraint $k = \arg\max_j \{z_{jt}\}$ has been also considered but simplified. Then z_{jt} is a binary decision variable that is only 1 if $j \in M$ has been visited in time $t \in T$. Under the condition that $j \in S$ and $S \subseteq M$, every k is part of the subtour and has been visited. But we found a more efficient way to find an optimal solution faster. If a violated constraint will not only be added for the violated time frame but also be added to all time frames of the test instance, then an average time saving of factor 1.5 can be expected. The traditional subtour elimination algorithm (12) mostly used in TSP related problems has been proven as inefficient.

Unfortunately, the consumed literature did not provide any exact implementation details even for the most critical part - namely the subtour elimination constraint. Based on a personal interpretation of the mathematical equation the first implementations and all further deviates have been implemented as best possible. Also, source code was not available - neither on request - from the paper authors.

4.1 Implementation of Subtour Elimination Constraints

When the solver finds a feasible solution the callback method checks its routes for subtours. For this the algorithm starts at the supplier ($i = 0$) and transits to a retailer j through an arc which is part of the tour ($y^t_{oj} \geq 0$). From this retailer j the tour proceeds through an arc to another retailer k ($y^t_{jk} = 1$). In this way the arcs forming the route are followed from retailer to retailer until the supplier is reached again. Subtours exist when not every served retailer ($z_{st} = 1$) has been visited by following these adjacent arcs. The retailers visited in this tour form a subtour but this subtour is not considered a real subtour because it contains the supplier.

From the remaining retailers which are served but have no connection to the supplier a retailer is selected as the starting point of the first real subtour. From there adjacent arcs are followed until this retailer is reached again. The visited retailers are then part of the first subtour. If there are still retailers which are served but not part of the first subtour and also are not connected to the supplier again one is selected as the starting point for the second subtour. This procedure is repeated until all served retailers are part of a subtour.

For every real subtour (subtour that does not contain the supplier) the following constraints are added to the model as lazy constraints:

$$\sum_{i \in S} \sum_{j \in S, j<i} y_{ij}^t \leq \sum_{i \in S} z_{it} - z_{kt}, t \in T \text{ for every } k \in S \tag{24}$$

where S is the set of retailers forming a subtour. Because similar subtours were likely to appear in multiple time frames the constraints (24) were not only constructed for the time frame t in which the specific subtour was found but for all $t \in T$.

5 Results

5.1 Algorithm Optimization Results

Experiments showed that subtle adjustments to some of the settings of the Gurobi solver sometimes resulted in noticeable change of the calculation time. Also experimenting with attributes of variables resulted in unexpected changes of performance.

With Gurobi the type attribute of variables can be defined as integer, binary or continuous. Variables also have attributes for upper and lower bound. The following two examples illustrate what was observed when experimenting with these attributes. Changing the attribute type of the y_{ij}^t (vertex i to j is part of the route at time t) variables from binary to integer with a lower and upper bound of 0 and 1 respectively, improved the performance of the program. Another example of the influence of variable attributes was observed with the x_{st} variables (delivered quantity). Initially the upper bound attribute of these variables was set to the maximum inventory level Us which is valid for the VMIR-OU and VMIR-ML policy. Decreased calculation time was observed when the upper bound attribute was set to infinity. The maximum value for x_{st} was restricted by the replenishment constraints (6) to (8) which were added to the Gurobi model.

Calculations have been performed for several different combinations of settings for the Gurobi solver. A comparison of the program performance with different settings for the number of used CPUs was made. A selection of instances was calculated using four and eight CPUs. The total time used for calculating the test set instances was lower when using eight CPUs. However, some instances were calculated faster with only four CPUs. A pattern relating the computation times to the instance size by number of retailers or time frames was not recognizable, although instances with a rather long computation time were more likely to perform better with eight CPUs.

To determine the influence of the branching priority settings of the solver, several computations were compared. For each computation over a set of problem instances a different variable group was prioritized. That means the instance set was solved setting the branching priority attribute of for example all z_{st} to 1 while for all other variables the attribute was 0 by default. Then the calculation was repeated for other variable groups, e.g., x_{st} or y_t with increased branching priority.

The conclusion from these experiments was that the branching priority did not have any noticeable impact on the overall performance of the solving process. The usage of different solving strategies by alteration of the MIPFocus parameter of the solver did not result in faster computation compared to the default setting. With the default setting the solver automatically balances between finding new feasible solutions and proving that the current solution is optimal. Neither emphasizing the first nor the later nor a focus on the bound improved the performance of the calculation. The increase of time the solver should spend for feasibility heuristics by raising the according parameter had a negative impact on the computation time.

5.2 Computation Results

Computation results have been calculated on a system with a Microsoft Windows 8.1 64bit operating system and a quad core Intel i7 3770k CPU with 16 GB of memory. The source code is written in C#.NET 4.0 and compiled with enabled code optimization. The MIP solving package is provided by Gurobi in version 6.0.4. Maximum running time is limited to 1800s (30 min). Source code is available on request to the authors.

The computation results are structured into two different benchmark categories. The first category compares the different implemented solutions discussed in Sect. 4 (Table 2). The second category shows the evolution of computation power (Table 3). The results are compared to the initial results from Archetti et al. (2007) and the newer result from (Coelho and Laporte 2014) (Tables 4 and 5).

The test runs in this paper have been executed with four different settings. At first the difference of computation time has been examined if different thread counts have been used (4T or 8T). The CPU in this test environment has four physical cores and the ability to use the Hyper-Threading technology from Intel which doubles the count of virtual cores. Other settings than four or eight threads has been proven to be not efficient. Lower counts don't use the full power of the machine and higher counts have the disadvantage of rescheduling overhead within the process execution. Between the two chosen settings no overall winner has been evaluated. For some test instances the setting with four threads performs better than using eight threads. And for some others the opposite is the case. But no prediction is possible which performs better according to retailer count or involved time frames. Regarding to all tests that have been done, the average computation time of the four thread count settings is slightly lower. The second setting was the introduction of the more efficient subtour elimination constraint (13) (SEE1 - Extended subtour elimination constraint (13) used/SEE0 - Original subtour elimination constraint (11) used). Implementation details of the new constraint are described in Sect. 4.1. This extension has been proven as more efficient. In most cases the computation time could be reduced. In some special cases the algorithm was able to find an optimal solution within the time limit for test instances, which are not solvable with the original subtour elimination constraint in time.

Unfortunately, the implementation in this paper was not able to produce similar or better results in case of computation time as the results from Coelho et al. Even after this study it is not evident why the implementation is worse. Reasons could be the implementation itself or the chosen solver package from Gurobi.

Table 2. Computation results for the self-implemented branch-and-cut algorithm

Instance	Costs	8T-SI E1		4T-SEE1		8T-SEE0		4T-SEE0	
		t [s]	G [%]	t [s]	G [%]	t [s]	G [%]	t [s]	G [%]
h3high1n10	4'970.62	0		0		0		0	
h3high1n15	5'713.84	0		0		0		0	
h3high1n20	7'353.82	1		1		1		1	
h3high1n25	8'657.70	2		1		1		1	
h3high1n30	12'635.55	232		324		1'007		850	
h3high1n35	11'984.69	28		6		21		35	
h3high1n40	14'006.60	*1'800	2.7%	*1'800	0.3%	*1'800	1.6%	*1'800	1.6%
h3high1n45	14'661.20	40		83		342		213	
h3high1n5	2'149.80	0		0		0		0	
h3high1n50	15'235.80	*1'800	0.8%	*1'800	0.6%	*1'800	0.5%	*1'800	1.8%
h3low1n10	2'167.37	0		0		0		0	
h3low1n15	2'236.53	0		0		0		0	
h3low1n20	2'793.29	1		1		1		1	
h3low1n25	3'309.64	1		1		1		1	
h3low1n30	3'918.76	229		199		1'032		529	
h3low1n35	3'694.48	7		5		18		13	
h3low1n40	4'263.43	*1'800	0.9%	*1'800	7.6%	*1'800	1.9%	*1'800	10.0%
h3low1n45	4'369.38	67		70		134		145	
h3low1n5	1'281.68	0		0		0		0	
h3low1n50	4'629.92	*1'800	8.2%	*1'800	1.3%	*1'800	4.9%	*1'800	5.6%
h6high1n10	8'870.15	1		1		4		3	
h6high1n15	12'118.83	2		1		2		12	
h6high1n20	14'702.95	329		141		134		159	
h6high1n25	15'581.47	16		7		51		75	5.0%
h6high1n30	23'184.00	*1'800	4.8%	*1'800	3.9%	*1'800	n/a	*1'800	
h6high1n5	5'942.82	0		0		0		0	
h6low1n10	4'499.25	1		1		1		2	
h6low1n15	5'462.68	1		1		2		2	
h6low1n20	6'490.18	466		128		353		159	
h6low1n25	7'095.86	92		35		138		69	
h6low1n30	8'319.59	*1'800	19.4%	*1'800	10.8%	*1'800	7.1%	*1'800	1.9%
h6low1n5	3'335.24	0		0		0		0	
Average [s]		385		369		439		225	
Solved		26 / 32		26 / 32		26 / 32			
								30	32

* = Not able to solve test instance to optimality within time limit

Table 3. Computation result comparison - Current implementation vs. Archetti vs. Coelho et al.

Instance	Costs	4T-SS E1		Archetti		Coelho et al.	
		t [s]	G [%]	t [s]	G [%]	t [s]	G [%]
h3high2n10	4'803.17	0		0		0	
h3high2n15	5'821.04	0		0		1	
h3high2n20	7'385.03	1		10		6	
h3high2n25	9'266.87	1		14		9	
h3high2n30	11'351.36	324		164		18	
h3high2n35	10'706.91	6		199		24	
h3high2n40	11'722.58	*1'800	2.7%	1'003		74	
h3high2n45	13'675.96	83		1'205		76	
h3high2n5	1'959.05	0		0		0	
h3high2n50	15'453.80	*1'800	0.8%	*1'800	n/a	148	
h3low2n10	2'510.13	0		0		2	
h3low2n15	2'506.21	0		0		0	
h3low2n20	2'799.90	1		12		7	
h3low2n25	3'495.97	1		25		7	
h3low2n30	3'737.11	199		84		13	
h3low2n35	3'796.80	5		173		27	
h3low2n40	4'166.95	*1'800	0.9%	1'500		74	
h3low2n45	4'226.82	70		1'133		38	
h3low2n5	1'176.63	0		0		0	
h3low2n50	4'919.75	*1'800	8.2%	*1'800	n/a	235	
h6high2n10	8'569.73	1		11		7	
h6high2n15	11'932.10	1		22		6	
h6high2n20	14'646.96	141		1'536		470	
h6high2n25	16'823.20	7		578		35	
h6high2n30	20'090.29	*1'800	4.8%	*1'800	n/a	802	
h6high2n5	5'045.91	0		0		1	
h6low2n10	5'236.98	1		7		16	
h6low2n15	5'494.74	1		39		10	
h6low2n20	6'082.54	128		*1'800	n/a	688	
h6low2n25	7'484.84	35		548		27	
h6low2n30	7'761.53	*1'800	19.4%	*1'800	n/a	1'190	
h6low2n5	2'722.33	0		0		1	
Average [s]		369		539		125	
Solved		26 / 32		27 / 32		32 / 32	

* = Not able to solve test instance to optimality within time limit

Table 4. Computation results for the self-implemented branch-and-cut algorithm

Instance	Costs	8T-SEE1		4T-SEE1		8T-SEE0		4T-SEE0	
		t [s]	G [%]	t [s]	G [%]	t [s]	G [%]	t [s]	G [%]
h3high2n10	4'803.17	0		0		0		0	
h3high2n15	5'821.04	1		0		1		1	
h3high2n20	7'385.03	1		1		1		1	
h3high2n25	9'266.87	85		69		89		85	
h3high2n30	11'351.36	8		50		31		14	
h3high2n35	10'706.91	3		2		3		4	
h3high2n40	11'722.58	128		187		570		456	
h3high2n45	13'675.96	338		119		262		244	
h3high2n5	1'959.05	0		0		0		0	
h3high2n50	15'453.80	702		1'191		1'250		422	
h3low2n10	2'510.13	0		0		0		0	
h3low2n15	2'506.21	0		0		0		1	
h3low2n20	2'799.90	1		1		1		1	
h3low2n25	3'495.97	60		48		126		262	
h3low2n30	3'737.11	67		11		41		35	
h3low2n35	3'796.80	23		3		3		3	
h3low2n40	4'166.95	90		257		1'119		783	
h3low2n45	4'226.82	428		303		232		349	
h3low2n5	1'176.63	0		0		0		0	
h3low2n50	4'919.75	246		462		386		408	
h6high2n10	8'569.73	1		0		1		0	
h6high2n15	11'932.10	2		2		10		11	
h6high2n20	14'646.96	16		5		79		46	
h6high2n25	16'823.20	1'352		1'432		1'204		*1'800	0.3%
h6high2n30	20'090.29	287		417		215		191	
h6high2n5	5'045.91	0		0		0		0	
h6low2n10	5'236.98	0		0		1		1	
h6low2n15	5'494.74	1		2		2		16	
h6low2n20	6'082.54	12		9		41		32	
h6low2n25	7'484.84	*1'800	0.9%	419		*1'800	1.1%	*1'800	1.0%
h6low2n30	7'761.53	711		182		437		240	
h6low2n5	2'722.33	0		0		0		0	
Average [s]		199		162		247		225	
Solved		31 / 32		32 / 32		31 / 32		31 32	

* = Not able to solve test instance to optimality within time limit

Table 5. Computation result comparison - Current implementation vs. Archetti vs. Coelho et al.

Instance	Costs	4T-SSE1		Archetti		Coelho et al.	
		t [s]	G [%]	t [s]	G [%]	t [s]	G [%]
h3high2n10	4'803.17	0		0		0	
h3high2n15	5'821.04	0		1		3	
h3high2n20	7'385.03	1		8		5	
h3high2n25	9'266.87	69		47		7	
h3high2n30	11'351.36	50		130		20	
h3high2n35	10'706.91	2		97		21	
h3high2n40	11'722.58	187		449		29	
h3high2n45	13'675.96	119		553		145	
h3high2n5	1'959.05	0		0		0	
h3high2n50	15'453.80	1'191		1'782		251	
h3low2n10	2'510.13	0		0		0	
h3low2n15	2'506.21	0		1		2	
h3low2n20	2'799.90	1		6		7	
h3low2n25	3'495.97	48		53		16	
h3low2n30	3'737.11	11		128		18	
h3low2n35	3'796.80	3		74		23	
h3low2n40	4'166.95	257		369		38	
h3low2n45	4'226.82	303		928		158	
h3low2n5	1'176.63	0		0		0	
h3low2n50	4'919.75	462		1'235		133	
h6high2n10	8'569.73	0		7		4	
h6high2n15	11'932.10	2		31		6	
h6high2n20	14'646.96	5		354		37	
h6high2n25	16'823.20	1'432		1'732		113	
h6high2n30	20'090.29	417		*1'800	n/a	197	
h6high2n5	5'045.91	0		0		1	
h6low2n10	5'236.98	0		4		2	
h6low2n15	5'494.74	2		22		8	
h6low2n20	6'082.54	9		282		52	
h6low2n25	7'484.84	419		1'710		222	
h6low2n30	7'761.53	182		*1'800	n/a	220	
h6low2n5	2'722.33	0		0		1	
Average [s]		162		425		54	
Solved		32 / 32		30 / 32		32 / 32	

* = Not able to solve test instance to optimality within time limit

6 Conclusions

The initial idea of this research project was to do experiments with a heuristic to solve a problem in the field of inventory routing. Probably a heuristic, ideally with problem instances for bench- marking, could be found and improved. However, the search in literature for such a heuristic was not successful, but problem instances for an inventory routing model could be found. The problem instances were created during the development of an exact branch-and-cut algorithm for solving an inventory routing problem by (Archetti et al. 2007).

With the existence of an optimal solution a heuristic could be benchmarked against it. For a benchmark to be valid the computing environment would have to be similar. Because the results by Archetti et al. were calculated some years ago and the algorithms implementation was not available, the work concentrated on re-implementing the branch-and-cut algorithm. Due to the limited time horizon for this research project the work concentrated on the exact algorithm and its implementation.

The algorithm was implemented using the Gurobi Optimizer software package. It was then run against the mentioned problem instances and the outcomes were compared to results available in literature. In general, we found the optimal solution faster, which is undoubtedly due to the better performance of the computer hardware and maybe, also to the advancement of the solver software (although not the same products were used than in other papers). However, for bigger instances the computation time significantly exceeded earlier results if they were at all solvable in the given time limit. It is not quite clear why this is the case. Probable causes could lie in the internals of the used solver software. We also assume that the detailed handling of the subtour elimination constraints could be different. The description of these constraints in the referred literature was not detailed enough for us to be clear on how exactly these were constructed and implemented.

Several experiments were made with solver specific settings and variable arguments. These tests showed that the performance of the calculation depends a lot on the configuration of the underlying solver software and on the specific problem instance.

References

Andersson, H., Hoff, A., Christiansen, M., Hasle, G., Løkketangen, A.: Industrial aspects and literature survey: combined inventory management and routing. Comput. Oper. Res. **37**(9), 1515–1536 (2010)

Applegate, D.L., Bixby, R.E., Chvátal, V., Cook, W.J.: The Traveling Salesman Problem: A Computational Study (Princeton Series in Applied Mathematics), Princeton University Press (2007)

Archetti, C., Bertazzi, L., Hertz, A., Speranza, M.G.: A hybrid heuristic for an inventory routing problem. INFORMS J. Comput. **24**(1), 101–116 (2012)

Archetti, C., Bertazzi, L., Laporte, G., Speranza, M.G.: A branch-and-cut algorithm for a vendor-managed inventory-routing problem. Transp. Sci. **41**(3), 382–391 (2007)

Bell, W.J., et al.: Improving the distribution of industrial gases with an on-line computerized routing and scheduling optimizer. Interfaces **13**(6), 4–23 (1983)

Bertazzi, L., Paletta, G., Speranza, M.G.: Deterministic order-up-to level policies in an inventory routing problem. Transp. Sci. **36**(1), 119–132 (2002). https://doi.org/10.1287/trsc.36.1.119.573

Bertazzi, L., Speranza, M.G.: Inventory routing problems with multiple customers. EURO J. Transp. Logist. **2**(3), 255–275 (2013). https://doi.org/10.1007/s13676-013-0027-z

Blumenfeld, D.E., Burns, L.D., Diltz, J.D., Daganzo, C.F.: Analyzing trade-offs between transportation, inventory and production costs on freight networks. Transp. Res. Part B: Methodol. **19**(5), 361–380 (1985)

Coelho, L.: Collection of test instances for the inventory routing problem (n.d.). http://www.leandro-coelho.com/instances/inventory-routing/

Coelho, L.C., Cordeau, J.-F., Laporte, G.: Thirty years of inventory routing. Transp. Sci. **48**(1), 1–19 (2013)

Coelho, L.C., Laporte, G.: The exact solution of several classes of inventory-routing problems. Comput. Oper. Res. **40**(2), 558–565 (2013)

Coelho, L.C., Laporte, G.: An optimised target-level inventory replenishment policy for vendor-managed inventory systems. Int. J. Prod. Res. **53**(12), 3651–3660 (2014). https://doi.org/10.1080/00207543.2014.986299

Dantzig, G., Fulkerson, R., Johnson, S.: Solution of a large-scale traveling-salesman problem. J. Oper. Res. Soc. Am. **2**(4), 393–410 (1954). http://www.jstor.org/stable/166695

Federgruen, A., Zipkin, P.: A combined vehicle routing and inventory allocation problem. Oper. Res. **32**(5), 1019–1037 (1984)

Fischetti, M., Gonzalez, J.J.S., Toth, P.: Solving the orienteering problem through branch-and-cut. INFORMS J. Comput. **10**(2), 133–148 (1998). https://doi.org/10.1287/ijoc.10.2.133

Gendreau, M., Laporte, G., Semet, F.: A branch-and-cut algorithm for the undirected selective traveling salesman problem. Networks **32**(4), 263–273 (1998). 10.1002/(sici)1097-0037(199812)32:4¡263::aid-net3¿3.3.co;2-h

Jitpaiboon, T.: The roles of information systems integration in the supply chain integration context-firm perspective. Supply Chain (2005)

Mitchell, J.E.: Branch-and-cut algorithms for combinatorial optimization problems. Mathematical Sciences (1999)

Padberg, M., Rinaldi, G.: A branch-and-cut algorithm for the resolution of large-scale symmetric traveling salesman problems. SIAM Rev. **33**(1), 60–100 (1991). https://doi.org/10.1137/1033004

Papadimitriou, C.H.: The Euclidean travelling salesman problem is NP-complete. Theor. Comput. Sci. **4**(3), 237–244 (1977). https://doi.org/10.1016/0304-3975(77)90012-3

Solyalı, O., Süral, H.: A branch-and-cut algorithm using a strong formulation and an a priori tour-based heuristic for an inventory-routing problem. Transp. Sci. **45**(3), 335–345 (2011). https://doi.org/10.1287/trsc.1100.0354

Iterated Local Search with Neighbourhood Reduction for the Pickups and Deliveries Problem Arising in Retail Industry

Hanyu Gu[1], Lucy MacMillan[2], Yefei Zhang[1](✉), and Yakov Zinder[1]

[1] School of Mathematical and Physical Sciences, University of Technology Sydney, 15 Broadway, Ultimo, NSW 2007, Australia
{Hanyu.Gu,yakov.zinder}@uts.edu.au, Ye.f.zhang@student.uts.edu.au
[2] Australian National Couriers, 29 Huntingwood Drive, Huntingwood, NSW 2148, Australia
Lucym@ancdelivers.com.au

Abstract. The paper studies a vehicle routing problem with simultaneous pickups and deliveries that arises in the retail sector, which considers a heterogeneous fleet of vehicles, time windows of the demands, practical restrictions on the drivers and a roster specifying the order of vehicle loading at the depot. The high competition in this industry requires that a viable optimisation approach must achieve a good balance of solution time, quality and robustness. In this paper, a novel iterated local search algorithm is proposed which dynamically reduces the neighbourhood so that only the most promising moves are considered. The results of computational experiments on real-world data demonstrate the high efficiency of the presented optimisation procedure in terms of computation time, stability of the optimisation procedure and solution quality.

Keywords: Vehicle routing problem · Iterated local search · Neighbourhood reduction

1 Introduction

This paper considers a vehicle routing problem with simultaneous pickups and deliveries (VRPSPD) which arises in the retail sector. The features of this problem include: a heterogeneous fleet of vehicles, time window for pickups and deliveries, open routes, restriction on shift length and loading roster at the depot. In spite of the practical importance of these features, few applications in the literature considered all of them simultaneously [6,9]. Furthermore, the objective of the considered problem is to maximise the number of allocations which is practically essential, but is rarely considered in the literature [9].

Supported by an Australian Government Research Training Program Scholarship.

B. Dorronsoro et al. (Eds.): OLA 2021, CCIS 1443, pp. 190–202, 2021.
https://doi.org/10.1007/978-3-030-85672-4_14

Since VRPSPD is NP-hard in the strong sense [3], the majority of the publications in this topic present various heuristics and metaheuristics [6,9]. In practice, a scheduler expects to produce a good solution within a short time limit, typically no more than one minute. In contrast, most research in the literature focuses more on solution quality. In this paper, an iterated local search [7] based optimisation algorithm is presented to achieve a satisfactory balance between solution quality and computation time.

The iterated local search algorithm (ILS) has been widely used to solve combinatorial optimisation problem [7]. It iteratively generates a sequence of local optimums. At each iteration a local search is performed on a problem-specific neighbourhood structure. A perturbation mechanism is employed to avoid local optimum and expand the search space. By allowing infeasible solutions in the designed neighborhood structure [5,12], ILS has been demonstrated to be much faster than the state-of-the-art for solving the Workforce Scheduling and Routing Problem, which is, from a practical application viewpoint, similar to the studied problem in this paper.

The most time-consuming component in ILS is the evaluation of potential moves in the local search procedure due to the large size of the neighbourhood of the current solution. It is also critical to select proper moves to increase the probability of converging to the global optimum. This paper presents a mechanism to reduce the neighbourhood dynamically, which makes the move evaluation faster, and at the same time direct search in the most promising part of the neighbourhood.

Contributions of this paper include

- development of a MIP model for a VRPSPD problem with many features from the retail sector
- introduction of neighbourhood reduction to speedup the ILS algorithm
- computational studies on real-world data

The remainder of the paper is organised as follows. Section 2 presents the problem formulation. Section 3 describes the proposed iterated local search. Section 4 presents the results for the computational experiments. Section 5 concludes the paper.

2 Problem Statement

The considered vehicle routing problem can be stated as follows. Let $G = \{L, A\}$ be a directed graph, where the set of vertices $L = \{0\} \cup C$ and $C = \{1, 2, ..., l\}$, the set of arcs $A = A_D \cup A_C$ and $A_D = \{(0, i)|i \in C\}$, $A_C = \{(i, j)|i \neq j, \forall i, j \in C\}$. Vertex 0 represents the depot and the remaining vertices represent the l customers. Each arc $(i, j) \in A$ has an associated travel time $t_{i,j}$.

The delivery to customer $i \in C$ is characterised by its weight w_i^d and volume v_i^d. The pickup from customer $j \in C$ is characterised by its weight w_i^p and volume v_i^p. For customer $i \in C$, the associated time window $[a_i, b_i]$ indicates the earliest and latest time when the driver can start the corresponding services, and let $p_i > 0$ be the service time required for the driver to complete the service.

Let T be the set of all vehicles. Each vehicle $i \in T$ is differed by its weight capacity W_i and volume capacity V_i. All vehicles $i \in T$ depart from the same single depot and are not required to return to depot after serving all allocated customers. The driver in each vehicle $i \in T$ finishes the shift after serving the last allocated customers. Due to the loading capacity of the depot, each vehicle $i \in T$ arrives at the depot at the specified starting time r_i with loading time δ_i. Furthermore, there exists an upper bound S_i on the shift time of the driver in vehicle $i \in T$, which is the length of time interval between the time when driver starts loading at the depot and the time when driver finishes the service of the last allocated customers.

Each customer $i \in C$ can be served only once, but not all vehicles are capable to serve certain customers. In this paper, two types of vehicles are considered, i.e., the one-man vehicle $T' \subset T$ and the two-men vehicle $T'' \subset T$. The customers are also classified as either one-man customer $C' \subset C$, or two-men customer $C'' \subset C$. The one-man customer can be served by all vehicles, while two-men customer can only be served by two-men vehicles.

The objective is to maximise the total number of allocated customer services while respecting all the constraints on drivers, vehicles and the depot.

Let x^i_{jk} be a binary variable indicating if customer j is the immediate predecessor of customer k in the route of vehicle i; η^i_j be a binary variable indicating if customer j is allocated to vehicle i; γ^i_j be a binary variable indicating if customer j is the first customer to visit after vehicle i departing from the depot; θ^i_j be a binary variable indicating if customer j is the last customer in the route of vehicle i. Denote the time when driver in vehicle i starts serving customer k by s^i_k; the weight of the vehicle when leaving customer j by y_j; the volume of the vehicle when leaving customer j by z_j. The considered problem can be formulated as follows:

$$J = \max \sum_{i \in T} \sum_{j \in C} \eta^i_j \tag{1}$$

subject to

$$\sum_{i \in T} \eta^i_j \leq 1, \qquad \forall j \in C \tag{2}$$

$$\sum_{j \in C} \gamma^i_j \leq 1, \qquad \forall i \in T \tag{3}$$

$$\gamma^i_j + \sum_{k \in C} x^i_{k,j} = \eta^i_j, \qquad \forall i \in T, j \in C \tag{4}$$

$$\theta^i_j + \sum_{k \in C} x^i_{j,k} = \eta^i_j, \qquad \forall i \in T, j \in C \tag{5}$$

$$a_j \leq s^i_j \leq b_j, \qquad \forall j \in C, i \in T \tag{6}$$

$$(r_i + \delta_i + t_{0,k})\gamma^i_k \leq s^i_k, \qquad \forall i \in T, k \in C \tag{7}$$

$$s_j^i + (p_j + t_{j,k})x_{j,k}^i + (a_k - b_j)(1 - x_{j,k}^i) \leq s_k^i, \quad \forall i \in T, \forall (j,k) \in A_C \tag{8}$$

$$p_j + s_j^i - r_i - (p_j + b_j - r_i)(1 - \theta_j^i) \leq S_i, \quad \forall j \in C, i \in T \tag{9}$$

$$\sum_{k \in C} w_k^d \eta_k^i \leq W_i, \quad \forall i \in T \tag{10}$$

$$y_k \leq W_i + (\max_{e \in T} W_e - W_i)(1 - \eta_k^i), \quad \forall i \in T, k \in C \tag{11}$$

$$\sum_{j \in C} w_j^d \eta_j^i - w_k^d + w_k^p - (\max_{e \in T} W_e - w_k^d + w_k^p)(1 - \gamma_k^i) \leq y_k, \quad \forall i \in T, k \in C \tag{12}$$

$$y_j - w_k^d + w_k^p - (\max_{e \in T} W_e - w_k^d + w_k^p)(1 - x_{j,k}^i) \leq y_k, \quad \forall i \in T, \forall (j,k) \in A_C \tag{13}$$

$$\sum_{k \in C} v_k^d \eta_k^i \leq V_i, \quad \forall i \in T \tag{14}$$

$$z_k \leq V_i + (\max_{e \in T} V_e - V_i)(1 - \eta_k^i), \quad \forall i \in T, k \in C \tag{15}$$

$$\sum_{j \in C} v_j^d \eta_j^i - v_k^d + v_k^p - (\max_{e \in T} V_e - v_k^d + v_k^p)(1 - \gamma_k^i) \leq z_k, \quad \forall i \in T, k \in C \tag{16}$$

$$z_j - v_k^d + v_k^p - (\max_{e \in T} V_e - v_k^d + v_k^p)(1 - x_{j,k}^i) \leq z_k, \quad \forall i \in T, \forall (j,k) \in A_C \tag{17}$$

$$\sum_{i \in T'} \sum_{k \in C''} \eta_k^i = 0 \tag{18}$$

$$x_{j,k}^i \in \{0,1\}, \quad \forall \{j,k\} \in A_C, i \in T \tag{19}$$

$$\eta_j^i \in \{0,1\}, \quad \forall i \in T, j \in C \tag{20}$$

$$\theta_j^i \in \{0,1\}, \quad i \in T, j \in C \tag{21}$$

$$y_j \geq 0, \quad \forall j \in C \tag{22}$$

$$z_j \geq 0, \quad \forall j \in C \tag{23}$$

where (7) and (3) guarantee that a vehicle either stays at the depot or visits exactly one customer; (4) and (5) ensure that each customer must have an immediate successor from the same route except for the last customer; together with (2) ensure that a customer is visited by at most one vehicle; the arrival time, loading time at the depot, travelling time between vertices, the time window are taken into account by (7), (8) and (6)–(8) respectively; the shift length, weight capacity, volume capacity are enforced by (9), (10)–(13), and (14)–(17) respectively; (6) and (8) eliminate subtours by virtue of $p_i > 0$.

3 ILS with Neighbourhood Reduction

A critical component of ILS is the design of proper neighbourhood structures. It has been demonstrated by many publications that permitting infeasible solutions in local search together with the use of an augmented objective function can significantly boost the performance of the meta-heuristics in the field of vehicle routing problem [1,2,8,12]. The neighbourhood structures considered in this paper are defined by the commonly used edge exchange operators, which allows the violation of the time window, shift length, weight and volume capacity constraints. However, the algorithm presented in this paper reduces the size of the neighbourhood by only allowing moves that lead to more allocations than the best known feasible solution. To be specific, let $J(s)$ be the number of allocated customers in a solution s which can be infeasible; $H(s, O)$ be the neighbourhood of a solution s induced by an edge exchange operator O permitting infeasible solution. The corresponding reduced neighbourhood is defined as

$$\widehat{H}(s, O) = \{s' \in H(s, O) | J(s') > J(s^*)\}$$

where s^* be the best known feasible solution. In the studied problem, it is permitted to have customers not allocated. Therefore, feasible solutions can be efficiently generated using simple heuristics (see Sect. 3.1 for more details). It should be noted that the reduced neighbourhood is dynamic since s^* can be updated in the iterative process of ILS. Since ILS can quickly find good solutions, the size of the reduced neighbourhood becomes significantly smaller after just a few iterations, which leads to faster convergence of the algorithm. Also, the solution process can be more stable because only solutions with more allocations are considered in the local search process.

The paper considers two edge exchange operators

- inter-route swap O_1: exchanges a sequence of up to two consecutive customers in a route with a sequence of up to two consecutive customers in another route; exchanges a sequence of up to two consecutive customers in a route with at most one unallocated customer;
- intra-route swap O_2: extract at most two consecutive customers from a route and insert it into a different position of the same route; reverse the order of a sequence of consecutive customers in the route.

It should be noted that O_2 cannot increase the number of allocated customers. Therefore, it is used mainly for repair infeasibility in the local search procedure.

In the local search procedure, the solution in the reduced neighbourhood is evaluated based on the augmented objective function

$$f(s) = J(s) - \alpha \times TW(s) - \beta \times WD(s) - \sigma \times Weight(s) - \psi \times Volume(s) \tag{24}$$

where $TW(s)$, $WD(s)$, $Weight(s)$, $Volume(s)$ are the total violation for constraints on time window, working duration, weight, volume corresponding to s and $\alpha, \beta, \sigma, \psi$ are non-negative weights for $TW(s)$, $WD(s)$, $Weight(s)$, $Volume(s)$. Furthermore, $TW(s)$, $WD(s)$, $Weight(s)$, $Volume(s)$ are computed by the technique used in [8,10,11].

The details of the local search procedure based on the reduced neighbourhood (NRS) are given in Algorithm 1.

Algorithm 1. NRS(s)

```
1: while TRUE do
2:    if Ĥ(s, O_1) == ∅ then return s* end if
3:    s' = s
4:    s = argmax_x{f(x)|x ∈ Ĥ(s, O_1)}
5:    if f(s') < f(s) then
6:       s' = s
7:       if s is feasible then s* = s end if
8:    else
9:       Break
10:   end if
11: end while
12: s = s'
13: if Ĥ(s, O_2) ≠ ∅ then s = argmax_x{f(x)|x ∈ Ĥ(s, O_2)} end if
14: if f(s') < f(s) then
15:    if s if feasible then s* = s end if
16: else
17:    s = s'
18: end if
19: return s
```

In this pseudocode, the input solution s is permitted to be infeasible. The edge exchange operator O_1 is applied until a local optimum is found under the reduced neighbourhood $\widehat{H}(s, O_1)$. Since the size of the reduced neighbourhood is related to the number of allocations in the current global optimum s^*, s^* is updated whenever a new global optimum is found (line 7 and 15). It should be noted that $\widehat{H}(s, O_1)$ is empty only if the current s^* has all customers allocated, which is also the global optimum. Following the strategy in [12], local search based on the edge exchange operator O_2 is performed for at most one iteration after the local optimum under O_1 is found (line 13). In line 13, $\widehat{H}(s, O_2)$ is empty only if s is a feasible solution. The output of NRS is either the input solution, or a solution with more allocated customers and higher augmented objective function value.

Algorithm 2. ILS with neighbourhood reduction (ILS-NR)

1: $s' = \text{INITIAL}()$
2: $s^* = s'$
3: $t = J(s^*)$
4: $h = 1$
5: **while** s^* has unallocated customers **and** $h \leq M$ **do**
6: $\alpha = \beta = \sigma = \psi = 1$
7: $e = 1$
8: **repeat**
9: $\bar{s} = s'$
10: $s' = \text{NRS}(s')$
11: **if** $f(\bar{s}) \neq f(s')$ **then** Update $\alpha, \beta, \sigma, \psi$ **end if**
12: $e++$
13: **until** $f(\bar{s}) == f(s')$ **or** s^* has unallocated customers **or** $e > E$
14: **if** $J(s^*) > t$ **then**
15: $t = J(s^*)$
16: $h = 1$
17: **end if**
18: $s' = \text{PERTURB}(h)$
19: $h++$
20: **end while**
21: **return** s^*

The ILS with neighbourhood reduction (ILS-NR) is now presented in Algorithm 2. It begins with the INITIAL procedure which generates a feasible solution for the problem (line 1). The details of INITIAL is given in Sect. 3.1. This solution is also the current best known solution s^* (line 2). It should be noted that the current best known s^* is a global variable and may be updated inside the NRS and PERTURB procedure.

After the call of the INITIAL procedure, the WHILE loop (line 5–20) is executed if the current best known solution s^* has at least one unallocated customer. The WHILE loop terminates if the current best known solution allocates all customers, or counter h exceeds M which is a parameter. Each iteration of the WHILE loop (line 5–20) attempts to find a solution with more allocations than the current best known solution s^* applying the local search procedure (line 8–13).

Each iteration of the local search (line 8–13) is an applications of NRS which aims to find a solution with a better value of the augmented objective function (24). The penalties for violation of corresponding constraints are updated to force the convergence to feasible solutions. Following [1,2,12], at the beginning of each iteration of the local search (8–13), the initial value for weights $\alpha, \beta, \sigma, \psi$ in the augmented objective function (24) are set to one (line 6). If NRS returns an improving solution, a weight is multiplied by $1 + \Delta$ if the corresponding constraint has a positive violation; otherwise the weight is divided by $1 + \Delta$. Δ is a parameter that controls the strength of the adjustment. This weight updating mechanism is effective in producing feasible solutions, which explains why O_2 is only applied for one iteration in NRS (line 10). Local search terminates if either the NRS procedure fails to obtain a solution with better value of the augmented objective function (24), the current best known solution s^* allocates all customers or the count e exceeds E which is a parameter.

3.1 INITIAL Procedure

The INITIAL procedure is a sweep heuristic [4] that constructs a feasible solution for the problem. First a list of customers is constructed based on the geographic coordinates of the customers. Then the customers are inserted to a route one by one until no customer can be inserted, in which case a new route is constructed. Since one-man vehicles can only serve one-man customers, whereas two-men vehicles can serve all-types of customers, the procedure constructs the routes for one-man vehicles first, then followed by the routes for two-men vehicles. When inserting a customer into the route, the procedure chooses the insertion position that respects all the constraints and gives the smallest increase in travel time. The procedure terminates until either no customers can be inserted into the vehicle's route, or all customers have been allocated.

3.2 PERTURB Procedure

The PERTURB procedure expands the search space by randomly perturbing the current best solution s^*. An unallocated customer is randomly chosen, and then inserted into a position among the routes which gives the largest value of (24) when $\alpha = \beta = \sigma = \psi = 1$. Then, two randomly selected sequences of consecutive customers are swapped between two randomly selected routes. This random swap will be performed multiple times which depends on the counter h in the pseudocode for the ILS-NR (Algorithm 2). To be specific, the number of swaps starts from one and increases by one each time when counter h in Algorithm 2 increase. The current best solution s^* may also be updated in this process.

4 Computational Study

This section presents the results of computational experiments aimed at the evaluation of the performance of ILS-NR. A total of 60 instances were provided by a transportation company working in the retail industry. Each instance represents the real-world situation on a particular day. The travel time from the location of the depot to each customer, and the travel time between the location of each customer are specified by a symmetric matrix. The time when driver arrives at the depot is specified by a roster and each driver can work for a maximum of 10 h. ILS-NR is implemented in c++, and compiled with g++ O3. The following settings are used throughout the experiments [12]:

Table 1. Comparison of performance between CPLEX, ILS-NR and CPLEX warm start

Instances	$\|C\|$	$\|T\|$	CPLEX			ILS-NR				CPLEX warm start			
			Obj	Gap(%)	Time(s)	Avg	Max	#Max	Time(s)	Input	Obj	Gap(%)	Time(s)
M-2017-07-23	30	3	27	10.88	9112	**28.00**	**28**	**30**	**0.13**	**28**	**28**	**0.00**	**3711**
M-2017-07-24	26	2	21	9.52	14257	21.93	22	28	0.10	22	22	9.09	21600
M-2017-07-25	14	2	**14**	**0.00**	**1**	**14.00**	**14**	**30**	**0.00**	**14**	**14**	**0.00**	0
M-2017-10-08	28	2	24	12.50	10288	24.63	26	1	0.17	26	26	3.85	21600
M-2017-10-09	22	2	21	4.76	21600	21.00	21	30	0.03	21	21	4.76	21600
M-2017-10-10	22	2	17	11.76	21600	17.00	17	30	0.07	17	17	11.76	11304
M-2017-10-16	34	2	26	19.99	21600	26.10	27	3	0.30	27	27	15.55	21600
M-2017-10-17	24	2	21	9.52	21600	21.30	22	10	0.10	22	22	4.55	21600
M-2017-10-21	34	2	24	29.17	21600	26.87	28	1	0.33	28	28	12.75	21600
M-2017-10-24	17	2	**17**	**0.00**	3	**16.90**	**17**	**27**	**0.00**	**17**	**17**	**0.00**	0
M-2017-10-30	37	2	27	29.63	21600	28.90	30	1	0.40	30	30	16.85	21600
M-2017-12-22	72	7	66	9.09	21600	69.43	70	13	3.47	70	70	2.86	21600
M-2017-12-23	70	5	59	18.64	21600	65.80	67	3	3.47	67	67	4.48	21600
M-2017-12-24	70	5	50	40.00	21600	57.47	59	1	3.20	59	59	18.64	21600
M-2017-12-25	70	5	52	25.00	21600	57.50	59	1	3.40	59	59	10.17	21600
R-2017-07-23	47	5	**47**	**0.00**	463	**47.00**	**47**	**30**	**0.00**	**47**	**47**	**0.00**	1
R-2017-07-24	65	3	48	14.58	21600	52.13	53	5	2.60	53	53	3.77	21600
R-2017-07-25	43	4	**42**	**0.00**	19472	**42.00**	**42**	**30**	**0.50**	**42**	**42**	**0.00**	0
R-2017-10-08	88	6	80	8.71	21600	**85.60**	**86**	**18**	**7.40**	**86**	**86**	**0.00**	**18582**
R-2017-10-09	63	4	54	5.56	21600	55.27	56	8	2.37	56	56	1.79	21600
R-2017-10-10	44	5	**44**	**0.00**	593	**44.00**	**44**	**30**	**0.00**	**44**	**44**	**0.00**	0
R-2017-10-16	72	5	64	9.37	21600	68.67	69	20	3.50	69	69	1.77	21600
R-2017-10-17	37	4	34	8.82	5084	35.93	36	28	0.37	36	36	2.78	21600
R-2017-10-21	60	5	55	5.45	21600	**58.00**	**58**	**30**	**1.80**	**58**	**58**	**0.00**	**1**
R-2017-10-24	53	6	**53**	**0.00**	**790**	**53.00**	**53**	**30**	**0.00**	**53**	**53**	**0.00**	**1**
R-2017-10-30	71	7	69	2.90	21600	**70.67**	**71**	**20**	**1.43**	**71**	**71**	**0.00**	**1**
R-2017-12-12	52	4	49	6.12	21600	**51.43**	**52**	**18**	**0.67**	**52**	**52**	**0.00**	**1**
R-2017-12-19	52	4	46	10.87	21600	**50.47**	**51**	**14**	**1.20**	**51**	**51**	**0.00**	0
R-2017-12-22	62	4	53	15.09	21600	57.03	58	3	2.47	58	58	5.17	21600
R-2017-12-23	70	5	63	9.52	21600	67.73	68	22	3.23	68	68	1.47	21600
R-2017-12-24	70	5	56	10.71	21600	**60.70**	**62**	**1**	**3.27**	**62**	**62**	**0.00**	**2**
R-2017-12-25	70	5	65	7.69	21600	**69.77**	**70**	**23**	**0.93**	**70**	**70**	**0.00**	**1**
T-2017-07-23	64	5	63	1.59	21600	**64.00**	**64**	**30**	**0.00**	**64**	**64**	**0.00**	**1**
T-2017-07-24	70	5	67	2.99	21600	**69.00**	**69**	**30**	**2.63**	**69**	**69**	**0.00**	**1**
T-2017-07-25	57	4	55	3.64	21600	**56.77**	**57**	**23**	**0.57**	**57**	**57**	**0.00**	0
T-2017-10-08	65	8	**65**	**0.00**	**3834**	**65.00**	**65**	**30**	**0.00**	**65**	**65**	**0.00**	**1**
T-2017-10-09	43	7	**43**	**0.00**	**31**	**43.00**	**43**	**30**	**0.00**	**43**	**43**	**0.00**	**1**
T-2017-10-10	46	5	**46**	**0.00**	**675**	**46.00**	**46**	**30**	**0.00**	**46**	**46**	**0.00**	0
T-2017-10-16	63	7	**63**	**0.00**	**6631**	**63.00**	**63**	**30**	**0.00**	**63**	**63**	**0.00**	**2**
T-2017-10-17	56	4	49	12.24	21600	52.53	53	16	1.43	53	53	3.77	13380
T-2017-10-21	76	4	58	8.62	21600	61.93	62	28	4.23	62	62	1.61	21600
T-2017-10-24	62	4	52	10.05	21600	55.33	56	10	2.30	56	56	1.79	21600
T-2017-10-30	36	5	**36**	**0.00**	**13**	**36.00**	**36**	**30**	**0.00**	**36**	**36**	**0.00**	0
T-2017-12-12	63	7	**63**	**0.00**	**1345**	**63.00**	**63**	**30**	**0.00**	**63**	**63**	**0.00**	**2**
T-2017-12-19	54	5	**54**	**0.00**	**923**	**54.00**	**54**	**30**	**0.00**	**54**	**54**	**0.00**	**1**
T-2017-12-22	91	7	75	18.67	21600	**88.73**	**89**	**22**	**7.47**	**89**	**89**	**0.00**	**7**
T-2017-12-23	70	5	63	11.11	21600	**69.93**	**70**	**28**	**0.67**	**70**	**70**	**0.00**	**1**
T-2017-12-24	70	5	63	9.52	21600	67.10	68	3	3.77	68	68	1.47	21600

(continued)

Table 1. (*continued*)

Instances	$\|C\|$	$\|T\|$	CPLEX			ILS-NR				CPLEX warm start			
			Obj	Gap(%)	Time(s)	Avg	Max	#Max	Time(s)	Input	Obj	Gap(%)	Time(s)
T-2017-12-25	70	5	64	9.37	21600	68.53	69	16	3.23	69	69	1.45	21600
T-2017-12-26	70	5	65	4.62	21600	**67.97**	**68**	**29**	**3.07**	**68**	**68**	**0.00**	**1**
A-2017-10-16	100	4	53	30.19	21600	61.73	3	2	4.90	63	63	9.52	21600
A-2017-12-22	100	7	76	22.37	21600	82.70	84	2	9.20	84	84	10.72	21600
B-2017-10-08	100	6	72	15.16	21600	79.60	80	18	8.70	80	80	3.65	21600
B-2017-10-16	100	5	71	15.49	21600	78.93	80	6	8.03	80	80	2.50	21600
B-2017-10-30	100	7	81	13.58	21600	86.93	88	5	9.40	88	88	4.55	21600
B-2017-12-22	100	4	58	41.79	21600	67.97	70	1	7.90	70	70	17.49	21600
C-2017-07-24	100	5	86	9.30	21600	92.83	93	25	8.77	93	93	1.65	21600
C-2017-10-16	100	7	96	2.08	21600	**97.97**	**98**	**29**	**6.97**	**98**	**98**	**0.00**	**5**
C-2017-10-21	100	4	67	20.47	21600	75.40	76	12	9.67	76	76	6.06	21600
C-2017-12-22	100	7	89	11.24	21600	**97.87**	**99**	1	**10.20**	**99**	**99**	**0.00**	**5**

- The maximum permissible number of consecutive unsuccessful attempts to improve the current best known solution (the parameter M in Algorithm 2) is computed as $|C| + \lambda|T|)$, where C is the set of all customers, T is the set of all vehicles, $\lambda = 10$.
- The maximum number of exchange operations in the perturbation is five.
- The parameter Δ for adjusting the weights (Sect. 3) is 0.5.

In addition, the maximum permissible iterations for local search (the paramter E in Algorithm 2) is 100. All computational experiments are conducted on a computer with Intel Xeon CPU E5-2697 v3 2.60 GHz and 8 GB RAM.

We first compare the performance of ILS-NR with CPLEX which solves the IP model in Sect. 2. Furthermore, we test the performance of CPLEX when the best solution from ILS-NR is used as a warm start. Both CPLEX and CPLEX with warm start have a time limit of 6 h and memory limit of 7.5GB RAM. Version 12.10 of CPLEX is used for all the tests. In Table 1, the groups titled "CPLEX" and "CPLEX warm start" contain results obtained by CPLEX and CPLEX with warm start. In these groups, the objective value, optimality gap, computational time are displayed in columns titled "Obj", "Gap(%)" and "Time(s)". The column titled "Input" in group "CPLEX warm start" displays the objective value of the warm start solution. ILS-NR is run 30 times on each instance with the average objective value ("Avg."), best objective value ("Max."), number of runs the best objective value is obtained ("#Max") and computation time ("Time(s)") being reported under the group "ILS-NR".

According to Table 1, CPLEX can prove optimality for 13 instances. With warm start, CPLEX can prove optimality for another 16 instances with significantly reduced CPU time. Among these 29 instances proved optimality by CPLEX, ILS-NR can find optimal solutions with high frequency (#Max) within 10.2 s. For 45 out of 60 instances, the average objective values produced by ILS-NR are better than the objective values produced by CPLEX which has a time limit of 6 h.

Table 2. Comparison of performance between ILS and ILS-NR

Instances	$\|C\|$	$\|T\|$	ILS			ILS-NR					
			Avg.	Max	Time(s)	Avg.	%	Max.	%	Time(s)	%
M-2017-07-23	30	3	28.00	28	0.17	28.00	0.00	28	0.00	**0.13**	**20.00**
M-2017-07-24	26	2	21.57	22	0.13	**21.93**	**-1.70**	22	0.00	**0.10**	**25.00**
M-2017-07-25	14	2	14.00	14	0.00	14.00	0.00	14	0.00	0.00	0.00
M-2017-10-08	28	2	24.13	25	0.17	**24.63**	**−2.07**	**26**	**−4.00**	0.17	0.00
M-2017-10-09	22	2	21.00	21	0.07	21.00	0.00	21	0.00	**0.03**	**50.00**
M-2017-10-10	22	2	16.97	17	0.03	**17.00**	**−0.20**	17	0.00	0.07	−100.00
M-2017-10-16	34	2	25.83	26	0.33	**26.10**	**−1.03**	**27**	**−3.85**	**0.30**	**10.00**
M-2017-10-17	24	2	21.03	22	0.10	**21.30**	**−1.27**	22	0.00	0.10	0.00
M-2017-10-21	34	2	26.23	28	0.27	**26.87**	**−2.41**	28	0.00	0.33	−25.00
M-2017-10-24	17	2	17.00	17	0.03	16.90	0.59	17	0.00	**0.00**	**100.00**
M-2017-10-30	37	2	28.47	29	0.40	**28.90**	**−1.52**	**30**	**−3.45**	0.40	0.00
M-2017-12-22	72	7	69.30	70	4.37	**69.43**	**−0.19**	70	0.00	**3.47**	**20.61**
M-2017-12-23	70	5	65.70	67	4.33	**65.80**	**−0.15**	67	0.00	**3.47**	**20.00**
M-2017-12-24	70	5	57.47	59	3.87	57.47	0.00	59	0.00	**3.20**	**17.24**
M-2017-12-25	70	5	57.37	58	4.00	**57.50**	**−0.23**	**59**	**−1.72**	**3.40**	**15.00**
R-2017-07-23	47	5	47.00	47	0.00	47.00	0.00	47	0.00	0.00	0.00
R-2017-07-24	65	3	51.80	52	2.93	**52.13**	**−0.64**	**53**	**−1.92**	**2.60**	**11.36**
R-2017-07-25	43	4	42.00	42	0.67	42.00	0.00	42	0.00	**0.50**	**25.00**
R-2017-10-08	88	6	85.50	86	16.10	**85.60**	**−0.12**	86	0.00	**7.40**	**54.04**
R-2017-10-09	63	4	55.07	56	2.90	**55.27**	**−0.36**	56	0.00	**2.37**	**18.39**
R-2017-10-10	44	5	44.00	44	0.00	44.00	0.00	44	0.00	0.00	0.00
R-2017-10-16	72	5	68.90	70	4.73	68.67	0.34	69	1.43	**3.50**	**26.06**
R-2017-10-17	37	4	36.00	36	0.43	35.93	0.19	36	0.00	**0.37**	**15.38**
R-2017-10-21	60	5	58.00	58	2.30	58.00	0.00	58	0.00	**1.80**	**21.74**
R-2017-10-24	53	6	53.00	53	0.00	53.00	0.00	53	0.00	0.00	0.00
R-2017-10-30	71	7	70.77	71	1.33	70.67	0.14	71	0.00	1.43	−7.50
R-2017-12-12	52	4	51.10	52	1.10	**51.43**	**−0.65**	52	0.00	**0.67**	**39.39**
R-2017-12-19	52	4	50.50	51	1.47	50.47	0.07	51	0.00	**1.20**	**18.18**
R-2017-12-22	62	4	56.67	58	9.67	**57.03**	**−0.65**	58	0.00	**2.47**	**74.48**
R-2017-12-23	70	5	67.77	68	4.07	67.73	0.05	68	0.00	**3.23**	**20.49**
R-2017-12-24	70	5	60.73	61	3.73	60.70	0.05	**62**	**−1.64**	**3.27**	**12.50**
R-2017-12-25	70	5	69.83	70	0.87	69.77	0.10	70	0.00	0.93	−7.69
T-2017-07-23	64	5	64.00	64	0.00	64.00	0.00	64	0.00	0.00	0.00
T-2017-07-24	70	5	69.00	69	3.20	69.00	0.00	69	0.00	**2.63**	**17.71**
T-2017-07-25	57	4	56.47	57	0.93	**56.77**	**−0.53**	57	0.00	**0.57**	**39.29**
T-2017-10-08	65	8	65.00	65	0.00	65.00	0.00	65	0.00	0.00	0.00
T-2017-10-09	43	7	43.00	43	0.00	43.00	0.00	43	0.00	0.00	0.00
T-2017-10-10	46	5	46.00	46	0.00	46.00	0.00	46	0.00	0.00	0.00
T-2017-10-16	63	7	63.00	63	0.00	63.00	0.00	63	0.00	0.00	0.00
T-2017-10-17	56	4	52.47	53	1.63	**52.53**	**−0.13**	53	0.00	**1.43**	**12.24**
T-2017-10-21	76	4	61.30	62	4.37	**61.93**	**−1.03**	62	0.00	**4.23**	**3.05**
T-2017-10-24	62	4	55.03	56	2.60	**55.33**	**−0.55**	56	0.00	**2.30**	**11.54**
T-2017-10-30	36	5	36.00	36	0.00	36.00	0.00	36	0.00	0.00	0.00
T-2017-12-12	63	7	63.00	63	0.00	63.00	0.00	63	0.00	0.00	0.00
T-2017-12-19	54	5	54.00	54	0.00	54.00	0.00	54	0.00	0.00	0.00
T-2017-12-22	91	7	88.97	89	15.00	88.73	0.26	89	0.00	**7.47**	**50.22**
T-2017-12-23	70	5	69.50	70	2.17	**69.93**	**−0.62**	70	0.00	**0.67**	**69.23**
T-2017-12-24	70	5	66.90	68	4.03	**67.10**	**−0.30**	68	0.00	**3.77**	**6.61**
T-2017-12-25	70	5	68.17	69	3.97	**68.53**	**−0.54**	69	0.00	**3.23**	**18.49**

(continued)

Table 2. (*continued*)

Instances	$\|C\|$	$\|T\|$	ILS			ILS-NR					
			Avg.	Max	Time(s)	Avg.	%	Max.	%	Time(s)	%
T-2017-12-26	70	5	67.87	68	12.83	**67.97**	**−0.15**	68	0.00	**3.07**	**76.10**
A-2017-10-16	100	4	61.23	62	5.97	**61.73**	**−0.82**	**63**	**−1.61**	**4.90**	**17.88**
A-2017-12-22	100	7	82.63	84	10.90	**82.70**	**−0.08**	84	0.00	**9.20**	**15.60**
B-2017-10-08	100	6	79.33	81	16.80	**79.60**	**−0.34**	80	1.23	**8.70**	**48.21**
B-2017-10-16	100	5	78.17	80	9.20	**78.93**	**−0.98**	80	0.00	**8.03**	**12.68**
B-2017-10-30	100	7	86.47	88	17.07	**86.93**	**−0.54**	88	0.00	**9.40**	**44.92**
B-2017-12-22	100	4	67.20	68	8.90	**67.97**	**−1.14**	**70**	**−2.94**	**7.90**	**11.24**
C-2017-07-24	100	5	92.50	93	12.00	**92.83**	**−0.36**	93	0.00	**8.77**	**26.94**
C-2017-10-16	100	7	98.00	98	10.33	97.97	0.03	98	0.00	**6.97**	**32.58**
C-2017-10-21	100	4	75.20	76	16.60	**75.40**	**−0.27**	76	0.00	**9.67**	**41.77**
C-2017-12-22	100	7	98.20	99	13.73	97.87	0.34	99	0.00	**10.20**	**25.73**
Average			56.19	56.70	4.05	56.33	−0.32	56.82	−0.31	2.67	17.61

To demonstrate the effectiveness of the neighbourhood reduction, Table 2 presents the computational results for ILS-NR and ILS without neighbourhood reduction. The performance of ILS-NR was measured against ILS by the percentage difference

$$\frac{X_{ILS} - X_{ILS-NR}}{X_{ILS}} \times 100 \tag{25}$$

where X can either be the average objective value (column "Avg."), best found objective value ("Max") or CPU time ("Time"); X_{ILS-NR} is the value obtained by ILS-NR and X_{ILS} is the value obtained by ILS. Therefore, a negative percentage difference indicates that ILS-NR is better with respect to the average objective value and best found objective value, while a positive percentage difference indicates that ILS-NR is better with respect to CPU time. For readers' convenience, the superior results produced by ILS-NR are shown in bold.

In Table 2, ILS-NR is faster than ILS on 41 out of 60 instances with an average difference of 17.61%, which clearly demonstrates the improvement on computation time due to neighbourhood reduction. In terms of stability, the average objective value produced by the ILS-NR outperforms the average objective value produced by ILS on 49 instances.

5 Conclusion

This paper considers a practical vehicle routing problem with simultaneous pickups and deliveries which arises in the retail sector. A novel neighbourhood reduction technique is introduced to enhance the performance of the state-of-the-art iterated local search algorithm. Computational experiments carried out on a set of real-world instances demonstrate the superior performance of the proposed algorithm in terms of computational time, solution quality and stability. The advantage of the proposed algorithm is more conspicuous for time-critical applications given the longest computation time among the test instances is just 10.2 s.

References

1. Cordeau, J.F., Gendreau, M., Laporte, G.: A tabu search heuristic for periodic and multi-depot vehicle routing problems. Netw.: Int. J. **30**(2), 105–119 (1997)
2. Cordeau, J.F., Laporte, G., Mercier, A.: A unified tabu search heuristic for vehicle routing problems with time windows. J. Oper. Res. Soc. **52**(8), 928–936 (2001)
3. Garey, M.R., Johnson, D.S.: Computers and Intractability, vol. 174. Freeman, San Francisco (1979)
4. Gillett, B.E., Miller, L.R.: A heuristic algorithm for the vehicle-dispatch problem. Oper. Res. **22**(2), 340–349 (1974)
5. Gu, H., Zhang, Y., Zinder, Y.: Lagrangian relaxation in iterated local search for the workforce scheduling and routing problem. In: Kotsireas, I., Pardalos, P., Parsopoulos, K.E., Souravlias, D., Tsokas, A. (eds.) SEA 2019. LNCS, vol. 11544, pp. 527–540. Springer, Cham (2019). https://doi.org/10.1007/978-3-030-34029-2_34
6. Koç, Ç., Laporte, G., Tükenmez, İ: A review on vehicle routing with simultaneous pickup and delivery. Comput. Oper. Res. 104987 (2020)
7. Lourenço, H.R., Martin, O.C., Stützle, T.: Iterated local search: framework and applications. In: Gendreau, M., Potvin, J.-Y. (eds.) Handbook of Metaheuristics. ISORMS, vol. 272, pp. 129–168. Springer, Cham (2019). https://doi.org/10.1007/978-3-319-91086-4_5
8. Nagata, Y., Bräysy, O., Dullaert, W.: A penalty-based edge assembly memetic algorithm for the vehicle routing problem with time windows. Comput. Oper. Res. **37**(4), 724–737 (2010)
9. Parragh, S.N., Doerner, K.F., Hartl, R.F.: A survey on pickup and delivery problems. J. für Betriebswirtschaft **58**(2), 81–117 (2008)
10. Vidal, T., Crainic, T.G., Gendreau, M., Prins, C.: A hybrid genetic algorithm with adaptive diversity management for a large class of vehicle routing problems with time-windows. Comput. oper. Res. **40**(1), 475–489 (2013)
11. Vidal, T., Crainic, T.G., Gendreau, M., Prins, C.: A unified solution framework for multi-attribute vehicle routing problems. Eur. J. Oper. Res. **234**(3), 658–673 (2014)
12. Xie, F., Potts, C.N., Bektaş, T.: Iterated local search for workforce scheduling and routing problems. J. Heurist. **23**(6), 471–500 (2017)

A Genetic Algorithm for the Three-Dimensional Open Dimension Packing Problem

Cong Tan Trinh Truong(✉), Lionel Amodeo, and F. Yalaoui

LOSI, University of Technology of Troyes, Troyes, France
{cong_tan_trinh.truong,lionel.amodeo,farouk.yalaoui}@utt.fr

Abstract. The three-dimensional Open Dimension Packing problem (3D-ODPP) is a real-world driven optimization problem that aims at the minimization of package volume in right-size packaging systems. The problem can be found in many industrial scenarios, such as e-commerce secondary packaging. The objective of the 3D-ODPP is to find out the length, width, and height of the cardboard box that can be used to pack a given set of or products so that the volume of the box is minimal. Many literature researches have focused on exact methods to deal with the 3D-ODPP. Despite the fact that the exact methods are capable of finding the global solution, their applications are very limited in terms of problem size and computational time because the 3D-ODPP is NP-hard in the strong sense. In addition, constructive and meta-heuristic methods for solving the 3D-ODPP have not been discussed frequently in the literature and remain a gap in the state-of-the-art.

This paper proposes a genetic algorithm that deals with the 3D-ODPP. The genetic process is to find out the packing sequence and the orientation of products. To construct the solution, a new greedy-search product placement algorithm is developed. This placement algorithm is used to determine the position where each product is placed and to calculate the volume of the package. Literature instances are tested and the obtained solutions are compared with that given by existing exact methods. The experiments show that the proposed algorithm has the capacity of solving the 3D-ODPP in a reasonable time and gives competitive solutions compared with the benchmark methods, especially for problems with many products.

1 Introduction

The three-dimensional Open Dimension Packing problem (*3D-ODPP*), one of the Cutting and Packing problems according to the typology of [13], is a real-world driven optimization problem that aims at the minimization of package volume in right-size packaging systems. The problem can be found in many industrial scenarios, such as e-commerce secondary packaging. The objective of the *3D-ODPP* is to find out the length, width, and height of the cardboard box

B. Dorronsoro et al. (Eds.): OLA 2021, CCIS 1443, pp. 203–215, 2021.
https://doi.org/10.1007/978-3-030-85672-4_15

that can be used to pack a given set of small items (or products) so that the volume of the box is minimal.

Many literature researches, such as [6,8,9,11] have focused on exact methods to deal with the *3D-ODPP*. These works are inspired by the mathematical model of [2] that deals with the mono container loading problem. Despite the fact that the exact methods like that in [6,8,9,11] are capable of finding the global solution, their applications are very limited in terms of problem size and computational time because the *3D-ODPP* is NP-hard in the strong sense. In addition, constructive and meta-heuristic methods for solving the *3D-ODPP* have not been discussed frequently in the literature. Therefore, solving large-sized *3D-ODPP* is still a gap in the state-of-the-art.

Genetic algorithms have been widely used in the literature for solving Cutting and Packing problems [7]. For the container loading problem, genetic algorithms are also used to deal with single and multi-objective problems [1,3,5,14]. However, meta-heuristic approaches are still not present much in the literature. This paper proposes a genetic algorithm called *"GA-ODP"* that deals with the *3D-ODPP*. The proposed method is inspired by the random-key biased genetic algorithm for solving 2D and 3D Bin Packing problem presented in [4]. The algorithm of [4] aims at minimizing the number of bins (container objects with fixed measurements) needed to pack a set of items. In order to construct a solution, three main decisions to be made are: the packing sequence indicating the order in which the items are packed; the orientation of each item; and the positions where the items are placed. [4] use a genetic algorithm to determine the orientation and the packing sequence of the items, then a constructive algorithm is applied to determine items' positions. The *GA-ODP* in this paper uses the same chromosome representation as that in [4]. However, the algorithm of [4] only deals with Bin Packing problems where bins' size is fixed, and the solutions are constructed based on the empty spaces rested inside the bins while for the *3D-ODPP*, all box measurements are variable, therefore, the encoding of item orientation, the placement strategy, and the fitness function of [4] cannot be used. This study proposes a new placement algorithm to construct the packing solutions. The new placement algorithm is based on a greedy search algorithm that finds the local optimum for each item placement in a packing sequence. This placement algorithm allows constructing a solution for any packing order and item orientation given by the genetic algorithm. No chromosome repairing is needed in the evolutionary process.

Literature test instances with different problem sizes are tested and the obtained results are compared with that given by existing method [11]. The experiments show that the proposed algorithm has the capacity of solving the *3D-ODPP* and gives competitive solutions compared with state-of-the-art exact methods while the computational time is much shorter and therefore, the proposed method has the capacity of dealing with larger problems.

The rest of this paper is organized as follows: the Sect. 2 represents the formulation of the problem with a Mixed integer nonlinear programming. The Sect. 3 introduces the proposed genetic algorithm that deals with the *3D-ODPP*. Computational experiments are shown in Sect. 4 and the conclusion of this study is in Sect. 5.

2 Problem Formulation

The *3D-ODPP* addressed in this study is a problem arises in an e-commerce secondary packaging system where a set of cuboid-shaped items is to be packed into a single cardboard box before being shipped to client. Given n items of cuboid-shaped with their fixed length (p), width (q), and height (r). Every item can be rotated in six possible orthogonal orientations inside the box. Knowing that the capacity of the system is much higher than the items volume, which means any set of items can be packed by the packaging system. As all the measurements of the box are variable, the purpose is to determine the length (L_b), width (W_b), and height (H_b) of the minimal volume packing box.

The mathematical model for solving the *3D-ODPP* is presented in [8]:
Parameters:

- n: number of items to be packed.
- p, q, r: the vectors indicating items' length, width and height, respectively.
- M: big number used in the model. $M = \left(\sum_{i=1}^{n} p_i\right)^3$

Variables:

- x_i, y_i, z_i ($i \in \{1...n\}$): Continuous variables indicating the coordinate of products.
- L_b, W_b, H_b: Continuous variables for length, width, height of the box, respectively.
- $o_{i,j}$ ($i \in \{1...n\}; j \in \{1...6\}$): Binary variables indicating weather the product i has orientation j. The orientations are defined as shown in Table 1.
- $a_{i,j}, b_{i,j}, c_{i,j}$ ($i, j \in \{1...n\}$): Binary variables indicating the "left-right", "front-behind", and "above-under" relative positions of products i and j. For example, if product i is on the left side of product j then $a_{2,3} = 1$, otherwise, $a_{2,3} = 0$. If there is at least one relative position between two product, then they are called "non-intersected".

Objective function:

$$Minimize \quad L_b \times W_b \times H_b \tag{1}$$

Subject to:

$$\sum_{j=1}^{6} o_{i,j} = 1 \quad \forall i \in \{1...n\} \tag{2}$$

$$a_{i,j} + a_{j,i} + b_{i,j} + b_{j,i} + c_{i,j} + c_{j,i} \geq 1 \quad \forall i, j \in \{1...n\}; i \neq j \tag{3}$$

$$x_i + p_i(o_{i,1} + o_{i,2}) + q_i(o_{i,3} + o_{i,4}) + r_i(o_{i,5} + o_{i,6}) \leq x_j + M(1 - a_{i,j}) \quad \forall i, j \in \{1...n\}; i \neq j \tag{4}$$

$$y_i + p_i(o_{i,3} + o_{i,5}) + q_i(o_{i,1} + o_{i,6}) + r_i(o_{i,2} + o_{i,4}) \leq y_j + M(1 - b_{i,j}) \quad \forall i, j \in \{1...n\}; i \neq j \tag{5}$$

$$z_i+p_i(o_{i,4}+o_{i,6})+q_i(o_{i,2}+o_{i,5})+r_i(o_{i,1}+o_{i,3}) \leq z_j+M(1-c_{i,j}) \quad \forall i,j \in \{1...n\}; i \neq j \tag{6}$$

$$L_b \geq x_i + p_i(o_{i,1} + o_{i,2}) + q_i(o_{i,3} + o_{i,4}) + r_i(o_{i,5} + o_{i,6}) \quad \forall i \in \{1...n\} \tag{7}$$

$$W_b \geq y_i + p_i(o_{i,3} + o_{i,5}) + q_i(o_{i,1} + o_{i,6}) + r_i(o_{i,2} + o_{i,4}) \quad \forall i \in \{1...n\} \tag{8}$$

$$H_b \geq z_i + p_i(o_{i,4} + o_{i,6}) + q_i(o_{i,2} + o_{i,5}) + r_i(o_{i,1} + o_{i,3}) \quad \forall i \in \{1...n\} \tag{9}$$

$$\max_{i\in\{1...n\}} r_i \leq \phi \leq \sum_{i=1}^{n} p_i \qquad \forall \phi \in \{L_b, W_b, H_b\} \tag{10}$$

$$\sum_{i=1}^{n} (p_i \times q_i \times r_i) \leq L_b \times W_b \times H_b \leq \sum_{i=1}^{n} p_i \times \left(\max_{i\in\{1...n\}} q_i\right) \times \left(\max_{i\in\{1...n\}} r_i\right) \tag{11}$$

The constraint (2) assures that an item can only have at most one orientation. The constraints (3) to (6) define the relative positions of items. The constraints (7) to (9) make sure all products are entirely placed inside the box. Finally, the constraints (10) to (11) show the upper and lower bounds of box length, width, height, and volume.

[6,8,9,11] use the logarithm transformation and piecewise linearization technique presented in [10,12] to solve the model with a liner solver, e.g. CPLEX.

Table 1. Item orientations.

Orientation	1	2	3	4	5	6
Among x-axis (l)	p	p	q	q	r	r
Among y-axis (w)	q	r	p	r	p	q
Among z-axis (h)	r	q	r	p	q	p

3 Genetic Algorithm

As mentioned in Sect. 1, the exact methods proposed by [6,8,9,11] are very limited by the problem size and require great computational power. Therefore, an heuristic approach would be necessary in many practical cases. This section describes how the genetic algorithm *GA-ODP* is applied to solve the *3D-ODPP*.

3.1 Solution Encoding and Decoding

In this genetic algorithm, an encoded solution (also known as a "chromosome") is an array made of $2n$ genes that contain the genetic information about items' packing sequence and orientation. The first n genes indicate the order in which the items are loaded into the box. Every genes of this part is a real number whose value is between 0 and 1. The second part of the chromosome includes n genes whose value is an entire number between 1 and 6 indicating the orientation of the items. The actual dimensions (l, w, h) of items among x, y, and z-axis corresponding to item orientations are as shown in Table 1.

Before constructing any solution, the chromosome must be decoded into the actual packing sequence and item orientation so that the constructing algorithm can turn them into item loading position and the box dimensions can be calculated (Fig. 1). The decoded genes are represented as two following vectors:

- **The Vector of Loading Sequence *(VLS)*:** a vector of size $(1 \times n)$ that is made of n elements whose value is an entire number between 1 and n, without repeating, indicating the order in which the items are loaded into the box. The *VLS* is obtained by sorting the first n genes in the ascending order. In other words, the corresponding item of the gene with smaller value among the first n genes will be packed earlier.
- **The Vector of Item Orientation *(VIO)*:** a vector of size $(1 \times n)$ whose elements are entire numbers within $\{1, 2, \ldots, 6\}$. It indicates the exact orientation of n corresponding items. As mentioned in Sect. 2, each item can have one out of the six possible orientations. This vector can be obtained by copying the second part of the encoded chromosome: $VIO_i = C_{2,i} \quad \forall i = 1...n$.

These vectors will be used as input arguments of the placement algorithm presented in Sect. 3.2.

3.2 Solution Construction

This section describes how to construct a solution from the decoded chromosomes and calculate the fitness of each solution. To construct a solution including items' orientation and loading position as well as the dimensions of the bounding box, the decoded chromosome of the solution will be used by a placement algorithm.

Placement Algorithm: The placement algorithm is based on the greedy search algorithm. It places the items one after the other by the specific order defined by the *VLS*, until there is no item left. All the possible position, including x, y, and z-coordinates, where the items can be placed are precalculated and stored in three lists: Px, Py, and Pz, respectively. At the beginning, every list has only one element, which equals to 0. This means the first item will always be placed at the origin of the coordinate system. Every time a new item is placed, at most one coordinate will be added into the each list. The new coordinates allow the following items can be placed next to the item that has just been placed. The placement algorithm will check all the combinations of x, y, and z-coordinates and find out the position for each item so that the increase of box volume caused by it's placement is minimal. Once all items are loaded, the dimensions of the bounding box will be calculated. This placement algorithm assures the feasibility of all solutions. The placement algorithm is shown in Algorithm 1.

Fitness Function: The quality of a solution is considered as the volume of the bounding box given by the placement algorithm. As the box's length, width, and height are L_b, W_b, and H_b, the fitn ess function is:

$$f = L_b \times W_b \times H_b \tag{12}$$

3.3 Evolutionary Process

A population is a set of individuals (chromosomes) created by a generator or by the evolutionary process. At the beginning, a set of n_p chromosomes are randomly generated to made up the initial population. The *GA-ODP* evolves the initial population through n_g generations to improve the solution and find out the best packing sequence as well as items' orientation for the given items. For each generation, the fitness value of every individual is computed, then the individuals are classified into two groups: the first group contains n_e elite individuals with highest fitness. The second group contains n_r "regular" individuals, which are the rest of the population. It is clear that $n_e + n_r = n_p$. To create the population of the next generation, new individuals are generated by the following operators:

- Copying: the chromosome of all the elite individuals are copied directly to the offspring's chromosome of the next generation without modification in the genes.
- Crossover: two individuals are selected from the population of the current generation. Their genes are mixed up by a specific operator to create the chromosome of a new offspring individual. Then the new individual is added to the population of the next generation.
- Mutation: Some individuals from the current population are selected to copy to the next generation but their genes will contain some random modifications created by the mutation operator.

Let n_c be the number of individuals created by the crossover operator, and n_m be the number of mutants, then, $n_r = n_c + n_m$. In other words, for any new generation, the population always has n_e individuals copied from the previous generation and n_r new individuals generated by the crossover and mutation operator. During both operators, there is a small possibility that some mutations appear on some genes and make the offspring's chromosome a little bit different before it is added to the population. Figure 3 illustrates the operations of generating a new population. The crossover and mutation operators are described as follows:

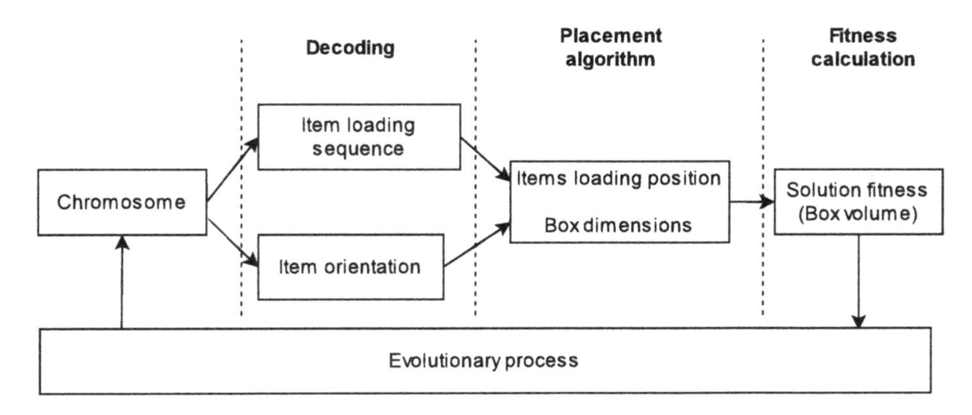

Fig. 1. Genetic algorithm

Copying: From the current population, the chromosome of all individuals with highest fitness will be copied to the offspring's chromosome of the population of the next generation. In other words, all elites of the current population will be copied to the next generation without modification.

Crossover: The *GA-ODP* uses a random-key crossover operator to generate new individuals. For each pair of parent individuals, a crossover vector of $2n$ random real numbers within $[0, 1]$ is generated. A given probability of crossover P_c is also used for this operator. For each number in the crossover vector, if the number is greater than P_c then the gene at the corresponding position of the first parent is copied to the offspring's chromosome. Otherwise, the gene from the second parent is copied to the offspring's chromosome. Figure 2 shows an example of crossover operator.

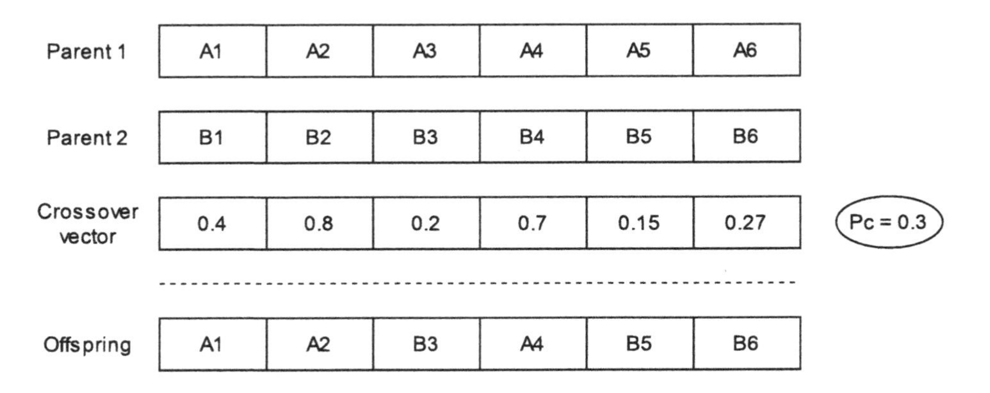

Fig. 2. Crossover operator

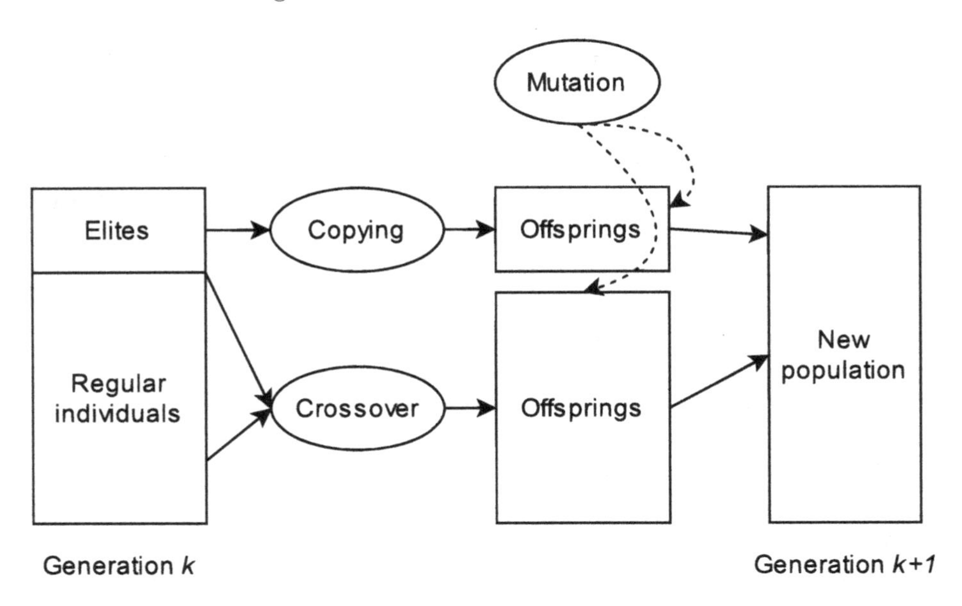

Fig. 3. Evolutionary process

As mentioned in Sect. 3.2, the placement algorithm guarantees that any chromosome can be decoded into a solution with no item intersection, therefore, no reparation process is needed for new individuals.

Mutation: The mutation operator is necessary to increase the diversity of the genetic pool and avoid premature convergence of the population. Let P_m be the probability of mutation, a mutation vector is created in the same way at the crossover vector above. If there is at least one element of the mutation vector is less than P_m, then there will be a mutation occurred on the gene at the corresponding position of the chromosome. Then the gene will be replaced by a new random gene whose value is between $[0, 1]$ if it is a packing sequence gene, and in $\{1, \ldots, 6\}$ if it is an item orientation gene.

Algorithm 1: Placement algorithm

```
Input: p, q, r, VLS, VIO
Output: L, W, H, x, y, z
Result: Items placement position and box dimensions
n = length of (p)
Calculate (l, w, h) corresponding to items' orientation VIO
Sort the items by order of VLS
Px = Py = Pz = {0}
Initialize the set of placed items: B = {}
for i ∈ {1...n} do
    V_min = M
    for x* ∈ Px do
        for y* ∈ Py do
            for z* ∈ Pz do
                Calculate V* when item i is placed at (x*, y*, z*)
                if (V* ≤ V_min) & (No intersection between item i and any item
                 in B) then
                    V_min = V*
                    x_i = x*
                    y_i = y*
                    z_i = z*
    Ajouter item i to B
    if (x_i + l_i) ∉ Px then
        Ajouter (x_i + l_i) au Px
    if (y_i + w_i) ∉ Py then
        Ajouter (y_i + w_i) au Py
    if (z_i + h_i) ∉ Pz then
        Ajouter (z_i + h_i) au Pz
Calculate L_b, W_b, and H_b
```

---

4 Computational Experiments

To see the efficiency of the proposed genetic algorithm, a set of test instances derived from [11] named *S1* will be tested. This set contains 10 test instances of homogeneous and heterogeneous packing items. The number of products is from 4 to 9 which is quite small in terms of problem size. To test the further performance of the proposed algorithm, a new set of 10 test instances, named *S2* is randomly generated. The products in *S2* are cuboid and heterogeneous. The size of the products is random number between 10 and 100 so that $p_i \geq q_i \geq r_i$ $\forall i \in \{S2.1, S2.2 \ldots S2.10\}$. The number of items of the instances is from 8 to 50.

The benchmark method is the mathematical model proposed in [11] resolved with the solver IBM CPLEX 12.8.0.

For the *GA-ODP*, different combinations of parameters are tested to find out the optimal configuration for the genetic algorithm and reduce the computational time. In this section, different combinations of GA parameters are tested to find out the combination that gives the best computational performance. n_g is selected from $\{50, 100, 200, 300\}$, $n_e \in \{0.1n_p, 0.2n_p, 0.3n_p\}$, $n_m \in \{0.1n_p, 0.2n_p, 0.3n_p\}$. As shown in [4], the population size n_p can significantly affect the calculation, and good results are obtained by indexing the population size into the problem size (or number of items). Therefore, in these experiments, the population size is also a function of problem size, the following values are tested to choose the best parameter: $n_p \in \{20n, 30n, 50n, 100n\}$. By testing all the combinations created by the parameters, the following configuration is chosen to be constant parameters for solving the test instances: $n_g = 200$, $n_p = 20n$, $n_e = 0.3n_p$, $n_m = 0.1n_p$, $P_c = 0.5$, and $P_m = 0.1$.

Next, the computational results obtained by using the best GA parameters are compared with them given by the benchmark method. The algorithms are programmed in C++ and the experiments are executed on an Intel Core i7-6820HQ CPU @2.70 GHz, Windows 7 PC with 32 GB of RAM. The improvement of the *GA-ODP* over the benchmark method is calculated as follows:

$$GAP_V = \frac{(V_2 - V_1)}{V_1} \tag{13}$$

$$GAP_t = \frac{(t_1 - t_2)}{t_2} \tag{14}$$

where V_1 and t_1 are box volume and computational time of the benchmark method, while V_2, t_2 are these values given by the proposed method.

Table 2 shows the test results and the computational time for ten test instances in *S1*. The row with header *"Solver"* shows the solutions given by the benchmark method, while the row with header *"GA-ODP"* shows the experimental results given by the proposed method. It can be seen that the solutions given by the *GA-ODP* is competitive with that given by the benchmark method. There are six out of ten test instances where the difference of box volume of both method are the same. For the other instances, the difference is not enormous, which is from about 1.2% to 3.8%. In terms of computational time, for the small problem (S1.1, S1.2, S1.7, and S1.8) where number of items is not greater than 5, the solver finds the solution faster than the *GA-ODP* (the Gap_t is negative). However, when there are more items (S1.3 to S1.6 and S1.9 to S1.10), the solver needs much more time to find out the solution, and the computational time increases quickly among the number of items. In the other hand, the computational time of the *GA-ODP* does not increase in the same function as that of the solver. In that way, the *GA-ODP* can solve the problems with more items in a reasonable time.

Table 2. Computational results for *S1*

Problem		S1.1	S1.2	S1.3	S1.4	S1.5	S1.6	S1.7	S1.8	S1.9	S1.10
# items		4	5	6	7	8	9	4	5	6	7
Solver	L	28	30	35	43	9	10	127	102	92	101
	W	26	28	28	28	8	8	57	95	81	89
	H	6	6	6	6	5	6	30	30	50	51
	V	4368	5040	**5880**	7224	360	480	217170	290700	**372600**	**458439**
	T(s)	**0.5**	**1.9**	4.2	32.1	1.6	3.4	**0.3**	**2.2**	22.1	215.2
GA-ODP	L	28	30	31	25	9	10	127	102	90	106
	W	26	28	16	24	8	8	57	95	85	88
	H	6	6	12	12	5	6	30	30	50	50
	V	4368	5040	5952	7200	360	480	217170	290700	382500	466400
	T(s)	1.84	2.8	**2.88**	**4.08**	**0.8**	**1.6**	1.68	2.56	**3.12**	**3.76**
	Gap_V	0	0	0.012	0.023	0	0	0	0	0.027	0.038
	Gap_t	-0.73	-0.32	0.46	6.87	1.0	1.13	-0.82	-0.14	6.08	56.23

The Table 3 shows the test results of instances of *S2* with more items. It can be seen that *GA-ODP* outperformed the benchmark method in terms of computational time in all instances. For the problems S2.9 and S2.10, the solver cannot find a solution within 3600 s while the *GA-ODP* finds the solutions in 113.45 and 1621.39 s, respectively. For other problem instances, the computational of the *GA-ODP* is also shorter than that of solver with 60% to 276% improvement.

Table 3. Computational results for *S2*

Problem		S2.1	S2.2	S2.3	S2.4	S2.5	S2.6	S2.7	S2.8	S2.9	S2.10
# items		8	9	10	11	12	13	14	15	20	50
Solver	V	265650	1170288	937020	713400	630336	888272	1656200	1132560	-	-
	T(s)	14.1	13.99	15.78	28.22	42.12	58.04	74.06	3600	3600	3600
GA-ODP	V	265650	1170288	948264	726241	642312	908702	1697605	1195983	2468840	3724660
	T(s)	**3.75**	**5.87**	8.22	**14.01**	**18.49**	**36.36**	**43.54**	**51.86**	**113.45**	**1621.39**
	Gap_V	0	0	0.012	0.018	0.019	0.023	0.025	0.056	-	-
	Gap_t	2.76	1.38	0.92	1.01	1.28	0.60	0.70	68.42	-	-

The test results show that the proposed algorithm has the capacity of giving competitive solutions compared with exact method in the literature. In the other hand, the proposed algorithm can deal with larger problem while its computational time is not exploding among with the problem size.

The computational experiments also show that for most of the case, minimizing the adjusted box volume can lead to the minimal actual box volume. In terms of computational times, the proposed algorithm is not exploding while number of items increases.

5 Conclusion

This paper has proposed a new genetic algorithm to deal with the 3D-ODPP. The proposed algorithm has shown the capability of solving the 3D-ODPP in a reasonable computational time while the given solutions are competitive with those given by exact methods in the literature. The proposed method has significant advantage in terms of computational time when solving problems with more items to be packed.

However, this work has not consider many practical constraints that often arise in real-world scenarios, such as item supporting, package balancing, weight distribution, etc. These constraints can be attached to the problem as the hard constraints or as a multi-objective optimization problem. Therefore, solving the 3D-ODPP by a genetic algorithm with the consideration of practical constraints will be an interesting subject for future researches.

References

1. Araujo, L., Ozcan, E., Atkin, J., Baumers, M., Tuck, C., Hague, R.: Toward better build volume packing in additive manufacturing: classification of existing problems and benchmarks (2015)
2. Chen, C., Lee, S.M., Shen, Q.: An analytical model for the container loading problem. Eur. J. Oper. Res. **80**(1), 68–76 (1995)
3. Gonçalves, J., Resende, M.: A parallel multi-population genetic algorithm for a constrained two-dimensional orthogonal packing problem. J. Comb. Optim. **22**(1), 180–201 (2011)
4. Gonçalves, J., Resende, M.: A biased random key genetic algorithm for 2d and 3d bin packing problems. Int. J. Prod. Econ. **145**, 500–510 (2013)
5. Gonçalves, J.F., Resende, M.G.: A parallel multi-population biased random-key genetic algorithm for a container loading problem. Comput. Oper. Res. **39**(2), 179–190 (2012)
6. Junqueira, L., Morabito, R.: On solving three-dimensional open-dimension rectangular packing problems. Eng. Optim. **49**(5), 733–745 (2017)
7. de Queiroz, T., Miyazawa, F., Wakabayashi, Y., Xavier, E.: Algorithms for 3d guillotine cutting problems: Unbounded knapsack, cutting stock and strip packing. Comput. Oper. Res. **39**, 200–212 (2012)
8. Truong, C., Amodeo, L., Yalaoui, F.: A mathematical model for three-dimensional open dimension packing problem with product stability constraints. In: Dorronsoro, B., Ruiz, P., de la Torre, J., Urda, D., Talbi, E.G., (eds) Optimization and Learning. OLA 2010. Communications in Computer and Information Science, **1173**, 241–251. Springer, Cham (2020). https://doi.org/10.1007/978-3-030-41913-4_20
9. Truong, C., Amodeo, L., Yalaoui, F., Hautefaye, J., Birebent, S.: A product arrangement optimization method to reduce packaging environmental impacts. IOP Conf. Ser. Earth. Environ. Sci. **463**, 012164 (2020)
10. Tsai, J.F., Li, H.L.: A global optimization method for packing problems. Eng. Optim. **38**(6), 687–700 (2006)
11. Tsai, J.F., Wang, P.C., Lin, M.H.: A global optimization approach for solving three-dimensional open dimension rectangular packing problems. Optimization **64**(12), 2601–2618 (2015)

12. Vielma, J.P., Nemhauser, G.L.: Modeling disjunctive constraints with a logarithmic number of binary variables and constraints. Math. Program. **128**(1–2), 49–72 (2011)
13. Wäscher, G., Haußner, H., Schumann, H.: An improved typology of cutting and packing problems. Eur. J. Oper. Res. **183**(3), 1109–1130 (2007)
14. Zheng, J., Chien, C., Gen, M.: Multi-objective multi-population biased random-key genetic algorithm for the 3-d container loading problem. Comput. Indust. Eng. **89**, 80–87 (2015)

Formulation of a Layout-Agnostic Order Batching Problem

Johan Oxenstierna[1,2(✉)], Louis Janse van Rensburg[3], Jacek Malec[1], and Volker Krueger[1]

[1] Department of Computer Science, Lund University, Lund, Sweden
johan.oxenstierna@cs.lth.se
[2] Kairos Logic AB, Lund, Sweden
[3] Thompson Institute, University of the Sunshine Coast, Sunshine Coast, QLD, Australia

Abstract. To date, research on warehouse order-batching has been limited by reliance on rigid assumptions regarding rack layouts. Although efficient optimization algorithms have been provided for conventional warehouse layouts with Manhattan style blocks of racks, they are limited in that they fail to generalize to unconventional layouts. This paper builds on a generalized procedure for digitization of warehouses where racks and other obstacles are defined using two-dimensional polygons. We extend on this digitization procedure to introduce a layout-agnostic minisum formulation for the Order Batching Problem (OBP), together with a sub-problem for the OBP for a single vehicle, the *single batch* OBP. An algorithm which optimizes the *single batch* OBP iteratively until an approximate solution to the OBP can be obtained, is discussed. The formulations will serve as the fundament for further work on layout-agnostic OBP optimization and generation of benchmark datasets. Experimental results for the digitization process involving various settings are presented.

Keywords: Order Batching Problem · Vehicle Routing · Warehouse digitization

1 Introduction

Order-picking is "the process of retrieving products from storage areas in response to a specific customer request" where "customer request" denotes a shipment order consisting of one or several products [1]. Order-picking is accountable for as much as 55% of all operating expenses in a warehouse and is considered an important process to optimize [2]. *Order-batching* is a common method with which to conduct order-picking. It means that each picker (vehicle) is set to pick a so-called *batch* of one or more orders [3]. As an optimization problem order-batching is known as the *Order Batching Problem* (OBP) [4] or the *Joint Order Batching and Picker Routing Problem* (JOBPRP) [5]. The *Picker Routing Problem* is a sub-problem of the OBP for one vehicle and is here treated as equivalent to the Traveling Salesman Problem (TSP) [6]. This paper follows

This work was partially supported by the Wallenberg AI, Autonomous Systems and Software Program (WASP) funded by the Knut and Alice Wallenberg Foundation.

B. Dorronsoro et al. (Eds.): OLA 2021, CCIS 1443, pp. 216–226, 2021.
https://doi.org/10.1007/978-3-030-85672-4_16

the convention that an "OBP" can include TSP optimization without having to include TSP optimization in the name of the problem (such as the JOBPRP) [4]. The *Picker Routing Problem* is henceforth referred to as TSP and the *Order Batching Problem,* which includes TSP optimization, as OBP. In the literature the OBP is usually formulated as a specific form of the more well-known Vehicle Routing Problem (VRP) [7], with two key amendments:

1. *Order-integrity*: In the OBP products in one order cannot be picked by more than one vehicle [8] whereas in the VRP this constraint is not used (there is no notion of a warehouse shipment "order" in the VRP) [7].
2. *Obstacle-layout:* We can observe two types of obstacle layouts (see Fig. 1): In the *conventional* layout, racks are laid out in a Manhattan style blocks. In the *unconventional* layout, racks or other obstacles can be freely placed (see Fig. 2. for examples). The *unconventional* layout includes the case when there are no racks or obstacles at all. All previous work on the OBP seems to require explicitly a conventional layout [5, 8–10], while the VRP does not have this requirement.

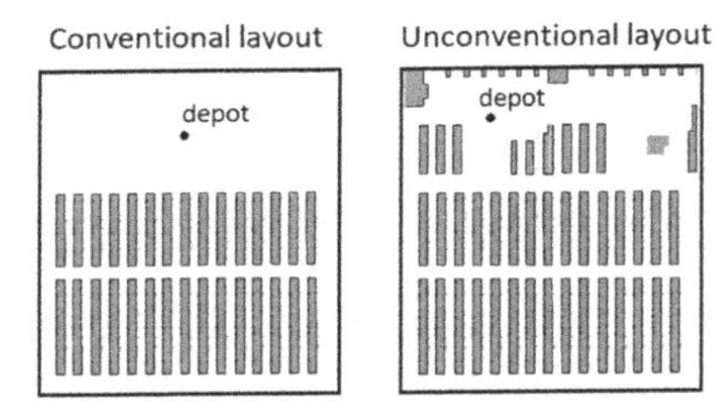

Fig. 1. Example of a *conventional* layout (left) with 30 *racks*, 16 *aisles* and 3 *cross-aisles.* Adding a single or a few irregular racks or other obstacles to the *conventional* layout renders it *unconventional.*

The aim of this paper is to formulate an OBP where orders and order-integrity are preserved, but where the layout is generalized towards any layout with or without polygonal obstacles. This is in line with a future research recommendation by Masae et al. [11]: "there is a strong need for developing [...] algorithms for [...] non-conventional warehouses". Below are some reasons for why this is important:

- It allows warehouses with unconventional layouts to formulate and optimize OBP's. This includes warehouses divided into zones where each zone has a conventional layout.
- It allows OBP optimization to be used as a tool with which to optimize warehouse layouts beyond conventional layouts.
- Problems in non-warehouse domains, such as agriculture, mining, road and aerial logistics to be explored as OBP's. The OBP is fundamentally similar to *batch processing* [12] where each process consists of constrained sub-processes (similar to *order-integrity*), and the Key Performance Indicator (KPI) depends on how well the

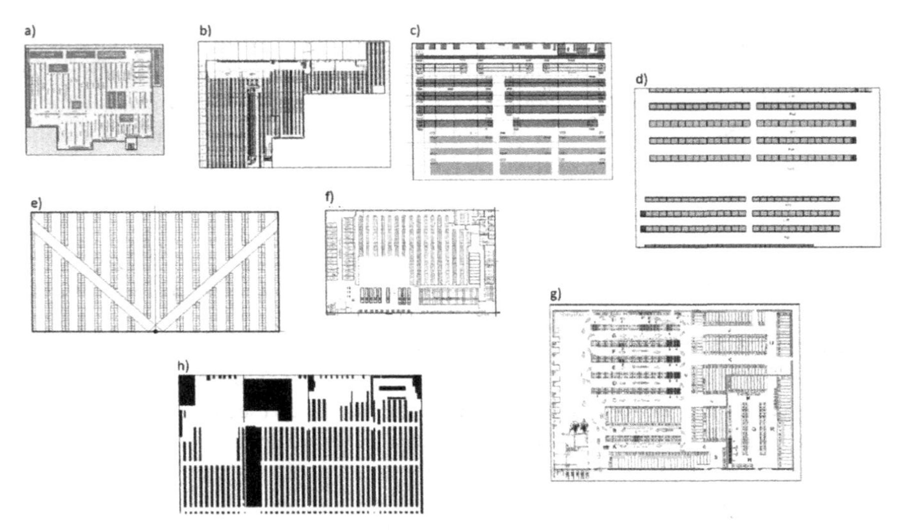

Fig. 2. Eight examples of unconventional warehouse layouts. (a) and (b) show cases where the layout has been built to fit within a non-rectangular outer wall. (e) is the so called "fishbone" layout.

sub-processes operate when they are combined. These types of broadened perspectives on the OBP can only be pursued if it is generalized beyond conventional layouts.

The paper continues with a literature review (Sect. 2), followed by the OBP formulation (Sect. 3). The formulation builds on a digitization process which generates the distances and shortest paths between all defined locations for a given warehouse [13]. The feasibility of the digitization process is examined in experiments involving various warehouse configurations (Sect. 4).

2 Literature Review

The OBP is a specific form of the Vehicle Routing Problem (VRP) [7] and a specific VRP-variant known as the *Steiner*-VRP [14]. A key feature of the *Steiner*-VRP is that multiple visits to the same location (representing a vertex in a graph) are allowed [5, 8, 10, 14]. OBP's and VRP's are known to be NP-hard [15, 16]. OBP's have been formulated using integer programming (e.g. [14]) or set-partitioning (e.g. [4]), with a heavy reliance on heuristics for a *conventional* warehouse layout. The conventional layout is modeled such that obstacles (racks) are arranged with parallel "aisles" (between racks) and parallel "cross-aisles" (between sections of racks) [9, 14]. Using such restrictive definitions for aisles and cross-aisles makes it possible to formulate heuristics that reduce the solution space of an OBP. Briant et al. [9], for example, use cutting planes and various relaxation heuristics to formulate an OBP which they then propose optimality bounds for. They use a conventional layout with 8 aisles and 3 cross-aisles, which corresponds to the size of the warehouse shown in Fig. 2(d).

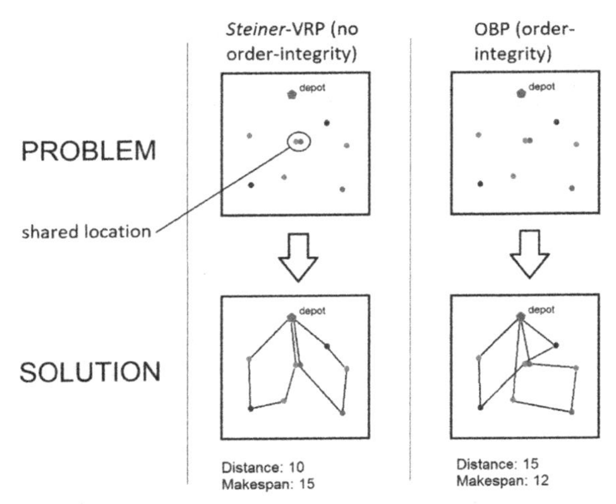

Fig. 3. A Steiner-VRP (left) plotted against the proposed layout-agnostic OBP in a setting without any obstacles. The dots denote products and the colors orders which the products belong to. The outlined green and red products in the middle share the same location. The difference between the Steiner-VRP and the OBP seen here is solely due to the *order-integrity* constraint. The vehicle distances may be longer in the OBP but the products which they are assigned to carry are more associated (by order color in this example). Order-integrity is used to e.g. reduce a later time-consuming sorting effort or to reduce pick-error i.e. the risk of the wrong product going into the wrong order. (Color figure online)

The conventional layout appears in formulations as "number of aisles" [8], "the cross-distance between two consecutive aisles" [4], "number of vertices in the subaisle" [14] or "intra-aisle distance" [17]. They are used as required inputs for OBP optimization. Some authors have called for formulations involving more layouts than the conventional layout [11, 18–21]. Without the conventional layout, however, it is a challenging task to effectively constrain an OBP solution space. This can for instance be exemplified in the scenario when there are no obstacles, and each order contains a single product. In that case the OBP is equivalent to a Steiner-VRP, and this problem has no yet proposed optimal solution [14]. Proposed OBP optimization algorithms for the conventional layout include dynamic programming [9], datamining [22], clustering [10] and meta-heuristics such as Tabu Search [23], Ant Colony Optimization [15] and Genetic Algorithms [24]. In the VRP research domain problem formulations are generally not concerned with obstacle layouts [25]. Instead the only requirement in a VRP is usually a cost matrix, providing the travel distance or time between all pairs of locations [7, 26]. In a VRP it is generally assumed that this cost matrix already exists, or that it is produced in a prior data collection process. In research on the OBP, on the other hand, plenty of attention is usually given to how to produce the cost matrix and how to define shortest paths or TSP's in an environment with obstacles. This can also be seen in some papers on VRP's that include obstacles (e.g. [27] and [28]). Concerning where vehicles begin and end their trips, most OBP papers assume that the origin and destination location is the same (usually this location is named *depot*). If this is not the case, the OBP is denoted *multi-depot* or a *Dial-A-Ride-Problem* (DARP) [21]. An example of this is when vehicles have one location where they drop off their picked orders, and where there are one or several locations where they can start their rides.

3 Problem Formulation

3.1 Preliminaries

The proposed OBP formulation is based on an undirected, symmetric and weighted graph. Without obstacles (racks or other) no graph is needed since distances between all pairs of locations in that case can be assumed to be Euclidean. Also, in the obstacle free case, the shortest path between any two locations can be assumed to be a single edge. With obstacles, however, shortest distances must be calculated based on the shortest paths that circumvent obstacles, and this is achieved here using the Floyd-Warshall graph algorithm [13, 29]. Concerning number of depots the below formulation assumes both an origin and a destination location for vehicles is formulated (but they can share the same coordinates).

First a set of locations is defined as $\mathcal{L} \subset \mathbb{R}^+ \times \mathbb{R}^+$. This set consists of different types of locations: $l_s \in \mathcal{L}$ is the starting (origin) location for all vehicles. $l_d \in \mathcal{L}$ is the destination location for all vehicles. $\mathcal{L_P} \subset \mathcal{L}$ is the set of product locations. $\mathcal{L_U} \subset \mathcal{L}$ is a union of sets of obstacles: $\mathcal{L_U} = \cup_i u_i, i \in \mathbb{N}^+$ where each u_i is a polygonal obstacle with a set of corner locations $u_i = \left\{l_i^1, l_i^2, \dots, l_i^k\right\} \subseteq \mathcal{L_U}, k \in \mathbb{N}^+$. All of the locations can thus be summarized as a union: $\mathcal{L} = \{l_s\} \cup \{l_d\} \cup \mathcal{L_P} \cup \mathcal{L_U}$. The products which are to be collected are defined as a set $\mathcal{P} = \{p_1, p_2, \dots, p_n\}, n \in \mathbb{N}^+$. Each product $p \in \mathcal{P}$ has a location $loc^p : \mathcal{P} \to \mathcal{L_P}$, weight $w^p : \mathcal{P} \to \mathbb{R}^+$ and volume $vol^p : \mathcal{P} \to \mathbb{R}^+$. *The unassigned orders* which are to be batched are defined as a subset of all possible combinations of products $\mathcal{O} \subset 2^{\mathcal{P}}$. The locations of the products in an order $o \in \mathcal{O}$ are defined as a function $loc^o : \mathcal{O} \to 2^{\mathcal{L_P}}$. Order weight and volume quantities are defined as $w^o : \mathcal{O} \to \mathbb{R}^+$ and $vol^o : \mathcal{O} \to \mathbb{R}^+$. $w(o) = \sum_{p \in o} w(p), vol(o) = \sum_{p \in o} vol(p)$. *Vehicles* are defined as $\mathcal{M} = \left\{(w, vol, k, id) | w, vol, id \in \mathbb{R}^+, k \in \mathbb{N}^+\right\}$ where w denotes weight capacity, vol denotes volume capacity, k denotes the maximum number of orders the vehicle can carry and id a unique identifier of a vehicle. The capacities of a single vehicle $m \in \mathcal{M}$ are provided using functions $w^m : \mathcal{M} \to \mathbb{R}^+$, $vol^m : \mathcal{M} \to \mathbb{R}^+$ and $k^m : \mathcal{M} \to \mathbb{N}^+$.

The digital model of the warehouse is represented as a graph with a set of *vertices* $\mathcal{V} = \{v_1, v_2, \dots, v_n\}, n = |\mathcal{L}|$. $\mathcal{V}$ consists of different types of vertices denoted as follows: $v_s \in \mathcal{V}$ is a starting (origin) vertex for vehicles, $v_d \in \mathcal{V}$ is a destination vertex for vehicles, $\mathcal{V}_{\mathcal{L_P}} \subset \mathcal{V}$ is a set of product location vertices and $\mathcal{V_U} \subset \mathcal{V}$ is a set of obstacle corner vertices. The union of all vertices, $\mathcal{V} = \{v_s\} \cup \{v_d\} \cup \mathcal{V}_{\mathcal{L_P}} \cup \mathcal{V_U}$, are defined similarly to the locations apart from one important difference: There may be several products in one location and there is one vertex per *product location*, not one vertex per product (this is to limit the size of the graph). To get a set of locations from a corresponding set of vertices the function $loc^{\mathcal{V}} : \mathcal{V} \to \mathcal{L}$ is used. To get a set of vertices from a set of locations is similarly provided by the function $v^{\mathcal{L}} : \mathcal{L} \to \mathcal{V}$.

The set of possible batches is defined as $\mathcal{B} \subset 2^{\mathcal{O}}, b \in \mathcal{B}, b \in 2^{\mathcal{O}}, b \neq \emptyset$. The locations of the products in the batch can be obtained using function $loc^b : \mathcal{B} \rightarrow 2^{\mathcal{L}_\mathcal{P}}. loc(b) = \cup_{o \in b} loc(o)$. Similarly, the vertices in the batch are $v^b : \mathcal{B} \rightarrow 2^{\mathcal{V}_{\mathcal{L}_\mathcal{P}}}. v(b) = \cup_{o \in b} v(loc(o))$. Batch weight and volume quantities are defined as $w^b : \mathcal{B} \rightarrow \mathbb{R}^+$ and $vol^b : \mathcal{B} \rightarrow \mathbb{R}^+$. The number of orders in a batch is defined as $k^b : \mathcal{B} \rightarrow \mathbb{N}^+$ or $|b|$.

The set of edges E is defined such that each edge is an ordered pair $e \in E = \{(i, j), i, j \in \mathcal{V}, \mathrm{i} \neq j\}$ where i is an origin and j a destination vertex. E excludes any edge which passes through the hull of any polygon in $\mathcal{U}$ (for details on how this can be achieved see [13]). Edges between adjacent corners in any polygon $u \in \mathcal{U}$ are not excluded in E. The edges and vertices are then used to construct the *symmetric undirected weighted graph* $G = (\mathcal{V}, E)$.

A *shortest paths distance matrix* $D : \mathcal{V} \times \mathcal{V} \rightarrow \mathbb{R}^+$ provides the minimum sum of edge distances between any two vertices in $\mathcal{V}$ without crossing any hull in $\mathcal{L}_\mathcal{U}$. Each edge cost $d_{loc(i),loc(j)} \in D$ (henceforth d_{ij}) is between two vertices, $i, j \in \mathcal{V}, i \neq j$. If there exists an unobstructed path between $loc(i)$ and $loc(j)$ (which does not go through any obstacle hull) the distance is Euclidean $\|loc(i) - loc(j)\|$. If obstacles must be bypassed to go from $loc(i)$ to $loc(j)$, however, the distance is a sum of Euclidean distances following the shortest path between them (without crossing obstacles). The Floyd-Marshall graph algorithm is used to compute these shortest paths and distances exactly [13].

The set of vertices, including origin and destination vertex, that have to be visited to pick a batch is defined as $\mathcal{V}_b = \{v_s\} \cup v(b) \cup \{v_d\}, b \in \mathcal{B}$. A function can then be built which provides *the sequence of vertex visits in a batch TSP solution (tour)*:

$$T^b : \mathcal{V}_b \rightarrow \{v_i\}_{i=1}^n, n = |\mathcal{V}_b|, \tag{1}$$

$$T(b)_i = \begin{cases} v_s & i = 1 \\ v_k & 1 < i < n \\ v_d & i = n \end{cases} \tag{2}$$

where $v_k \in v(b)$ and i gives the sequence of visits. *The distance of a batch TSP solution (tour)* is similarly provided in a function:

$$D^b : T(b)_i \rightarrow \mathbb{R}^+, i \in \mathbb{N}^+, i \leq |T(b)|. \tag{3}$$

$$D(b) = \sum d_{T(b)_i T(b)_j}, i, j \in \mathbb{N}^+, j = i + 1, i < |T(b)| \tag{4}$$

Note D^b could be renamed D^{T^b} to clarify that the distance of a batch is computed over a certain *tour* to visit all the products in the batch. $\mathcal{V}$, E, G and D are assumed to be produced in a digitization preprocessing step and the computational effort at this stage is assumed to not be included in subsequent OBP optimization. Out of $\mathcal{V}$, E, G and D only D is needed as input for OBP optimization assuming vehicles are capable of finding the shortest path between any two locations on their own. $\mathcal{V}$, E, G are also needed for directions on how to follow the shortest path, and if visualizations of edges are sought, both of which are arguably important in an industrial OBP optimization service. One example of a visualization of G and a small OBP optimization instance can be seen in Fig. 4 below:

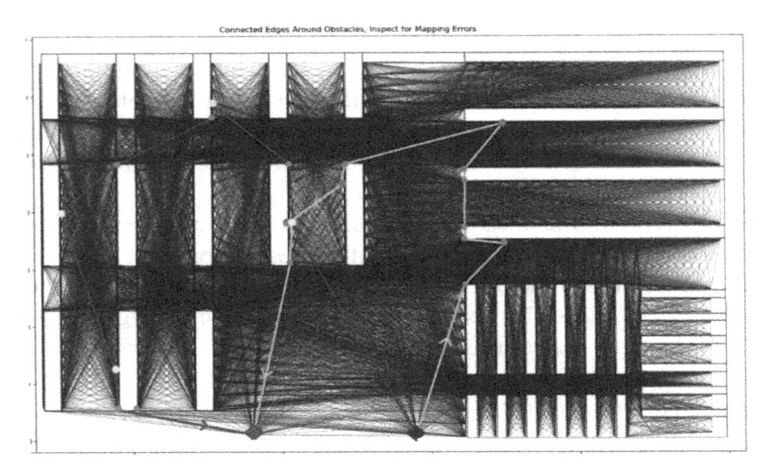

Fig. 4. Visualization of the digital graph (G) of a warehouse, and an example OBP with two orders, two vehicles and vehicle capacity of one order. Each blue line is an edge $e \in E$ that connects two vertices (i, j), $i, j \in \mathcal{V}$. The white hulls are racks (obstacles) laid out in an "unconventional" way and no edges pass through them. The orange vertices show a subset of $\mathcal{V}_{\mathcal{U}}$ and the green and yellow vertices along the racks are the sought products in $\mathcal{V}_{\mathcal{L}_{\mathcal{P}}}$ (where color indicates which order it belongs to). Note one of the products is visited by both vehicles. At the bottom the origin and destination, v_s and v_d can be seen (blue and red respectively). The OBP solution is here shown as the red and lime edges following the shortest paths between v_s, the yellow or green vertices and v_d (the two tours are obtained using T^b above). (Color figure online)

3.2 General OBP Formulation

A set-partitioning formulation with an exponential number of binary variables is used to formulate the layout-agnostic general OBP. The binary decision variable x_{mb} is used to indicate whether batch $b \in \mathcal{B}$ is assigned to vehicle $m \in \mathcal{M}$ ($x_{mb} = 1$ if m is assigned to b, $x_{mb} = 0$ otherwise). The binary decision variable x_{mo} is used to indicate whether order $o \in \mathcal{O}$ is assigned to vehicle $m \in \mathcal{M}$ ($x_{mo} = 1$ if m is assigned o, $x_{mo} = 0$ otherwise). The binary decision variable x_{ml} is used to indicate whether vehicle m visits location $l \in \mathcal{L}_{\mathcal{P}}$ ($x_{ml} = 1$ if m visits l, $x_{ml} = 0$ otherwise).

$$min \sum_{b \in \mathcal{B}} D(b) x_{mb}, m \in \mathcal{M} \tag{5}$$

$s.t.$

$$\sum_{m \in \mathcal{M}} x_{mo} = 1, \forall o \in \mathcal{O} \tag{6}$$

$$\sum_{l \in loc(o)} x_{ml} \geq x_{mo}, \forall o \in \mathcal{O}, m \in \mathcal{M} \tag{7}$$

$$q(b) \leq q(m) x_{mb}, b \in \mathcal{B}, q \in \{w, vol, k\}, m \in \mathcal{M} \tag{8}$$

The optimization aim of the OBP (5) is to assign batches to vehicles such that the sum of the distances of all batches is minimized. (6) ensures that each unassigned order is assigned to exactly one vehicle (*order-integrity*). (7) ensures that every product location in every order assigned to a vehicle is visited at least once. This inequality is what renders the OBP a general *Steiner*-VRP. (8) ensures capacity of vehicles is never exceeded.

3.3 Single Batch OBP Formulation

The general OBP formulation is problematic to work with due to the large number of possible combinations of vehicles and batches. Below is a proposal for a more tractable problem where the aim is to find a batch for an already selected vehicle. After vehicle m has been selected the aim is to assign as many orders as possible to it while keeping batch distance at a minimum:

$$\underset{b \in \mathcal{B}}{argmin}\, D(b) \tag{9}$$

$$\exists q(q(b) + q(o) \geq q(m)), \forall o \in \mathcal{O}, o \notin b, q \in \{w, vol, k\} \tag{10}$$

where $k(o)$ (i.e. the number of orders in an order) is 1. The aim in the *single batch* OBP (9) is to, for a given vehicle m, find a single batch b with the minimal batch distance. Constraints (6), (7), and (8) from the general OBP still apply (for the given vehicle). Constraint (10) is further added to ensure that the number of orders in the batch is as large as possible (for all unassigned orders there exists a weight, volume or number of orders quantity which will exceed vehicle capacity if the order is added to the batch). Without this maximization of number of orders an optimization algorithm would always create a batch with just a single order because this would produce the minimal batch distance. The single batch OBP formulation is a specific version of the so called *minimum cost maximal knapsack packing problem* (MCMKP) if distance is treated as "profit" and number of orders as knapsack "weight" (according to the definition by [30]). Note in the formulation here batch "weight" and "volume" are not included in the maximization since this would impose decision making over the importance of the different quantities (which one is most important to maximize while not exceeding vehicle capacity). The intention of the single batch OBP formulation is to provide the means with which to build an efficient *single batch* OBP optimization algorithm. This algorithm can then be used to produce one batch at a time within an algorithm which optimizes the general OBP, as proposed in Algorithm 1 below:

Algorithm 1. Iteration of single Batch OBP optimization.

1. total_OBP_cost ← 0
2. **WHILE** $\mathcal{O}$:
3. m ← select_vehicle($\mathcal{M}$)
4. b ← single_batch($\mathcal{O}, m, D$)
5. total_OBP_cost += $D(b)$
6. $\mathcal{O}$ ← remove_batched($\mathcal{O}, b$)

Algorithm 1 runs with the assumption that there are always enough vehicles to choose from, and it creates single batches until there are no more unassigned orders left. The total cost is expressed in the TSP tour distances of the batches $D(b)$. After a batch has been created its orders are removed from $\mathcal{O}$.

4 Experimental Results

This section evaluates the computational effort and memory requirement needed to generate the datastructures used by the formulation in Sect. 3. The only datastructure needed for OBP optimization is the distance matrix D, but graph G, including shortest paths between all locations are also included (Table 1).

Table 1. Experimental results for the digitization of distances and shortest paths.

Name	Number of locations	Number of polygons	Memory allocated		CPU time (s)
			Disk (mb)	RAM (mb)	
c6953_B01	781	0	14.4	78	541
c7561_B02	288	16	2.7	65	268
c0543_B03	760	1	14.2	79	501
c9109_AYD	218	5	0.7	62	195
c3495_BER	579	69	61	83	620
c0054_JUL	752	42	12.2	81	763
c3401_DAD	1384	296	377	385	5616
c9543_ARA	4037	234	520	574	26028
c2456_CAG	6491	306	1219	1290	65232

CPU used: Intel Core i7-4710MQ 2.5 GZ 4 cores, 8GB of RAM

Computational time and memory requirement grows fast with number of locations in the digitization procedure. The largest instance included 6491 defined locations and required 18 hours of *CPU-time*. Please note the computation only has to be run once (and re-run if the obstacle layout is changed in the warehouse). Once the graph has been generated, distances and shortest paths can be queried quickly by pre-allocating them in Random Access Memory (RAM), which is why RAM usage is also a relevant parameter. "Number of locations", denoted as $|\mathcal{L}|$ in Sect. 3, and the number of products in each defined location, varies depending on precision sought in the digitization process. For example, the warehouse denoted c9543_ARA, holds around 40000 products, but there are only 4037 defined locations. Each location in that case represents the products within an area of around 3 m^2 on the horizontal axis and 5 shelf levels on the vertical axis, with a total of around 10 products represented by every defined location. Clearly, a faster digitization process would be achieved if more products were mapped to the same locations, but then the digital model would be less precise. The tradeoff between memory and CPU-time on the one hand, and digitization precision on the other, is an interesting topic left for future work.

5 Conclusion

This paper set out to formulate an Order Batching Problem (OBP) that does not depend on the way in which racks or other obstacles are laid out in the warehouse. A digitization procedure to generate necessary datastructures was first described. A minisum set-partitioning formulation with an exponential number of binary variables was introduced for the layout-agnostic OBP. A more tractable version of the OBP, the *single batch* OBP, was additionally formulated where the aim is to find a single batch for an already specified vehicle. Experiments evaluating CPU-times and memory footprints for generating necessary datastructures was presented. In ensuing work new layout agnostic OBP optimization algorithms and benchmark instances will be introduced.

References

1. de Koster, R., Le-Duc, T., Roodbergen, K.J.: Design and control of warehouse order picking: a literature review. Eur. J. Oper. Res. **182**(2), 481–501 (2007). https://doi.org/10.1016/j.ejor.2006.07.009
2. Jiang, X., Zhou, Y., Zhang, Y., Sun, L., Hu, X.: Order batching and sequencing problem under the pick-and-sort strategy in online supermarkets. Proc. Comput. Sci. **126**, 1985–1993 (2018). https://doi.org/10.1016/j.procs.2018.07.254
3. Sharp, G.P., Gibson, D.R.: Order batching procedures. Eur. J. Oper. Res. **58**, 57–67 (1992)
4. Gademann, N., van de Velde, S.: Order batching to minimize total travel time in a parallel-aisle warehouse. IIE Trans. **37**(1), 63–75 (2005). https://doi.org/10.1080/07408170590516917
5. Valle, C.A., Beasley, B.A.: Order batching using an approximation for the distance travelled by pickers. Eur. J. Oper. Res. **284**(2), 460–484 (2019)
6. Ratliff, H., Rosenthal, A.: Order-picking in a rectangular warehouse: a solvable case of the traveling salesman problem. Oper. Res. **31**, 507–521 (1983)
7. Cordeau, J.-F., Laporte, G., Savelsbergh, M., Vigo, D.: Vehicle routing. In: Transportation, Handbooks in Operations Research and Management Science, vol. 14, pp. 195–224 (2007)
8. Bozer, Y.A., Kile, J.W.: Order batching in walk-and-pick order picking systems. Int. J. Prod. Res. **46**(7), 1887–1909 (2008). https://doi.org/10.1080/00207540600920850
9. Briant, O., Cambazard, H., Cattaruzza, D., Catusse, N., Ladier, A.-L., Ogier, M.: An efficient and general approach for the joint order batching and picker routing problem. Eur. J. Oper. Res. **285**(2), 497–512 (2020). https://doi.org/10.1016/j.ejor.2020.01.059
10. Kulak, O., Sahin, Y., Taner, M.E.: Joint order batching and picker routing in single and multiple-cross-aisle warehouses using cluster-based tabu search algorithms. Flex. Serv. Manuf. J. **24**(1), 52–80 (2012). https://doi.org/10.1007/s10696-011-9101-8
11. Masae, M., Glock, C.H., Grosse, E.H.: Order picker routing in warehouses: a systematic literature review. Int. J. Prod. Econ. **224**, 107564 (2020). https://doi.org/10.1016/j.ijpe.2019.107564
12. Chang, P.-Y., Damodaran, P., Melouk, S.: Minimizing makespan on parallel batch processing machines. Int. J. Prod. Res. **42**(19), 4211–4220 (2004). https://doi.org/10.1080/00207540410001711863
13. van Rensburg, L.J.: Artificial intelligence for warehouse picking optimization - an NP-hard problem. M.Sc., Uppsala University (2019)
14. Valle, C.A., Beasley, J.E., da Cunha, A.S.: Optimally solving the joint order batching and picker routing problem. Eur. J. Oper. Res. **262**(3), 817–834 (2017). https://doi.org/10.1016/j.ejor.2017.03.069

15. Li, J., Huang, R., Dai, J.B.: Joint optimisation of order batching and picker routing in the online retailer's warehouse in China. Int. J. Prod. Res. **55**(2), 447–461 (2017). https://doi.org/10.1080/00207543.2016.1187313
16. Psaraftis, H., Wen, M., Kontovas, C.: Dynamic vehicle routing problems: three decades and counting. Networks **67**, 3–31 (2015). https://doi.org/10.1002/net.21628
17. Bué, M., Cattaruzza, D., Ogier, M., Semet, F.: A two-phase approach for an integrated order batching and picker routing problem. In: Dell'Amico, M., Gaudioso, M., Stecca, G. (eds.) A View of Operations Research Applications in Italy. ASS, vol. 2, pp. 3–18. Springer, Cham (2019). https://doi.org/10.1007/978-3-030-25842-9_1
18. Bortolini, M., Faccio, M., Ferrari, E., Gamberi, M., Pilati, F.: Design of diagonal cross-aisle warehouses with class-based storage assignment strategy. Int. J. Adv. Manuf. Technol. **100**(9), 2521–2536 (2019). https://doi.org/10.1007/s00170-018-2833-9
19. Fumi, A., Scarabotti, L., Schiraldi, M.: The effect of slot-code optimization in warehouse order picking. Int. J. Bus. Manag. **5**, 5–20 (2013). https://doi.org/10.5772/56803
20. Gue, K.R., Meller, R.D.: Aisle configurations for unit-load warehouses. IIE Trans. **41**(3), 171–182 (2009). https://doi.org/10.1080/07408170802112726
21. Henn, S.: Algorithms for on-line order batching in an order picking warehouse. Comput. Oper. Res. **39**(11), 2549–2563 (2012). https://doi.org/10.1016/j.cor.2011.12.019
22. Chen, M.-C., Wu, H.-P.: An association-based clustering approach to order batching considering customer demand patterns. Omega **33**(4), 333–343 (2005). https://doi.org/10.1016/j.omega.2004.05.003
23. Henn, S., Wäscher, G.: Tabu search heuristics for the order batching problem in manual order picking systems. Eur. J. Oper. Res. **222**(3), 484–494 (2012). https://doi.org/10.1016/j.ejor.2012.05.049
24. Cergibozan, Ç., Tasan, A.: Genetic algorithm based approaches to solve the order batching problem and a case study in a distribution center. J. Intell. Manuf. 1–13 (2020). https://doi.org/10.1007/s10845-020-01653-3
25. Braekers, K., Ramaekers, K., Nieuwenhuyse, I.V.: The vehicle routing problem: state of the art classification and review. Comput. Ind. Eng. **99**, 300–313 (2016). https://doi.org/10.1016/j.cie.2015.12.007
26. Pillac, V., Gendreau, M., Guéret, C., Medaglia, A.L.: A review of dynamic vehicle routing problems. Eur. J. Oper. Res. **225**(1), 1–11 (2013). https://doi.org/10.1016/j.ejor.2012.08.015
27. Mansouri, M., Lagriffoul, F., Pecora, F.: Multi vehicle routing with nonholonomic constraints and dense dynamic obstacles (2017). https://doi.org/10.1109/IROS.2017.8206195
28. Bochtis, D.D., Sørensen, C.G.: The vehicle routing problem in field logistics part I. Biosyst. Eng. **104**(4), 447–457 (2009). https://doi.org/10.1016/j.biosystemseng.2009.09.003
29. Santis, R.D., Montanari, R., Vignali, G., Bottani, E.: An adapted ant colony optimization algorithm for the minimization of the travel distance of pickers in manual warehouses. Eur. J. Oper. Res. **267**(1), 120–137 (2018). https://doi.org/10.1016/j.ejor.2017.11.017
30. Furini, F., Ljubić, I., Sinnl, M.: An effective dynamic programming algorithm for the minimum-cost maximal knapsack packing problem. Eur. J. Oper. Res. **262**(2), 438–448 (2017). https://doi.org/10.1016/j.ejor.2017.03.061

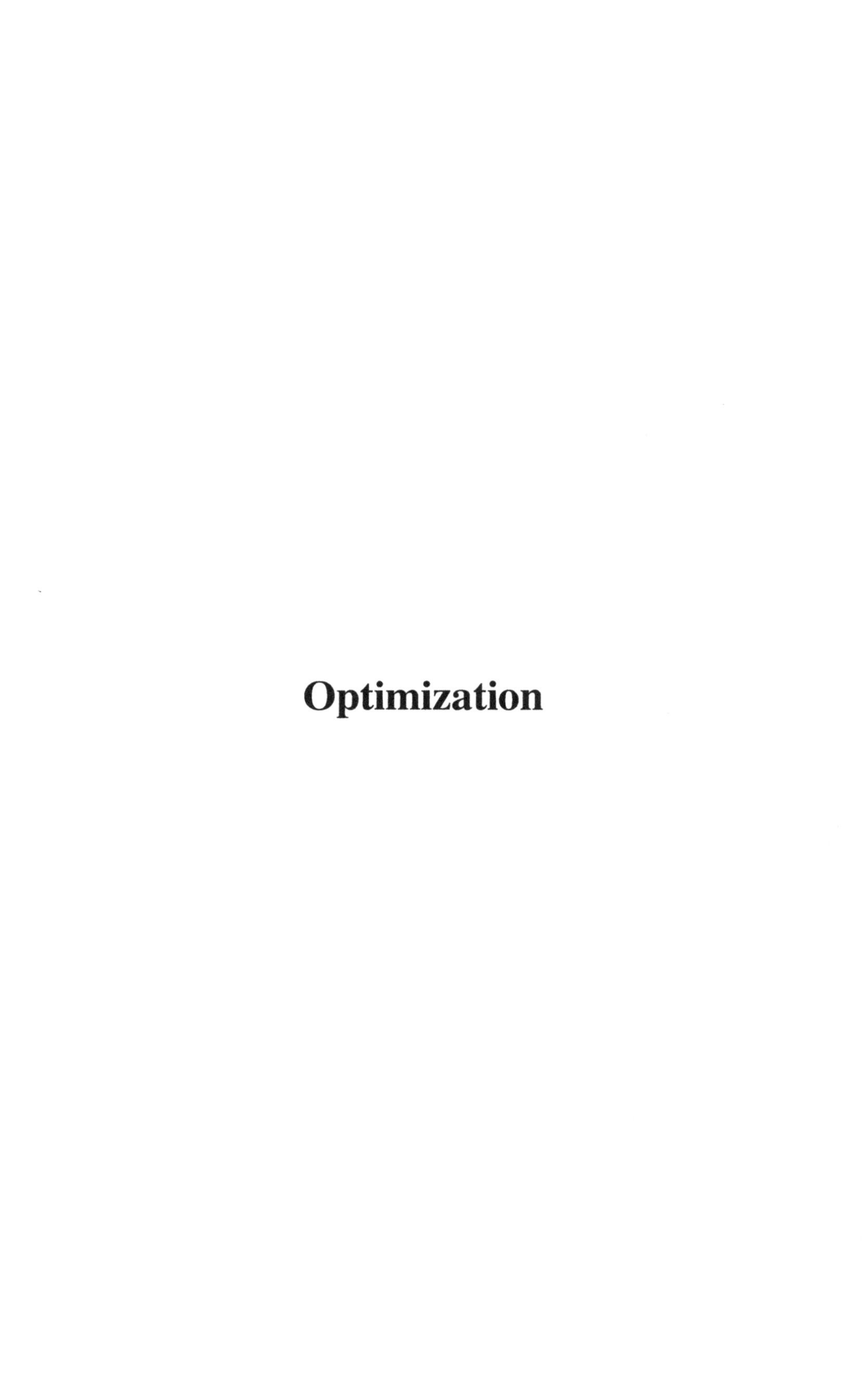

Optimization

Neighborhood Enumeration in Local Search Metaheuristics

Michiel Van Lancker(✉), Greet Vanden Berghe, and Tony Wauters

Department of Computer Science, CODeS, KU Leuven, Leuven, Belgium
michiel.vanlancker@cs.kuleuven.be

Abstract. Neighborhood enumeration is a fundamental concept in the design of local search-based metaheuristics. It is often the only principle of intensification present in a metaheuristic and serves as the basis for various metaheuristics. Given its importance, it is surprising that academic reporting on enumeration strategies lacks the necessary information to enable reproducible algorithms. One aspect of neighborhood enumeration in particular has been under the radar of researchers: the order in which neighbors are enumerated. In this paper, we introduce a versatile formalism for neighborhoods which makes explicit enumeration order and we analyse the impact of enumeration order on the outcome of search procedures with a small set of benchmark problems.

Keywords: Enumeration order · Local search · Neighborhoods · Metaheuristics

1 Introduction

Metaheuristics have gained a somewhat ambiguous reputation over the years. On the one hand they are lauded for their useful characteristics in practical applications: metaheuristics are problem-independent, general optimization algorithms. They are not only capable of being reused over a wide variety of problems, but many are also anytime algorithms which maintain a valid solution throughout the entire search process. Furthermore, they can be implemented in a highly configurable fashion, enabling automated algorithm design and parameter tuning. This results in algorithm templates that can be instantiated and automatically tailored to solve specific problems or instances. On the other hand, metaheuristics research has not yet reached the scientific rigor found in other fields, with many researchers tending to focus on algorithmic efficiency – or worse, novelty – rather than algorithmic understanding. This has led to a large variety of algorithms which differ only slightly from one another or are identical except for the terminology used [6].

While big steps have been made – especially during the last two decades – to transform the field into a more academic one with rigorous scientific discipline built on formalized concepts, many publications continue to operate in the sphere of problem-solving rather than algorithmic understanding. This resulted

B. Dorronsoro et al. (Eds.): OLA 2021, CCIS 1443, pp. 229–240, 2021.
https://doi.org/10.1007/978-3-030-85672-4_17

in many metaheuristics, but few insights. Nevertheless, efforts are underway to mature the discipline. Notable examples of this are (i) the endorsement by the Journal of Heuristics of the view that nature should no longer serve as an explicit inspiration for "novel" metaheuristics, (ii) the recognition of the need for white-box algorithm implementations, preferably described in a purely functional style [8], (iii) the call for rigorous evaluation and testing practices, and (iv) the active promotion of a view of what metaheuristics research ought to be [7].

In this paper we zoom in on one specific component of metaheuristics: the concept of local search neighborhoods. We argue that a gap exists between common theoretical neighborhood definitions and how they are implemented in practice. In other words: we argue that neighborhoods are not implemented according to the white-box principle, preventing algorithm reproducibility and standardized evaluation.

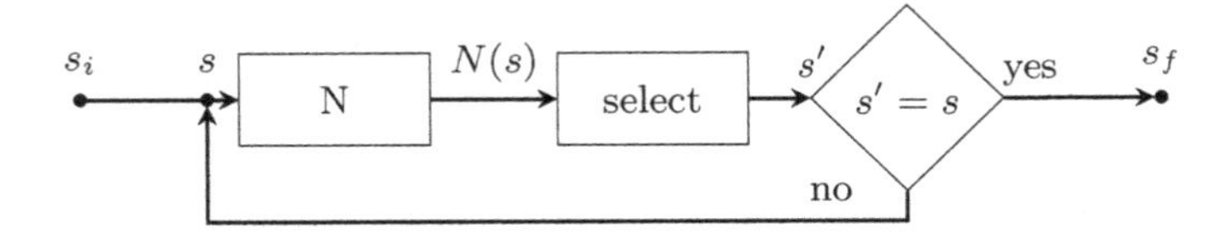

Fig. 1. Iterative improvement consists of repeatedly applying an improving operation to the solution.

Many optimization techniques can be considered instantiations of the iterative improvement-scheme (II-scheme), the distinction between which results from the interaction between their instantiating components. The II-scheme itself is straightforward: starting from an incumbent solution the search process consists of a series of iterations, where in each iteration a selection function selects an alternative solution of better quality than the incumbent solution. If a better solution is found, it replaces the incumbent solution. This process is repeated until no improving solution can be found.

In local search metaheuristics, a set of alternative solutions – called the neighborhood of the incumbent solution – is constructed by making a set of small modifications to the incumbent solution. Most, if not all, metaheuristics can be mapped to the II-scheme shown in Fig. 1. The difference between various metaheuristics yet again results from the differing interactions between their constituent components. Which components to consider and how to combine them is the responsibility of the (human) algorithm designer. Design choices which require some thought include how to generate a neighborhood of the incumbent solution, which solution to select from the neighborhood and how to compare solutions. Good neighborhood design is important when it comes to the efficiency of a local search (meta)heuristic. Choosing an appropriate selection criterion is equally important, as it strongly determines the behavior of the search and can have a dramatic impact on runtime.

Given the importance of these two design questions, it is fair to assume that reporting on metaheuristic algorithms should include complete information concerning which choices were made and, ideally, why. However, at present the opposite situation is the case: many publications concerning metaheuristics do not report neighborhood specifications to the level of detail required to facilitate reproducibility. Most obvious is the lack of information concerning how operators in a neighborhood are enumerated. This information is crucial if an order-dependent selection criterion is used and, indeed, virtually all deterministic selection criteria are order-dependent. A second, more subtle issue is the lack of information concerning which operators are a priori included in a neighborhood.

Our contributions in this paper are threefold. First, we introduce a formalism for the concept of a local search neighborhood, which makes explicit the enumeration order. Second, we analyze the effect of enumeration order on the outcome of a search procedure through a series of computational experiments. Third and finally, we provide several examples of the expressiveness of the proposed formalism.

The remainder of this paper is structured as follows. Beginning with the concept of iterative improvement, Sect. 2 introduces neighborhoods and selection methods and provides a brief overview of how neighborhood enumeration is commonly reported in metaheuristics research. In Sect. 3 we introduce a formalism for neighborhood enumeration. Section 4 then analyzes the effect of enumeration order on the outcome of a search procedure on a set of benchmark instances. Several examples demonstrating the flexibility of the formalism are given in Sect. 5. Section 6 then concludes the paper.

2 Iterative Improvement, Neighborhoods and Selection

In this section we review the relationship between iterative improvement, neighborhoods and selection criteria. In doing so we identify a gap between the commonly used definitions for the aforementioned concepts and the components required to implement the II-scheme, resulting in an incomplete algorithm specification. The section ends with a brief analysis of how neighborhood enumeration is currently reported on in the academic literature.

To approach local search metaheuristics as instantiations of iterative improvement, strict definitions are required for the instantiating components. Consider the II-scheme shown in Fig. 1. It is clear that an instantiation of the scheme is determined by three factors, namely: a neighborhood generation function N, a neighbor selection function *select* and a condition to test whether or not the search has ended. Since we are only interested in improvement methods, the ending condition can be excluded from the analysis and thus the behavior of a deterministic II-procedure is dependent on only two functions: the neighborhood function $N : S \rightarrow \mathcal{P}(S)$ and the selection function $select : \mathcal{P}(S) \rightarrow S$. As is clear from its type, the neighborhood-function must map the incumbent solution to a set of alternative solutions, resulting in the common introductory definition of a neighborhood [2,9]:

Definition 1. *A neighborhood function is a mapping $N : S \rightarrow \mathcal{P}(S)$ which assigns to each solution $s \in S$ a set of solutions $N(s) \subseteq S$. The members of $N(s)$ are called neighbors of s.*

In the context of local search however, a different definition is sometimes used to more adequately capture the notion of operators and locality. A neighborhood is defined in terms of a relation – the local search operator – on S:

Definition 2. *The R-neighborhood $N_R(s)$ of solution $s \in S$ is the neighborhood defined by the relation R on S, $N_R(s) = \{s' \in S : sRs'\}$.*

The second component of the II-scheme is a selection function, which returns a single neighbor from the neighborhood it receives as input. We refrain from giving a general definition of selection criteria, but note that any selection criterion must be a function of type *select* : $\mathcal{P}(S) \rightarrow S$ and we shall examine how well two of the most popular selection criteria adhere to this definition.

The first criterion we will consider is the *argmin* selection criterion (Eq. 1), which selects the best solution from the neighborhood. Next is the *firstmin* selection criterion (Eq. 2), which selects the first improving solution from the neighborhood. More formal definitions of both criteria are as follows:

$$\underset{s' \in N(s)}{\text{argmin}}\, c(s') := \{s' \mid \forall s'' \in N(s) : c(s') \leq c(s'')\} \tag{1}$$

$$\underset{s_i \in N(s)}{\text{firstmin}}\, c(s_i) := \{s_i \in N_{\downarrow}(s) \mid \forall s_j \in N_{\downarrow}(s) : i \leq j\} \tag{2}$$

$$\text{where } N_{\downarrow}(s) := \{s' \in N(s) \mid c(s') \leq c(s)\} \tag{3}$$

Note that this definition of *argmin* denotes a set of solutions instead of a single solution: if multiple solutions have the best objective value, all of these solutions will be returned. As such the definition specifies a function of type $\mathcal{P}(S) \rightarrow \mathcal{P}(S)$ and a modification, a tie-breaker, is needed to acquire the required type. Common tie-breakers are to select the first, the last or a random solution from the set of most improving solutions. Only the first two of these tie-breakers are deterministic and both of these are order-dependent.

For *firstmin*, the impact of order is obvious. To be able to return the first improving neighbor an order must be imposed on neighborhood N. In the worst case all solutions in the neighborhood are improving and thus each possible ordering of N will return a different solution. It follows that the neighborhood enumeration order must be known to achieve a full specification of a single iteration in the II-scheme. While the effect of enumeration order on the outcome of a single iteration is generally fairly limited, this is less so when considering the entire II-scheme. Since every iteration starts from the outcome of the previous iteration, the effect of an enumeration order compounds throughout the whole search.

Given the effect of enumeration order on the outcome of a search procedure, it is somewhat surprising that most publications do not contain any information about it. Many publications only describe neighborhoods in terms of their local search operator. A notable exception is [5], which does not only mention the use of a random enumeration order, but also publishes the complete source code of its implementation.

Finally, let us examine some open-source implementations of metaheuristics and see how they implement neighborhood enumeration. The following two implementations serve as an example: the Java Metaheuristics Search Framework(JAMES) [4] and the suite of metaheuristic frameworks PARADISEO [3]. In JAMES it is possible for users to implement custom neighborhoods through a neighborhood- and operator-interface, but imposing orders on neighborhood sets through an interface is not possible and must be programmed from scratch by the user. When querying the full neighborhood, an eagerly constructed list of operators is returned. In PARADISEO, users can implement custom neighborhoods in a similar fashion, though here order *is* made explicit by means of an iterator-interface. Querying the full neighborhood returns a lazy iterator over the neighborhood. Furthermore, neighborhoods can be linked together into new neighborhoods.

Before continuing with the next section, we end this section with an example of issues arising when neighborhood definitions are incomplete. We will illustrate these issues by considering the *TwoOpt*-operator for the Traveling Salesperson Problem (TSP). Let $C = \{c_1, \ldots, c_n\}$ be a set of points on the Euclidean plane representing cities and let $d : C \times C \rightarrow \mathbb{N}$ be the distance between two cities. Then, the goal of the TSP is to find the shortest tour which visits each city once. Let permutation $\pi \in \Pi$ represent a tour through all cities in C and let $I_\pi = \{1, \ldots, n\}$ be the index set of π. Element $\pi_i \in \pi$, where $i \in I_\pi$, represents the ith visited city in the tour. The objective value $c(\pi)$ is computed with (Eq. 4).

$$c(\pi) = \sum_{i \in I_\pi \setminus \{n\}} d(\pi_i, \pi_{i+1}) + d(\pi_n, \pi_1) \tag{4}$$

Applying the *TwoOpt*-operator to a solution for the TSP equals swapping two edges in the tour, or equivalently, inverting a subsequence of the solution representation π. The operator takes as input the current tour and two indices $i, j \in I_\pi$. To implement a function to generate the *TwoOpt* neighborhood, a double for loop is typically used. A naive implementation would generate neighbors for all possible pairs $(i, j) \in I_\pi^2$. This is however redundant: *TwoOpt* is a symmetric operator, thus a more efficient implementation would only generate neighbors for the pairs (i, j) for which $i < j$, as these are sufficient to cover the whole neighborhood. Aside from redundancy, which is unwanted but not problematic, if it is unclear which moves are included in the neighborhood and which are not, any order-dependent selection function can cause diverging search outcomes for two neighborhoods that "look" the same.

3 Neighborhood Enumeration

The previous section provided an introduction to how common definitions of neighborhoods, selection criteria and local optima are not sufficiently exact from an implementation perspective and how this in turn results in an incomplete algorithm specification. As suggested by the *TwoOpt*-example, there are two pieces of information missing from Definition 2: how many (i.e. which) solutions belong to a neighborhood and the order in which these solutions are visited. In this section we present an alternative definition of a neighborhood function, which makes the aforementioned information concrete. The purpose of the definition being introduced is to capture the structure of a local search neighborhood in such a way that the required implementation steps become clear.

Consider the neighborhood $N_M(s) \subseteq S$. For all $s_\phi \in N_M(s)$ we know that we can move from s to s_ϕ. Let $m_\phi : s \mapsto s_\phi$ be the function representing the move from s to s_i. There are $|N_M(s)|$ such functions, one for each $s_\phi \in N_M(s)$. Thus we can define the neighborhood as $N_M(s) = \{m_i(s)\}_{i \in \Phi}$, where Φ is an index over $N_M(s)$. Note that if we provide a constructor function $M : \Phi \to (S \to S)$, we can construct function $m_\phi : S \to S$ by evaluating $M(\phi)$. Given an iterator T over Φ, the first neighbor in the neighborhood can be generated as follows: take the first element ϕ from the iterator, call constructor M to construct move m_ϕ, and apply $m_\phi(s)$. To generate subsequent neighbors, take the next element from T and repeat the process until all elements from T have been consumed. The neighborhood can then be defined as:

Definition 3. *A neighborhood $N_M(s, T)$ is the set of solutions constructed by applying each function $m_\phi : S \to S$ for each $\phi \in T$ to s, where T is an iterator over (a subset of) Φ_M, the parameter space of operator $M : \Phi_M \to (S \to S)$. As T is ordered, a neighborhood enumeration is uniquely defined by the triple (s, M, T).*

This definition results in several extra design questions concerning the parameter space used in a neighborhood. While neighborhood design typically only considers the choice of operator, now two more design choices must be made: *which operator parameters should be included in a neighborhood* and *in what order they should be generated.* In the next two sections we take a more detailed look at what options are available regarding these choices.

3.1 Parameter Spaces

When considering operators, we make three observations: First, the parameter space Φ_M of operator M is dependent on the solution representation. Second, it is dependent on functional properties of its operator. Third, any subset of the parameter space can be used to generate a neighborhood.

Consider the TSP and three operators defined in Table 1. All three operators are quadratic and, since solution representation π is unconstrained, each operator can take any $(i, j) \in I_\pi^2$ as input, where I_π^2 is the Cartesian product of I_π. However, depending on the operator, we can eliminate some elements from I_π^2. For example, we know that the *Swap*- and *TwoOpt*-operators are symmetric operators and thus parameter combinations (i, j) and (j, i) will construct the same moves. Furthermore, for all three operators it is the case that no matter the state of the incumbent solution, parameter (i, i) will construct the identity move.

Table 1. Definitions of the Swap, 2opt and Shift operators and their respective parameter spaces.

Operator	Parameter space	Neighbor relation
Swap	$(i, j) \in I_\pi^2 : i < j$	$\pi'_i = \pi_j \wedge \pi'_j = \pi_i$
TwoOpt	$(i, j) \in I_\pi^2 : i < j$	$\forall k \in [0, j - i] : \pi'_{i+k} = \pi_{j-k}$
Shift	$(i, j) \in I_\pi^2 : i \neq j$	$\pi'_j = \pi_i$
		$\pi'_k = \begin{cases} \forall k \in [i+1, j] : \pi_{k-1}, & \text{if } i < j \\ \forall k \in [j, i-1] : \pi_{k+1} & \text{otherwise} \end{cases}$

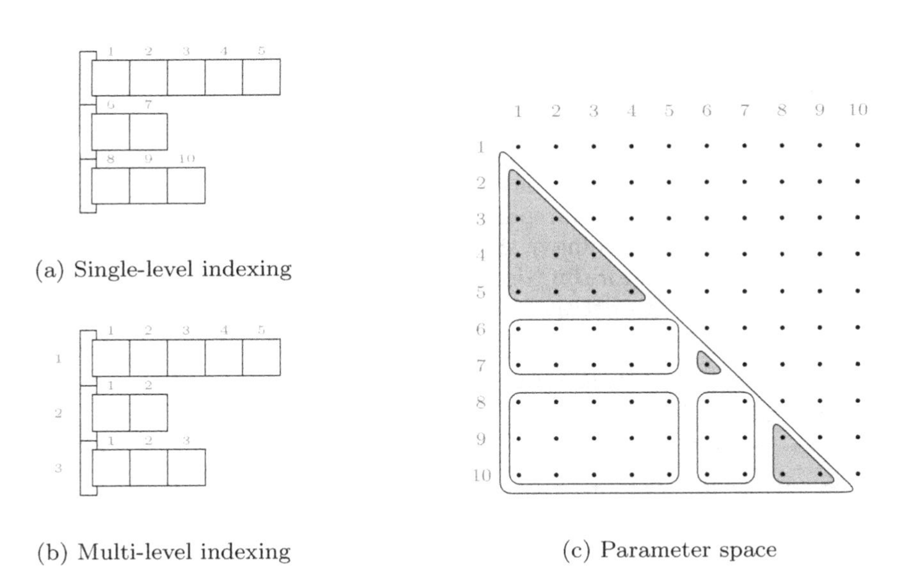

(a) Single-level indexing

(b) Multi-level indexing

(c) Parameter space

Fig. 2. The interpretation of operators and their respective parameter spaces is dependent on the indexing system used.

The importance of the chosen solution representation and index set becomes obvious when we consider more complex solution representations. Instead of permutation π, consider an ordered set of permutations ρ. To implement a neighborhood for this structure, we require an index set to base our parameter space on. Looking at Fig. 2 it is clear that multiple options are available. We can use a single-level, linear index – like we did for permutation π – where every position in the representation is represented by a single integer: its position in the overall element order. Alternatively, a multi-level index can be used, where every position in the representation is represented by two integers: the position of the permutation in the set and the position within the permutation. Figure 2c illustrates the correspondence between the parameter spaces of a symmetric operator using single-level and multi-level indexing. In light grey is the parameter space based on the single-level index. In dark grey are parameters corresponding to moves that operate inside a permutation of the set of permutations, using the multi-level index. Similarly, in white are the parameters corresponding to inter-permutation moves when using the multi-level index.

3.2 Enumeration Order

The final step is to impose an order on the defined parameters. Given a set of parameters of size n, there are $n!$ ways to impose an order. However, some of these orders are more interesting than others. Of special interest are those that follow particular patterns, which can usually be efficiently implemented as an iterator which generates the parameter sequence lazily. Some of these patterned sequences can be interpreted as prioritizing certain moves: consider the *TwoOpt* operator for the TSP and assume that we are using the $firstmin$ selection function. If *TwoOpt* moves are enumerated according to the scheme $(1,2),(1,3),(1,4),\ldots,$ the beginning position of the subsequence is considered more important than that of the end. Similarly scheme $(2,1),(3,1),(4,1),\ldots$ deems the end position more important. Finally, scheme $(1,2),(2,3),(3,4),\ldots$ prioritizes moves corresponding to shorter subsequence inversions. Such semantic distinctions can help algorithm designers gain insights into the behavior and performance of their algorithms.

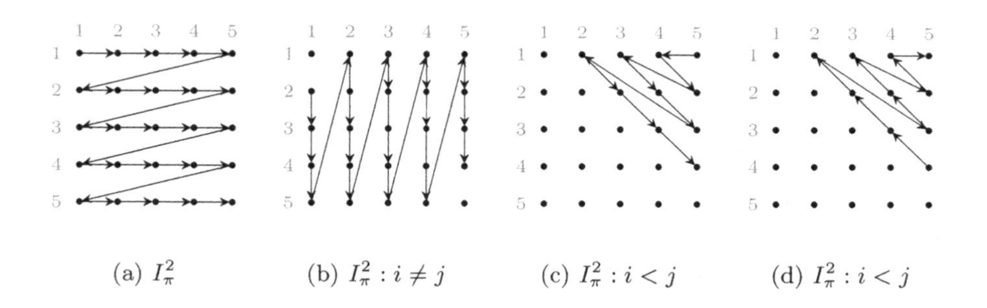

(a) I_π^2 (b) $I_\pi^2 : i \neq j$ (c) $I_\pi^2 : i < j$ (d) $I_\pi^2 : i < j$

Fig. 3. Various iterators over I_π^2.

Four iterators for quadratic operators are shown in Fig. 3 which differ in terms of their parameters included, order and direction. Figure 3a illustrates an iterator over the full parameter space – the Cartesian product I_π^2 – ordered along the rows. Figure 3b is ordered along the columns and eliminates parameters $(i,i) \in I_\pi^2$. Figures 3c and 3d are both ordered along the diagonals and eliminate parameters $(i,j) \in I_\pi^2$ for which $i \geq j$, but they differ in the direction they take.

4 Experimental Evaluation

To evaluate the influence of enumeration order on search procedures we consider a search procedure to be a program of type $solve : S \rightarrow S$. This program takes an initial solution s_i and returns a local optimum as final solution s_f. We refer to the change induced on s_i by *solve* as $\Delta_s = |C| - |e_c| - 1$, where $|C|$ is the number of cities and $|e_c|$ is the number of edges s_i and s_f have in common. In a similar fashion, we refer to the difference between the objective value of s_i and s_f as $\Delta_v = c(s_f) - c(s_i)$ and its runtime as Δ_t.

Table 2. The set of algorithm design parameters considered when experimentally evaluating enumeration order.

Constructive	Select	Operator	Order	Direction
Random	Argmin	Swap	Column	Forward
Greedy	Firstmin	TwoOpt	Row	Reverse
	Rolling	Shift	Diagonal	

To study the impact of enumeration order on the search we compare Δ_s, Δ_v and Δ_t for *solve* procedures instantiated with different design parameters. Table 2 lists these design parameters. As the first three columns do no influence enumeration order, they can be considered design parameters resulting in different "contexts" in which the effect of enumeration order is evaluated. These parameters serve to broaden the scope of our analysis. All of the included design parameters have been defined in earlier sections of this paper, except for the selection function *rolling*. This selection function is an adapted version of $firstmin$. Whereas $firstmin$ begins from scratch in the next iteration after selecting the first improving neighbor $s_i = m_i(s)$, *rolling* will continue enumerating from its current position. The last two columns determine enumeration order. Three different iterators are used as parameter **Order**, each of which can be used in two **Directions**, resulting in six enumeration order. Every configuration is tested on 42 TSP instances from TSPLIB. All algorithms and experiments are implemented in the Julia programming language for technical computing [1] and run in a single-core-per-run configuration on an Intel(R) Xeon(R) CPU E5-2650 v2 @ 2.60GHz machine with 16 cores. A complete description of the experimental setup and data is available online[1].

[1] https://github.com/Michiel-VL/Neighborhood_Enumeration_Data.

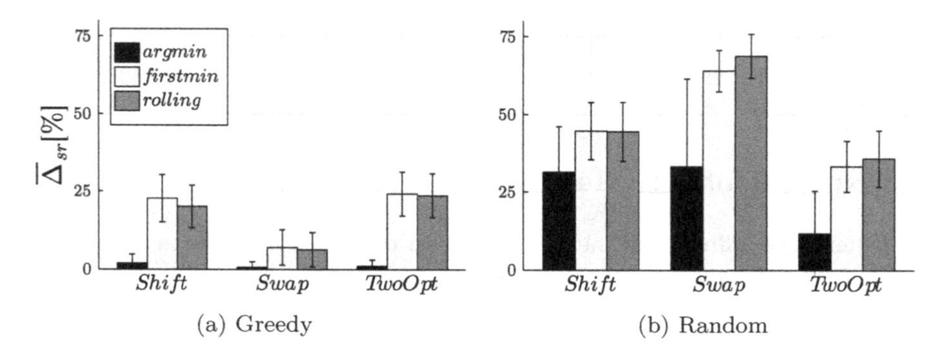

(a) Greedy (b) Random

Fig. 4. The relative difference between the final solutions should be zero if order had no influence.

First, we examine the effect of enumeration order on the solution state. If no such effect were to exist, then the final solutions of the six runs for a given context and instance should be identical, independent of the enumeration order parameters. To measure if there is an effect of enumeration order on the solution state, we compute the mean relative pairwise distance $\overline{\Delta}_{sr}$ between the set of final solutions of a given context and instance. Figure 4 is given for each of the 18 contexts. It is clear that the enumeration order does have an influence on the search outcome. Even for *argmin* selection, which is just barely order-dependent, $\overline{\Delta}_{sr}$ is fairly large, suggesting that the effect compounds quickly over the iterations of a search procedure.

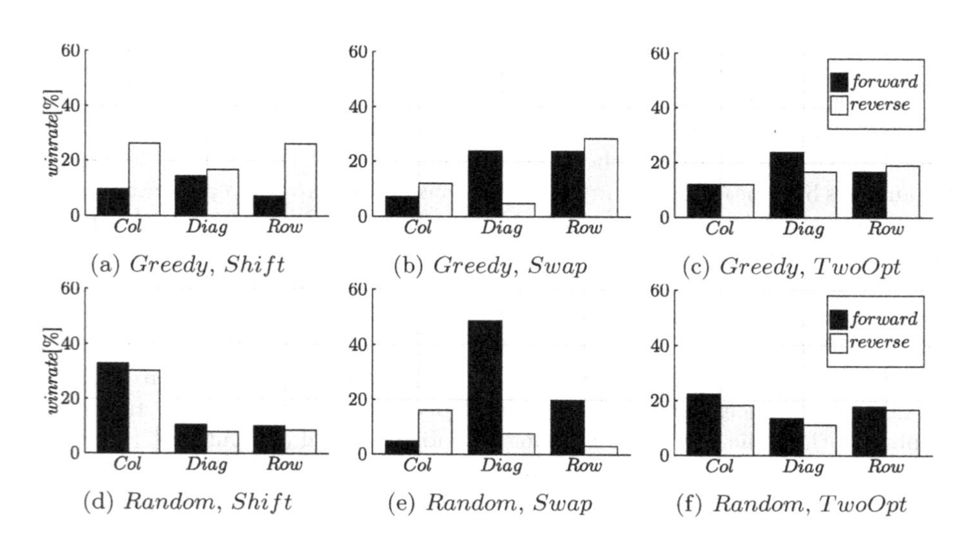

(a) *Greedy*, *Shift* (b) *Greedy*, *Swap* (c) *Greedy*, *TwoOpt*

(d) *Random*, *Shift* (e) *Random*, *Swap* (f) *Random*, *TwoOpt*

Fig. 5. The numbers of wins for different orders and operators.

Figure 5 shows the relative number of wins per enumeration order for different constructive heuristics and local search operators. While enumeration order does seem to affect the winrate, the results are inconclusive as to which order should be preferred for a given operator or constructive heuristic.

5 Discussion

Modeling a local search neighborhood as the combination of an operator with its own parameter space and an iterator over this parameter space has several advantages. First, it renders explicit the enumeration order to explore the neighborhood, which we have shown has an impact on the search outcome. Furthermore it is modular, as operator, set of parameters and order are completely separable implementation-wise. This not only enables easy reuse of code but it is also expressive, offering a range of neighborhood structures at virtually no cost.

It is also possible to encode structural properties of the problem in the neighborhood. As shown in Sect. 3, parameter spaces based on structured index sets can be used to distinguish between different parts of a solution representation. By opening up a neighborhood's structure through its parameter space, it is possible to use a wide variety of known algorithms to construct parameter spaces and reuse these over various neighborhoods.

Given a set of neighborhood definitions, new neighborhoods can be constructed in an algorithmic manner. Using function composition, operators can be composed into new operators and through the Cartesian product and disjoint union, various enumeration structures are available. Furthermore, given that in many programming languages iterators are denoted by a data structure that is composable in various ways – like filtering, linking or zipping – the definition as a whole is very expressive and enables concise descriptions of algorithms like Variable Neighborhood Descent and concepts such as path relinking or higher-order neighborhoods.

Note that defining a neighborhood as a triple $(s, M, T(\Phi_M))$ replaces the nested for-loops in many neighborhood implementations with a single foreach-loop. This triple separates three different neighborhood-design concerns that are typically entangled in code: local-search operators, neighborhood size and enumeration order. This enables algorithm designers not only to reuse operator, parameter space and enumeration order implementations for multiple neighborhoods, but it also leads to a more descriptive way of handling neighborhoods, enabling swift development and automated algorithm configuration.

6 Conclusion

In this paper we introduced a novel definition for neighborhoods aimed at formalizing their implementation. Defining local search neighborhoods in terms of a parametrized local search operator and an iterator over the parameter space of the operator leads to an expressive, composable definition which can be readily used during implementation. The iterator makes explicit two algorithm design considerations that are typically overlooked: in what order should neighbors be generated and which neighbors should be included in a neighborhood. Furthermore, by basing the operator parameter spaces on the indexing mechanism of a solution representation, significant parts of neighborhood design can be automatically derived from a solution representation. Finally, many enumeration orders can be efficiently implemented as a lazy sequence and therefore neighborhoods can be generated lazily.

While we only considered unconstrained problem representations, it would be interesting to look at constrained problems to examine how particular types of constraints affect the use of the definition, as complex constraints could prevent efficient iterator implementations. Though interesting, this primarily concerns implementation efficiency rather than formalization and thus lay outside the scope of this paper.

Acknowledgements. Research supported by Data-driven logistics (FWO-S007318N). Editorial consultation provided by Luke Connolly, KU Leuven.

References

1. Bezanson, J., Edelman, A., Karpinski, S., Shah, V.B.: Julia: a fresh approach to numerical computing. SIAM Rev. **59**(1), 65–98 (2017). https://doi.org/10.1137/141000671
2. Blum, C., Roli, A.: Metaheuristics in combinatorial optimization: overview and conceptual comparison. ACM Comput. Surv. (CSUR) **35**(3), 268–308 (2003)
3. Cahon, S., Melab, N., Talbi, E.G.: Paradiseo: a framework for the reusable design of parallel and distributed metaheuristics. J. Heuristics **10**(3), 357–380 (2004)
4. De Beukelaer, H., Davenport, G.F., De Meyer, G., Fack, V.: JAMES: an object-oriented java framework for discrete optimization using local search metaheuristics. Softw. Pract. Exp. **47**(6), 921–938 (2017). https://onlinelibrary.wiley.com/doi/abs/10.1002/spe.2459
5. Mecler, J., Subramanian, A., Vidal, T.: A simple and effective hybrid genetic search for the job sequencing and tool switching problem. Comput. Oper. Res. **127**, 105153 (2020). https://doi.org/10.1016/j.cor.2020.105153, http://www.sciencedirect.com/science/article/pii/S0305054820302707
6. Sörensen, K.: Metaheuristics-the metaphor exposed. Int. Trans. Oper. Res. **22**(1), 3–18 (2015)
7. Swan, J., et al.: Towards metaheuristics "in the large" (2020). arXiv preprint: arXiv:2011.09821
8. Swan, J., et al.: A research agenda for metaheuristic standardization (2015)
9. Talbi, E.G.: Metaheuristics: From Design To Implementation, **74**. John Wiley & Sons, Hoboken (2009)

Cryptographic Primitives Optimization Based on the Concepts of the Residue Number System and Finite Ring Neural Network

Andrei Tchernykh[1,2,3](✉), Mikhail Babenko[2,4], Bernardo Pulido-Gaytan[1], Egor Shiryaev[4], Elena Golimblevskaia[4], Arutyun Avetisyan[2], Nguyen Viet Hung[5], and Jorge M. Cortés-Mendoza[3]

[1] CICESE Research Center, Ensenada, BC, Mexico
chernykh@cicese.mx, lpulido@cicese.edu.mx
[2] Ivannikov Institute for System Programming, Moscow, Russian Federation
mgbabenko@ncfu.ru, arut@ispras.ru
[3] South Ural State University, Chelyabinsk, Russia
kortesmendosak@susu.ru
[4] North-Caucasus Federal University, Stavropol, Russian Federation
eshiriaev@ncfu.ru
[5] Le Quy Don Technical University, Hanoi, Vietnam
hungnv@lqdtu.edu.vn

Abstract. Data encryption has become a vital mechanism for data protection. One of the main challenges and an important target for optimization is the encryption/decryption speed. In this paper, we propose techniques for speeding up the software performance of several important cryptographic primitives based on the Residue Number System (RNS) and Finite Ring Neural Network (FRNN). RNS&FRNN reduces the computational complexity of operations with arbitrary-length integers such as addition, subtraction, multiplication, division by constant, Euclid division, and sign detection. To validate practical significance, we compare LLVM library implementations with state-of-the-art, high-performance, portable C++ NTL library implementations. The experimental analysis shows the superiority of the proposed optimization approach compared to the available approaches. For the NIST FIPS 186-5 digital signature algorithm, the proposed solution is 85% faster, even though the sign detection has low efficiency.

Keywords: Residue number system · Finite ring neural network · Encryption · High-performance · Cryptographic primitives

1 Introduction

Security becomes commonplace in all modern computing areas and affects many fields, including casual people communication, Internet of Things (IoT), analytics, self-learning systems, cloud computing, etc. Advanced cryptographic algorithms provide key mechanisms for data confidentiality, integrity, authentication, non-repudiation, etc.

B. Dorronsoro et al. (Eds.): OLA 2021, CCIS 1443, pp. 241–253, 2021.
https://doi.org/10.1007/978-3-030-85672-4_18

The cryptographic primitives are usually complex in terms of computational overhead and memory usage. They are designed based on mathematical theory, elliptic curves, Neural Networks (NNs), etc.

The high performance of cryptographic algorithms is important for numerous reasons. The principal one is the computational cost in terms of execution time. They can be executed by conventional computers, accelerated computing servers, and specialized hardware devices. In many cases, they are implemented as software components.

Many approaches are used to optimize encryption operations. Neuromorphic computing is concerned with emulating the neural structure and operation of the human brain. The main goals are to create a device that can extract better features, learn, recognize, classify, acquire new information, and even make a logical inference.

For instance, a single-chip prototype of the BrainScaleS 2, Intel Labs designed Loihi, and IBM's TrueNorth neuromorphic systems provide a proof-of-concept of a spiking neural network application to learn neurons and synapses [13, 14]. They include a hundred thousand neurons, each of which can communicate with thousands of others.

A Residue Number System (RNS) can achieve both fast computation and low power consumption. It is parallel, adaptable, and fault-tolerant, meaning it can produce results after components are failed [9, 10]. These properties allow for the successful development of cybersecurity systems [11–18].

RNS is a number system that represents integers by the remainders of division by several pairwise coprimes, called moduli. The arithmetic is called multi-modular arithmetic. It is widely used for computation with arbitrary length integers, for instance, in cryptography. It provides faster computation than with the usual numeral systems, even when converting between numeral systems is taken into account. By decomposing a large integer into a set of smaller integers, a large calculation is performed as a series of smaller calculations that can be performed independently and in parallel. The number of parallel elementary processes equals the number of RNS moduli.

In this paper, we propose a new optimization method RNS&FRNN of operations with arbitrary-length integers based on RNS and Finite Ring Neural Network (FRNN).

This paper is organized as follows. Section 2 describes the main concept of modular arithmetical operations. Section 3 introduces modular logical operations. Section 4 presents the scaling of RNS numbers by RNS base extension and introduces RNS&FRNN optimization method. Section 5 focuses on the experimental analysis. The conclusions and future work are discussed in the last Sect. 6.

2 Modular Arithmetical Operations

2.1 Addition, Subtraction, Multiplication, and Division

In the RNS, arithmetic operations are performed on each residue, according to the following general formula:

$$X \circ Y \overset{RNS}{\rightarrow} \left(|x_1 \circ y_1|_{p_1}, |x_2 \circ y_2|_{p_2}, \ldots, |x_n \circ y_n|_{pn} \right), \quad (1)$$

where $\{p_1, p_2, \ldots, p_n\}$ is a moduli set of pairwise coprime numbers. "$\circ$" denotes the operation of addition, subtraction, or multiplication.

Integer numbers X and Y are defined in RNS as tuples $(x_1, x_2, \ldots, x_n)$ and $(y_1, y_2, \ldots, y_n)$, where x_i represents the remainder of the division of X by pi, defined by $x_i = |X|_{p_i}$.

However, an additional restriction is imposed on the multiplication operation, which follows from the Chinese Remainder Theorem (CRT): $X \cdot Y < P$, where $P = \prod_{i=1}^{n} p_i$.

Integer division can be performed by various methods [2–4]. The most reliable algorithm is based on the scaling method. In this case, a dividend is an arbitrary number in the range $[0, P)$, and a divisor is any factor of $P = p_1 \cdot p_2 \cdot \ldots \cdot p_n$.

This division is similar to dividing by numbers belonging to a certain limited set, which is faster than dividing by an arbitrary divisor (2).

$$X = \left\lfloor \frac{X}{p_1} \right\rfloor \cdot p_1 + x_1, \tag{2}$$

where X is the dividend, and p_1 is the divisor.

The dividend is represented by the residues $X \overset{RNS}{\to} (x_1, x_2, \ldots, x_n)$, and the divisor is one of the moduli p_i. x_i is the residue of the division. In the first step of scaling, it is necessary to subtract the residue from the dividend (3):

$$X' \overset{RNS}{\to} \left(x_1', x_2', \ldots, x_n'\right) = \left(\left|x_1 - |x_i|_{p_1}\right|_{p_1}, \left|x_1 - |x_i|_{p_2}\right|_{p_2}, \ldots, \left|x_n - |x_i|_{p_n}\right|_{p_n}\right). \tag{3}$$

In the second step, the division of X' by p_i is carried out directly by (4):

$$\left\lfloor \frac{X}{p_1} \right\rfloor \overset{RNS}{\to} \left(-, \left|x_1' \left|p_1^{-1}\right|_{p_2}\right|_{p_2}, \ldots, \left|x_n' \left|p_1^{-1}\right|_{p_n}\right|_{p_n}\right), \tag{4}$$

where $\left|p_1^{-1}\right|_{p_i}$ is the multiplicative inversion of p_i.

At the end of the second stage, the residue x_i modulo p_i remains unknown, which can be found using the base extension (Sect. 4).

2.2 Euclidean Division

Euclidean division is carried out using the approximate division method. The essence of the approximate method for calculating the positional characteristic to compare and restore the positional notation of the numbers in RNS. It is based on the relative values of the numbers to the full range determined by CRT.

We have:

$$X = \left|\sum\nolimits_{i=1}^{n} \frac{P}{p_i} \left|P_i^{-1}\right|_{p_i} x_i \right|_P, \tag{5}$$

where $P = \prod_{i=1}^{n} p_i$, p_i is the RNS moduli, $\left|P_i^{-1}\right|_{p_i}$ is the multiplicative inversion of P_i relative to p_i, $P_i = \frac{P}{p_i}$.

If we divide the left and right sides of (5) by the constant P corresponding to the range of numbers, we obtain an approximate value

$$F(X) = \left|\frac{X}{P}\right|_1 = \left|\sum\nolimits_{i=1}^{n} k_i x_i\right|_1, \tag{6}$$

where $k_i = \frac{\left|P_i^{-1}\right|_{p_i}}{p_i}$ and $|x|_1$ is fractional part real number x.

The result is obtained after summing and discarding the integer part of the number while maintaining the sum fractional part.

The fractional value $F(X) = \left|\frac{X}{P}\right|_1 \in [0, 1)$ contains both information about the value of the number and its sign [1]. If $\left|\frac{X}{P}\right|_1 \in \left[0, \frac{1}{2}\right)$, then the number x is positive, and $F(X)$ is equal to the value of x divided by P. Otherwise, x is a negative number, and $1-F(X)$ shows the relative value of the number x [5].

There are several methods of calculating $F(X)$ [6–8]. The method of integer division X/Y can be described by an iterative scheme, which is performed in two stages.

In the first stage, the search for the highest degree of 2^i is carried out when approximating the quotient with a binary series.

In the second stage, the approximation series is refined. To get a range larger than P, you can choose the value $P = P' \cdot_{p_n+1}$, i.e., it is necessary to extend the RNS base by adding a redundant modulus. To avoid this base extension, a computationally complex operation, it is necessary to compare the current results of iteration i with previous values of iteration $i - 1$, and not dividends with intermediate divisors. This will satisfy the condition $0 < Y < P - 1$.

Known dividing algorithms determine the quotient based on the iteration $X' = X - Q_1 \cdot D$, where X and X' are the current and the next dividend respectively, D is the divisor, Q_1 is the quotient that is generated at each iteration from the full range of the RNS, and is not selected from a small set of constants.

In this method, the quotient is determined based on the iteration $r_i = X - B2^i$, where X is some divisible, B is the divisor, and 2^i is a member of the approximating series of the quotient. A comparison of the algorithms shows that the dividend in all iterations does not change, and the divisor is multiplied by a constant, which significantly reduces computational complexity.

The above method is easily modified in RNS using the approximate method of comparing modular numbers. In the iterative division process on a weighted number system, to search for the highest degree of a series of approximations of a quotient and to refine the approximating series, the dividend is compared with doubled divisors or with the sum of the members of the series.

The application of this idea for RNS can lead to an error in the division process. When the dynamic range is overflowed, the recovered number goes beyond the working range. For example, if the RNS moduli are $p_1 = 2, p_2 = 3, p_3 = 5,$ and $p_4 = 7$, then the range is $P = 2 \cdot 3 \cdot 5 \cdot 7 = 210$.

Suppose, during recovery, we got the number $X = 220$. In RNS $X = 220 = (0, 1, 0, 3)$. The range P is exceeded by the number 10, which in the RNS is $(0, 1, 0, 3)$. When using relative values, the number $X = 220$ is expressed as $X' = 10$, which is not true.

To overcome this difficulty, it is necessary to compare the current iteration values with the previous ones in the RNS. It allows to correct determining a larger or smaller number. The overflow of the dynamic range in the RNS can be used to make the decision "more – less".

In the first iteration, the dividend is compared with the divisor, and at the other iterations, the doubled values of the divisors $q_i Y < q_{i+1} Y$ are compared. In each new iteration, the current value is compared with the previous one.

The number of iterations required depends on the divisible and divisor values. Successive application of this operation leads to the formation of a sequence of integers $Yq_1 < \ldots < Yq_n > Yq_{n+1}$.

Let the case $Yq_n > Yq_{n+1}$ be fixed at $n + 1$ iterations, which corresponds to an overflow of the RNS range, i.e., $Yq_{n+1} > P$ and $X < Yq_{n+1}$. This completes the process of generating interpolation of the quotient by a binary series or by a set of constants in the RNS.

The process of approximating the quotient can be carried out by comparing only doubled neighboring approximate divisors. An important issue when implementing the function $F()$ is the accuracy of the coefficients.

It should also be noted that the number of characters in the fractional part should be twice as much as the number of characters in the RNS range. The modular numbers' division based on the approximate method of comparing numbers consists of the following steps (see Algorithm 1).

In this case, when the divisor has the minimum value and the dividend has the maximum, the threshold Δ_i is more than zero. It reduces the number of iterations when dividing a large divisible and a small divisor.

Algorithm 1. Euclidean division in RNS.

Input: $X \xrightarrow{RNS} (x_1, x_2, \ldots, x_n)$, $Y \xrightarrow{RNS} (y_1, y_2, \ldots, y_n)$, $F(X)$.

Output: $\omega = \left\lfloor \frac{X}{Y} \right\rfloor$, $\gamma = |X|_Y$.

Step 1. We calculate the approximate values of the divisible $F(X)$ and the divisor $F(Y)$ and compare them. If $F(X) < F(Y)$, then the division process ends and the quotient $\left\lfloor \frac{X}{Y} \right\rfloor = 0$. If $F(X) = F(Y)$, then the division process ends, and the quotient is equal to unity. If $F(X) > F(Y)$, then a higher degree 2^k is searched for by approximating the quotient with a binary code.

Step 2. We select the constant 2^k (the highest power of the series), multiply it by the divisor $F_1(X) = X2^k$ and introduce it into the comparison scheme. The constants $2^j \bmod p_i$, where $i = \overline{1,n}$, $1 \leq j \leq \log_2 P$ are previously stored in the memory.

Step 3. We find $\Delta_i = F(X) — F_1(Y)$. If in the sign digit Δ_i is "1", then the corresponding degree of the series is discarded, if it is "0", then in the adder of the quotient we add the value of a member of the series with this degree, that is 2^k.

Step 4. We find $F_1(Y)$, and check the term of the series with a degree 2^{k-1}.

Step 5. We find $\Delta_2 = \Delta_1 — F_1(Y)$ and perform the actions in accordance with paragraph 4.

Step 6. Similarly, we check all the remaining members of the series of the pre-zero degree. The resulting residue $\Delta_i = \Delta_{i-1} - F_{i-1}(Y) \approx 0$.

---

3 Modular Logical Operations

The operations of determining the sign of a number and comparing numbers can be performed using the approximate method discussed in Sect. 2.2.

The value obtained by (6) is a Positional Characteristic (PC). The determination of the sign detection is reduced to the PC, which consists of comparing the PC with the half value p_i, where $i = \overline{1, n}$. Thus, the sign of the number determines the following relation: if $\left|\frac{X}{P}\right|_1 < \frac{1}{p_i}$, then the number is positive; if $\left|\frac{X}{P}\right|_1 > \frac{1}{p_i}$, then the number is negative (Algorithm 2).

Algorithm 2. Sign detection in RNS.

Input: $X \xrightarrow{RNS} (x_1, x_2, \ldots, x_n), (p_1, p_2, \ldots, p_n), P, p_j, (P_1, P_2, \ldots, P_n)\ j$, for $i = \overline{1, n}$.
Output: $sign(X)$
1. **for** i **to** n **do**: $k_i = |P_i^{-1}|_{p_i}$
2. $F = 0$
3. **for** i **to** n **do**: $F \mathrel{+}= x_i \cdot k_i$
4. $F = F - (\boldsymbol{int})F$
5. **if** $F > 1/p_j$: **return 1**
6. **else**: **return 0**

The algorithm of sign detection in RNS is more complex in comparison with the algorithm of sign detection in the binary number system.

The numbers comparison can be viewed as calculating the PC of both numbers (X and Y) and comparing them:

if $\left|\frac{X}{P}\right|_1 - \left|\frac{Y}{P}\right|_1 = 0$, then $X = Y$; if $\left|\frac{X}{P}\right|_1 - \left|\frac{Y}{P}\right|_1 > 0$, then $X > Y$; if $\left|\frac{X}{P}\right|_1 - \left|\frac{Y}{P}\right|_1 < 0$, then $X < Y$. The implementation of the number comparison can provide an output $s = 0$ if $X = Y$, if $X > Y$ then $s = 1$, and $s = -1$ when $X < Y$.

4 Scaling RNS Numbers by Base Extension

Scaling an RNS number allows choosing a divisor that can simplify the division operation before dividing by p_i. The scaling depends on the coefficient K. The general formula for scaling the remainder is as follows:

$$x_i' = \left|\overline{x_{k_i}} \cdot K_i\right|_{p_i} \tag{7}$$

Thus, it is required to define two variables: x_k and K_i.

$$\overline{x_{k_i}} = |x_i - x_{n+1}|_{p_i}, \tag{8}$$

where $x_{n+1} = r_X$. r_X is a rank of a number, and in the RNS, it is reduced to the following calculations:

$$r_X = \frac{\sum_{i=1}^{n-1} x_i \cdot B_i}{P}, \tag{9}$$

where $B_i = \left|P_i^{-1}\right|_{p_i}$ is RNS modulo. Having defined $\overline{x_{k_i}}$, we need to calculate K_i, which is defined as $K_i = \left|K_i^{p_i-2}\right|_{p_i}$.

The base extension of the number in the RNS for neuromorphic computing can be effectively performed using the following procedure. The calculation of the new residue is based on the rank of the number, which can be defined as:

$$r_X = \left|\sum\nolimits_{i=1}^{n} x_i B_i\right|_{p_n}, \tag{10}$$

where $B_i = P_i \cdot \left|P_i^{-1}\right|_{p_i}$ is an orthogonal basis. Considering that, based on CRT and orthogonal basis, the number X in the base system $p_1, p_2, \ldots, p_{n-1}$ can be written as

$$X = \sum\nolimits_{i=1}^{n} a_i B_i - x_j P \tag{11}$$

Substituting (10) in (11), we obtain the following:

$$X = \left|\sum\nolimits_{i=1}^{n-1} x_i |B_i|_{p_j} + x_j\left(p_j - |P|_{p_j}\right)\right|_{p_j} \tag{12}$$

Based on the above, for the base extension, it is necessary to calculate the rank of the number x_j in the base system $p_1, p_2, \ldots, p_{n-1}$ according to the expression (11) and find the remainder x_j by (12).

The proposed optimization method for the base extension is characterized by calculations for small modulo p_n. However, when compared with the CRT method, it simplifies the calculation with the large modulo P, and then the calculation with p_j. The residue of the number on the base extension is obtained by optimization method RNS&FRNN of RNS operations based on FRNN.

The constants of the expressions (11) and (12) can be calculated in advance. They determine the network structure. FRNN presented in Fig. 1 works as follows.

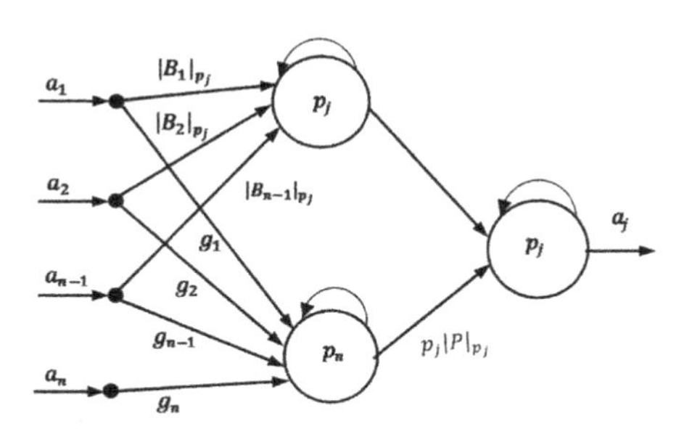

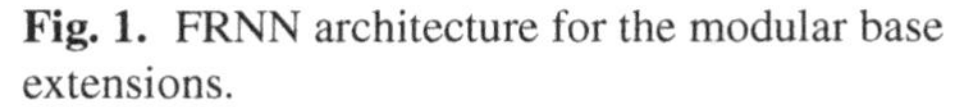

Fig. 1. FRNN architecture for the modular base extensions.

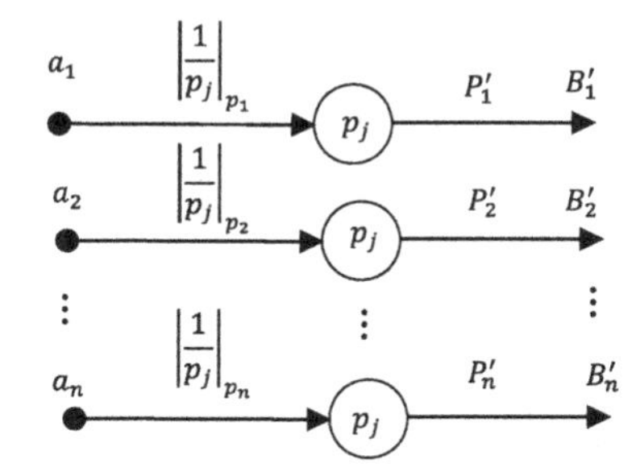

Fig. 2. The architecture of FRNN recalculation of the base extension.

The network input receives the modular values $x_1, \ldots, x_n$. In the first stage, a modular neural network modulo p_n by weighted summation of the modular values of the number

$x_1 \div x_n$ with coefficients $g_1 \div g_n$ calculates the rank of the number r_A. Then the modular network modulo p_j calculates the value $\sum_{i=1}^{n-1} a_i |B_i|_{p_{n+1}}$. In the second stage, $x_j = |X|_{p_j}$ is calculated using the computational model (11).

Each set of moduli of the modular code is characterized by an orthogonal basis, due to which, for the base extension, it is necessary to recalculate the basis $B_i^{'}$, $i = \overline{1, n+1}$. To recalculate them, the input data are: orthogonal basis B_i, $i = \overline{1, n}$, the moduli $p_1, p_2, \ldots, p_n$ and the values of the extended modulo p_j. Since $P_i^{'} = P'/p_i$ and P_i are coprime, we can calculate the orthogonal basis of the extended system as follows

$$B_i^{'} \equiv \frac{P'}{p_i} \cdot \left| P_i^{'-1} \right|_{p_i} \tag{13}$$

To calculate it on a NN basis, it is necessary to calculate two constants: $\left| \frac{1}{P_j} \right|_{pi}$. and $P_i^{'} = \frac{P'}{p_i}$. Thus, the NN architecture can be presented as following (Fig. 2).

The proposed algorithm has lower computational complexity compared to the known methods. However, the method involves multiplying pre-calculated constants. These constants are usually known in advance.

5 Experimental Results

We perform experimental analysis on CPU 2.7 GHz Intel Core i5, RAM 8 GB 1867 MHz DDR3, macOS High Sierra version 10.13.6 operating system. We use NTL, a high-performance, portable C++ library version 11.4.3, and LLVM's OpenMP runtime library version 10.0.0. RNS moduli are generated as a sequence of decreasing consecutive coprime numbers starting from $p_1 = 32,749, \ldots, p_{285} = 29,789$, and $L = \lceil \log_2 P \rceil$. One million random values of X and Y are generated using RandomBnd() function, an NTL routine for generating pseudo-random numbers. Execution time T of arithmetic and logical operations are measured in microseconds (μs). The number of threads is four. The results are presented in Table 1.

First, we measure the relative performance of each operation independently. The speedup of RNS&FRNN is between 9,954 and 25,888 for the addition, 12,348 and 31,385 for the subtraction, 13,193.9 and 318,203 for multiplication, 15,353.5 and 140,290 for division by constant, and 17,815.5 and 40,359.7 for Euclid division, varying n and L. RNS sign detection performance is between 4.5 and 15 times lower.

Now, let us compare the performance of NIST FIPS 186–5 digital signature algorithm with two implementations. It is based on the operation of multiplying the point of an elliptic curve over $GF(q)$ by a scalar, the most time-consuming operation, where q is a prime number.

Different approaches for computing the elliptic scalar multiplication are introduced. Well-known Montgomery approach is based on the binary method, where scalar multiplication is defined to be the elliptic point resulting from adding value to itself several times. It performs addition and doubling in each iteration.

Let us evaluate the mathematical expectation of the number of additions and doubling.

Table 1. Execution time of operations on NTL 11.4.3 (binary) and RNS&FRNN (RNS) (μs).

(a) Addition, Subtraction, and Multiplication

n	L	Addition			Subtraction			Multiplication		
		Binary	RNS	Binary/RNS	Binary	RNS	Binary/RNS	Binary	RNS	Binary/RNS
15	225	99,548	10	**9,954.8**	105,076	8	13,134.5	118,745	9	**13,193.9**
30	450	110,852	10	11,085.2	126,619	8	15,827.4	194,619	9	21,624.3
45	675	103,665	8	12,958.1	111,137	8	13,892.1	198,589	10	19,858.9
60	900	108,377	10	10,837.7	116,266	8	14,533.3	322,731	8	40,341.4
75	1,124	113,044	8	14,130.5	115,830	9	12,870	392,779	10	39,277.9
90	1,349	114,060	8	14,257.5	120,409	8	15,051.1	510,666	8	63,833.3
105	1,573	116,498	9	12,944.2	123,482	10	**12,348.2**	604,474	9	67,163.8
120	1,797	168,430	9	18,714.4	180,615	10	18,061.5	727,589	9	80,843.2
135	2,021	167,513	8	20,939.1	179,552	8	22,444	827,077	8	103,384.6
150	2,245	172,927	8	21,615.9	185,494	9	206,10.4	973,639	10	97,363.9
165	2,469	172,716	9	19,190.7	218,787	8	27,348.4	1,140,607	9	126,734.1
180	2,693	180,369	9	20,041.0	231,800	8	28,975	1,328,500	8	166,062.5
195	2,917	186,132	9	20,681.3	199,568	10	19,956.8	1,397,494	9	155,277.1
210	3,140	186,433	9	20,714.8	211,051	8	26,381.4	1,602,832	8	200,354.0
225	3,364	187,804	9	20,867.1	209,095	9	23,232.8	1,757,143	9	195,238.1
240	3,587	201,887	8	25,235.9	221,684	9	24,631.6	1,936,657	8	242,082.1
255	3,810	201,556	8	25,194.5	243,480	10	24,348	2,117,587	8	264,698.4
270	4,033	233,000	9	**25,888.9**	241,572	8	30,196.5	2,208,706	9	245,411.8
285	4,256	215,689	10	21,568.9	282,472	9	**31,385.8**	2,545,628	8	**318,203.5**

(b) Division by constant, Euclid division, and Sign detection

n	L	Division by constant			Euclid division			Sign detection		
		Binary	RNS	Binary/RNS	Binary	RNS	Binary/RNS	Binary	RNS	Binary/RNS
15	225	122,828	8	**15,353.5**	171,928	8	21,491.0	1	9	0.11
30	450	168,685	9	18,742.8	182,879	9	20,319.9	1	8	0.13
45	675	145,610	9	16,178.9	178,155	10	**17,815.5**	1	9	0.11
60	900	174,282	8	21,785.3	201,592	10	20,159.2	1	9	0.11
75	1,124	198,819	8	24,852.4	183,151	9	20,350.1	1	9	0.11
90	1,349	220,280	9	24,475.6	191,398	8	23,924.8	1	9	0.11
105	1,573	244,787	9	27,198.6	194,943	8	24,367.9	1	9	0.11
120	1,797	319,813	8	39,976.6	251,513	8	31,439.1	1	8	0.13
135	2,021	334,435	9	37,159.4	252,916	9	28,101.8	1	15	**0.07**
150	2,245	362,685	8	45,335.6	266,925	9	29,658.3	1	10	0.10
165	2,469	407,955	9	45,328.3	262,714	8	32,839.3	1	9	0.11
180	2,693	439,295	10	43,929.5	282,383	8	35,297.9	1	9	0.11
195	2,917	451,525	9	50,169.4	287,426	8	35,928.3	1	9	0.11
210	3,140	461,168	9	51,240.9	283,955	9	31,550.6	2	10	0.20
225	3,364	486,675	10	48,667.5	285,086	8	35,635.8	1	8	0.13
240	3,587	504,493	9	56,054.8	332,445	10	33,244.5	1	9	0.11
255	3,810	537,938	10	53,793.8	331,538	9	36,837.6	1	10	0.10
270	4,033	1,262,615	9	**140,290.6**	363,237	9	**40,359.7**	1	10	0.10
285	4,256	553,609	9	61,512.1	355,031	10	35,503.1	2	9	**0.22**

Doubling can be expressed as:

$$\frac{1}{2^{\lceil \log_2 q \rceil}} \sum_{i}^{\log_2 q-1} i \cdot 2^i = \frac{(\lceil \log_2 q \rceil - 2)2^{\lceil \log_2 q \rceil} + 2}{2^{\lceil \log_2 q \rceil}} \approx \lceil \log_2 q \rceil - 2 \tag{14}$$

Addition can be expressed as:

$$\frac{1}{2^{\lceil \log_2 q \rceil}} \sum_{i=0}^{\lceil \log_2 q \rceil} i \cdot C^i_{\lceil \log_2 q \rceil} = \frac{\lceil \log_2 q \rceil}{2^{\lceil \log_2 q \rceil}} \cdot 2^{\lceil \log_2 q \rceil - 1} = \frac{\lceil \log_2 q \rceil}{2}, \tag{15}$$

where $C^a_b = \frac{b!}{(b-a)! \cdot a!}$.

Using the projective Jacobian coordinates for the case when $Z \neq 1$ and $a = -3$, it takes 16 multiplications to add points, and 8 multiplications to double a point.

Statistical analysis of the algorithm demonstrates that the mathematical expectation of number of modular multiplications is about

$$\frac{\lceil \log_2 q \rceil}{2} \cdot 16 + (\lceil \log_2 q \rceil - 2) \cdot 8 = 16 \log_2 q - 16 \tag{16}$$

The execution time of the modular multiplication can be estimated as a sum of one multiplication and one addition; hence, $T_{Bin} = (16\lceil \log_2 q \rceil - 16)(M_{Bin} + A_{Bin})$, where M_{Bin} is the execution time of the multiplications and A_{Bin} is the execution time of the addition.

To assess the RNS implementation of the algorithm, first, we consider the RNS to binary T_C and binary to RNS T_E conversion times (Table 2).

The modular multiplication of an elliptic curve point by a scalar in RNS requires one multiplication, one addition, $n(n-1)/4$ divisions by a constant, and one operation for determining the sign of a number, where n is the number of moduli.

The execution time of the RNS implementation can be estimated as

$$T_{RNS} = (16\lceil \log_2 q \rceil q - 16)\left(M_{RNS} + A_{RNS} + \frac{n(n-1)}{4} DC_{RNS} + S_{RNS}\right) + 2(T_C + T_E),$$

where M_{RNS} is the execution time of multiplication of two numbers in RNS, A_{RNS} is the execution time of addition in RNS, DC_{RNS} is the execution time of the division by constant in RNS, S_{RNS} is the execution time of the sign detection, T_C is the time of binary to RNS conversion, and T_E is the time of RNS to binary conversion.

Thus, for $q = 2^{511} - 1$, $n = 75$, T_{Bin} and T_{RNS} are estimated, in the worst case, as $T_{Bin} = (16 \cdot 511 - 16) \cdot (392,779 + 113,044) = 4,127,515,680$, $T_{RNS} = (16 \cdot 511 - 16) \cdot (10 + 8 + 75 \cdot (75 - 1)/4 \cdot 8 + 9) + 2 \cdot (384,687,100 + 685,935,110) = 2,232,040,738$. Therefore, $T_{Bin}/T_{RNS} \approx 1.85$ times.

Table 2. Time of Binary to RNS (T_C) and RNS to Binary (T_E) conversion (μs).

n	L	T_C	T_E
15	225	43,685,000	58,829,340
30	450	109,762,100	152,454,240
45	675	192,411,600	290,528,220
60	900	309,272,700	464,671,140
75	1,124	384,687,100	685,935,110
90	1,349	484,623,300	952,097,660
105	1,573	595,553,500	1,251,265,520
120	1,797	713,753,000	1,604,888,270
135	2,021	815,894,900	1,995,859,970
150	2,245	944,469,700	2,440,457,340
165	2,469	1,085,540,400	2,962,540,610
180	2,693	1,264,208,500	3,524,798,450
195	2,917	1,424,684,400	4,180,765,530
210	3,140	1,648,081,600	4,692,365,130
225	3,364	1,713,357,600	5,268,070,980
240	3,587	1,882,562,600	6,117,228,220
255	3,810	2,049,367,600	6,824,793,510
270	4,033	2,242,481,300	7,592,480,470
285	4,256	2,469,894,800	8,680,054,900

6 Conclusion

We propose an optimization of six encryption operations: addition, subtraction, multiplication, division by constant, Euclid division, and sign detection with integers of arbitrary length based on modular arithmetic and finite ring neural networks.

We show that they provide significant advantages in comparison with long arithmetic implemented in NTL. The higher benefits of RNS&FRNN are derived for the multiplication of large numbers. RNS shares them into smaller numbers that can be performed independently and in parallel without carries between them. We demonstrate that the proposed solution is 85% faster than NIST FIPS 186-5 digital signature algorithm, even, calculation of the sign detection operation is inefficient.

The structure of the non-positional operations of RNS, fault tolerance, and parallelism can be well suited for neuromorphic systems. However, the proposed approach does not provide an efficient implementation of logical operations, such as comparison, number sign determination, etc., which we will study in future work.

Acknowledgments. This work was partially supported by the Ministry of Education and Science of the Russian Federation (Project 075–15-2020–788).

References

1. Krasnobayev, V.A., Yanko, A.S., Koshman, S.A.: A Method for arithmetic comparison of data represented in a residue number system. Cybern. Syst. Anal. **52**(1), 145–150 (2016). https://doi.org/10.1007/s10559-016-9809-2
2. Ruchkin, V., Romanchuk, V., Sulitsa, R.: Clustering, restorability and designing of embedded computer systems based on neuroprocessors. In: 2013 2nd Mediterranean Conference on Embedded Computing (MECO), pp. 58–61 (2013). https://doi.org/10.1109/MECO.2013.6601318
3. Vinogradov, I.M.: Elements of Number Theory. Courier Dover Publications (2016)
4. Yu, D.-J., Hu, J., Tang, Z.-M., Shen, H.-B., Yang, J., Yang, J.-Y.: Improving protein-ATP binding residues prediction by boosting SVMs with random under-sampling. Neurocomputing **104**, 180–190 (2013). https://doi.org/10.1016/j.neucom.2012.10.012
5. Hu, J., Li, Y., Yan, W.-X., Yang, J.-Y., Shen, H.-B., Yu, D.-J.: KNN-based dynamic query-driven sample rescaling strategy for class imbalance learning. Neurocomputing **191**, 363–373 (2016). https://doi.org/10.1016/j.neucom.2016.01.043
6. Babenko, M., et al.: Positional characteristics for efficient number comparison over the homomorphic encryption. Program. Comput. Softw. **45**(8), 532–543 (2019). https://doi.org/10.1134/S0361768819080115
7. Tchernykh, A., et al.: Scalable data storage design for non-stationary IoT environment with adaptive security and reliability. IEEE Internet Things J. **7**(10), 10171–10188 (2020). https://doi.org/10.1109/JIOT.2020.2981276
8. Burgess, N.: Scaling an RNS number using the core function. In: Proceedings 2003 16th IEEE Symposium on Computer Arithmetic, pp. 262–269 (2003). https://doi.org/10.1109/ARITH.2003.1207687
9. Tchernykh, A., et al.: Performance evaluation of secret sharing schemes with data recovery in secured and reliable heterogeneous multi-cloud storage. Cluster Comput. **22**(4), 1173–1185 (2019). https://doi.org/10.1007/s10586-018-02896-9
10. Miranda-López, V., Tchernykh, A., Babenko, M., Avetisyan, A., Toporkov, V., Drozdov. A.Y.: 2Lbp-RRNS: two-levels RRNS with backpropagation for increased reliability and privacy-preserving of secure multi-clouds data storage. IEEE Access **8**, 199424–199439 (2020). https://doi.org/10.1109/ACCESS.2020.3032655
11. Babenko, M., Shiriaev, E., Tchernykh, A., Golimblevskaia, E.: Neural network method for base extension in residue number system. In: Bychkov, I., Tchernykh, A., Feoktistov, A. (eds.) ICCS-DE 2020- 2nd International Workshop on Information, Computation, and Control Systems for Distributed Environments, Irkutsk, Russia, 6–7 July 2020, vol. 2638, pp. 9–22. CEUR-WS (2020). http://ceur-ws.org/Vol-2638/paper1.pdf
12. Babenko, M., Tchernykh, A., Golimblevskaia, E., Hung, N.V., Chaurasiya, V.K.: Computationally secure threshold secret sharing scheme with minimal redundancy. In: Bychkov, I., Tchernykh, A., Feoktistov, A. (eds.) ICCS-DE 2020- 2nd International Workshop on Information, Computation, and Control Systems for Distributed Environments, Irkutsk, Russia, 6–7 July 2020, vol. 2638, pp. 23–32. CEUR-WS (2020). http://ceur-ws.org/Vol-2638/paper2.pdf
13. Davies, M., et al.: Loihi: a neuromorphic manycore processor with on-chip learning. IEEE Micro **38**(1), 82–99 (2018). https://doi.org/10.1109/MM.2018.112130359

14. DeBole, M.V., et al.: TrueNorth: accelerating from zero to 64 million neurons in 10 years. Computer **52**(5), 20–29 (2019). https://doi.org/10.1109/MC.2019.2903009.
15. Babenko, M., et al.: RNS number comparator based on a modified diagonal function. Electronics **9**, 1784 (2020). https://doi.org/10.3390/electronics9111784
16. Miranda-Lopez, V., et al.: Weighted two-levels secret sharing scheme for multi-clouds data storage with increased reliability. In: 2019 International Conference on High Performance Computing & Simulation (HPCS), pp. 915–922. IEEE (2019). https://doi.org/10.1109/HPCS48598.2019.9188057
17. Babenko, M., Deryabin, M., Tchernykh, A.: The accuracy estimation of the interval-positional characteristic in residue number system. In: 2019 International Conference on Engineering and Telecommunication (EnT), pp. 1–5. IEEE (2019). https://doi.org/10.1109/EnT47717.2019.9030549
18. Kucherov, N., Babenko, M., Tchernykh, A., Kuchukov, V., Vashchenko, I.: Increasing reliability and fault tolerance of a secure distributed cloud storage. In: The International Workshop on Information, Computation, and Control Systems for Distributed Environments (2020) https://doi.org/10.47350/ICCS-DE.2020.16.

Investigating Overlapped Strategies to Solve Overlapping Problems in a Cooperative Co-evolutionary Framework

Julien Blanchard[1(✉)], Charlotte Beauthier[2], and Timoteo Carletti[1]

[1] Department of Mathematics and naXys institute, University of Namur, Namur, Belgium
{julien.blanchard,timoteo.carletti}@unamur.be
[2] Cenaero Research Center, Gosselies, Belgium
charlotte.beauthier@cenaero.be

Abstract. Cooperative co-evolution is recognized as an effective approach for solving large-scale optimization problems. It breaks down the problem dimensionality by splitting a large-scale problem into ones focusing on a smaller number of variables. This approach is successful when the studied problem is decomposable. However, many practical optimization problems can not be split into disjoint components. Most of them can be seen as interconnected components that share some variables with other ones. Such problems composed of parts that overlap each other are called overlapping problems. This paper proposes a modified cooperative co-evolutionary framework allowing to deal with non-disjoint subproblems in order to decompose and optimize overlapping problems efficiently. The proposed algorithm performs a new decomposition based on differential grouping to detect overlapping variables. A new cooperation strategy is also introduced to manage variables shared among several components. The performance of the new overlapped framework is assessed on large-scale overlapping benchmark problems derived from the CEC'2013 benchmark suite and compared with a state-of-the-art non-overlapped framework designed to tackle overlapping problems.

Keywords: Large-scale global optimization · Evolutionary algorithms · Cooperative co-evolution · Overlapping problem

1 Introduction

Nowadays, many real-world optimization problems arising in engineering and sciences deal with a large number of variables [7]. They present challenging

The present research benefited from computational resources made available on the Tier-1 supercomputer of the Fedération Wallonie-Bruxelles, infrastructure funded by the Walloon Region under the grant agreement n°1117545.

B. Dorronsoro et al. (Eds.): OLA 2021, CCIS 1443, pp. 254–266, 2021.
https://doi.org/10.1007/978-3-030-85672-4_19

characteristics making them hard to efficiently optimize. They are commonly solved by means of metaheuristics such as evolutionary algorithms or swarm intelligence [3]. However, the standard metaheuristics are not suitable to solve such large-scale global optimization (LSGO) problems because they suffer from the curse of dimensionality, i.e. their performance deteriorates when increasing the number of variables [1]. In this context, new approaches relying on the "divide-and-conquer strategy" have been proposed. They divide the initial LSGO problem into smaller ones which focus on smaller groups of variables. The latter are optimized in a round-robin fashion with a standard metaheuristic with the aim of producing the solution of the initial problem. This framework has been introduced by Potter and De Jong [9]. They designed a cooperative co-evolutionary (CC) approach to optimize LGSO problems by means of a genetic algorithm. Following this promising approach, the CC strategy have been embedded in many other metaheuristics such as evolutionary programming [6], particle swarm optimization [2] and differential evolution [11].

In any case, the efficiency of this approach is highly dependent on the performed decomposition. The latter depends on the characteristics of the objective function in terms of separability. A function is separable if the influence of any variable on the function value depends only on itself [18]. In this case, any decomposition that reduces the dimensionality is efficient in the CC framework. Other functions can be classified as additively separable [8] if they can be written as:

$$f(x) = \sum_{i=1}^{m} f_i(x_i), \tag{1}$$

where x_i $(i = 1, \ldots, m)$ are mutually exclusive k_i-dimensional decision vectors of f_i, x is the n-dimensional decision vector of the function f and m is the number of independent components such that $k_1 + \cdots + k_m = n$. In this way, the influence, of any variable in a component, on the function value depends only on other variables of the same component. Therefore, an ideal decomposition would divide the initial problem such that each subproblem focuses on one component given in Equation (1). The main challenge is thus to identify these components. It can be done by using the differential grouping strategy [8,16].

However, separable and partially separable problems are not representative of most LSGO problems arising in real-world optimization applications. Most of them incorporate several components that usually interact with each other. For example, the supply chain design and optimization [4] involves several components such as suppliers, manufacturers and distributors that interact with each other through a variety of transportation and delivery methods. Such interconnected problems are often referred as overlapping problems [17] because they are composed of parts that overlap others. In other words, each component involves multiple variables and some of them are shared with one or several other components. This kind of function is very challenging and standard CC algorithms fail to optimize them efficiently. Indeed, most of them rely either on random grouping [18] or on intelligent decomposition methods based on interaction identification [8]. The former simply completes several random decompositions in order to try catching linked variables in a same component but does not explicitly consider the interaction structure. The latter assigns all the linked variables in a single group and therefore does not reduce the dimensionality when dealing with overlapping problems. Two exceptions are the decomposition based on spectral clustering introduced in [5] and the decomposition specially designed for overlapping problems introduced in [15]. The latter breaks the linkage at shared variables between components in order to reduce the problem dimensionality, even for overlapping problems. It will be further discussed in Sect. 2.2.

In addition to the above methods, other CC strategies considering subsets that overlap each other have also received some attention. They raise some questions related to the exchange of information between components and related to the construction of the complete n-dimensional solution. In [14], non separable problems are decomposed into overlapping subproblems on the basis of a statistical variable interdependence learning scheme. The exchange of information is ensured by a periodically updated global solution (built on the basis of subproblem cores) used as shared memory. In [13], an overlapping decomposition covering the set of variables is predetermined. Compete and sharing strategies are implemented to choose the representative variables and share them among components. In [12], overlapping is not used to facilitate the decomposition but to overlap influential variables and evolves them in several components.

Some of these algorithms claim to tackle overlapping problems but do it with non-overlapped strategies [5,15]. Others, although based on overlapped strategies, do not explicitly claim to be able to tackle overlapping problems [12–14]. One may obviously think that the best way to optimize them in a CC framework is to do it with overlapped strategies. Nevertheless, to the best of the authors' knowledge, there are no research studies in that way. This paper introduces such a strategy and compare it with the non-overlapped approach specially designed for overlapping problems in [15]. The paper is organized as follows: Sect. 2 briefly describes the CC framework and the recursive differential grouping. Section 3 introduces the new strategy to split LSGO problems into overlapping subproblems and the overlapped CC framework that manages the exchange of information between subproblems. Experimental settings and results analysis are given in Sect. 4. Finally, findings and perspectives are discussed in Sect. 5.

2 Related Work

2.1 Cooperative Co-evolutionary Algorithms

The first attempt to optimize a LSGO problem with an evolutionary algorithm by means of a divide-and-conquer strategy was presented in 1994 [9]. Since then, this new approach, called cooperative co-evolution, has been widely studied [7]. The classical structure of this framework is described as follows:

1. *Decomposition*: Split the n-dimensional decision vector into some smaller disjoint subcomponents;

2. *Optimization*: Optimize each subcomponent with a standard evolutionary algorithm for a fixed number of iterations in a round-robin strategy;
3. *Combination*: Merge the solutions from each subcomponent to build the n-dimensional solution.

Throughout the optimization stage, the individuals in each subcomponent need to be evaluated with the n-dimensional function. For this purpose, they are completed with the variables of the *context vector*. The latter is a n-dimensional vector that contains information from all the subcomponents. Typically, it is composed of the variables of the current best solutions in each subcomponent and it is updated each time a better solution is found in a subcomponent.

2.2 Recursive Differential Grouping

In a CC framework, the decomposition should ideally be performed in such a way that there is no interaction between variables from different subcomponents. For additively separable problems, it can be uncovered with the Differential Grouping (DG) strategy [8,16]. In particular, the Recursive Differential Grouping (RDG) that benefits, as stated by its name, from recursive interaction detections between subsets of variables, relies on the following result [15,16]:

Theorem 1. *Let $f : \mathbb{R}^n \rightarrow \bar{\mathbb{R}}$ be an objective function; X_1 and X_2 be two mutually exclusive subsets of decision variables: $X_1 \cap X_2 = \emptyset$. If there exist a candidate solution $x^\star$ and sub-vectors a_1, a_2, b_1, b_2 such that*

$$f_{1,1}(x^\star) - f_{2,1}(x^\star) \neq f_{1,2}(x^\star) - f_{2,2}(x^\star) \tag{2}$$

where, $f_{i,j}(x^\star)$ is the function value obtained when replacing, in $x^\star$, the variables of X_1 with a_i and the variables of X_2 with b_j $(i, j = 1, 2)$, then there is some interaction between the decision variables in X_1 and X_2.

In practice, all the variables of $x^\star$, a_1 and b_1 are set to the lower bounds l of the search space. The variables of a_2 are set to the upper bounds u and those of b_2 are set to the mean $\bar{m}$ of the lower bounds and the upper bounds. Furthermore, equation (2) is not directly employed since the inequality may be the results of computational round-off errors instead of interaction detection, as expected. Thus, the following quantities are computed

$$\Delta_1 = f_{1,1}(x^\star) - f_{2,1}(x^\star),\ \Delta_2 = f_{1,2}(x^\star) - f_{2,2}(x^\star),\ \lambda = |\Delta_1 - \Delta_2| \tag{3}$$

and some interaction is detected when λ is greater than a threshold ϵ (see [16] for further details). Eventually, the success of the RDG algorithm relies on the recursive use of Theorem 1 to identify variables in X_2 that interact with those of X_1. Indeed, if any interaction between X_1 and X_2 is detected using Equation (3), the set X_2 is divided into two nearly equally-sized groups G_1 and G_2. Then, the interaction between X_1 and G_1 and X_2 and G_2 is checked. The process is repeated until all single variables in X_2 that interact with X_1 are identified.

In brief, the complete RDG algorithm can be presented as follows: (1) determine all the variables that interact with a selected variable x_i using the above recursive strategy and put them in a set X_1; (2) identify variables that interact with X_1 and add them to X_1, repeat the process until no more variable is added to X_1; (3) select another variable that is yet to be classified and return to step (1). Note that this approach would set all the variables of an overlapping problem into a single group. In [15], this issue was solved by slightly modifying the step (2) by imposing a condition on the size of X_1. In this new approach called RDG3, the step (2) is repeated: (a) until no more variable is added to X_1 or (b) until X_1 contains more than ϵ_n variables, where ϵ_n is fixed to a predetermined value.

3 Proposed Algorithm

The newly proposed algorithm aims to tackle LSGO overlapping problems within an overlapped CC framework. The fact that it has to deal with subcomponents that share several variables raises new challenges. The first one is to perform an accurate decomposition that detects overlapping variables efficiently and share them among several subcomponents. It can be achieved by using the modified approach of the RDG strategy presented in Sect. 3.1. The second challenge concerns the management of overlapping variables during the optimization, in particular for function evaluations. It will be discussed in Sect. 3.2.

3.1 Overlapped Recursive Differential Grouping

The main idea of the newly proposed decomposition strategy is to relax the grouping by identifying variables that make the link between several components in interconnected problems and share them among these components. For example, in the interaction graph presented in Fig. 1, three components can be identified:

$$S_1 = \{x_1, x_2, x_3, x_4\}, S_2 = \{x_3, x_4, x_5, x_6, x_7\} \text{ and } S_3 = \{x_7, x_8, x_9\}. \quad (4)$$

In each of them, interaction between variables are plentiful while there is no direct interaction between variables from distinct components, i.e. $\forall\ i, j(i \neq j)$, $k, l(k \neq l)$such that $x_i \in S_k \backslash S_l$ and $x_j \in S_l \backslash S_k$, x_i does not interact with x_j. Using the RDG3 strategy to decompose such a problem will break the linkage at shared variables and will lead to the decomposition illustrated in Fig. 1a. The latter might not be the optimal one since x_3 and x_4 (resp. x_7) are not optimized with x_5, x_6 and x_7 (resp. x_8 and x_9) while they are strongly connected. The new strategy, called *Overlapped RDG* (ORDG), is aimed to allow some overlapping between subcomponents to prevent from breaking these important linkages. It will produce the decomposition proposed in Fig. 1b.

The ORDG strategy is presented in Algorithm 1. It is very closed to the RDG algorithm except for the instructions in the "else" statement at line 12. In particular, the instruction at line 5 recursively identifies variables in X_2 that interact with X_1. They are added to X_1 to constitute the set $X_1^\star$ (see Algorithm 2).

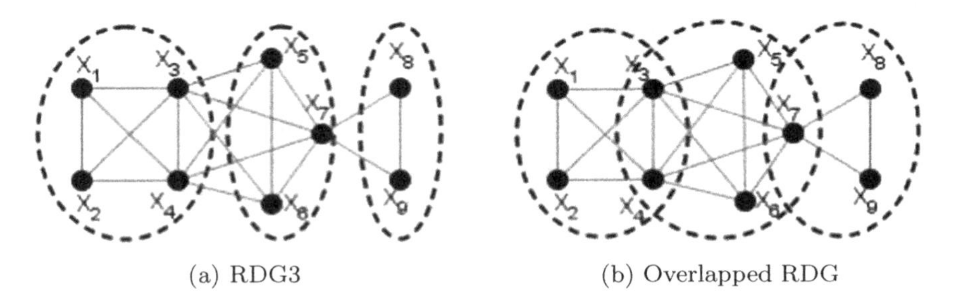

(a) RDG3 (b) Overlapped RDG

Fig. 1. The two obtained decompositions for an overlapping problem using RDG3 and Overlapped RDG strategies respectively.

Algorithm 1: Overlapped Recursive Differential Grouping

```
1  seps = {}, nonseps = {};
2  Set all the variables of x_l to the lower bounds, compute f_ = f(x_l) ;
3  X1 = {x1}, X2 = {x2, ..., xn} ;
4  while X2 ≠ {} do
5      X1* = R_Inter(X1, X2, f_, f) ;
6      if |X1*| = |X1|
          // For RDG3, the if would be: if |X1*| = |X1| or |X1*| > ε_n
7      then
8          if |X1| = 1 then seps = seps ∪ X1 ;
9          else nonseps = nonseps ∪ X1 ;
10         X1 = {xj} s.t. j ≤ i ∀xi ∈ X2;
11         X2 = X2 \ {xj} ;
12     else
           // For RDG3, the else statement would only contains the
              following instructions: X1 = X1*,  X2 = X2 \ X1 ;
13         if |X1| = 1 then
14             X1 = X1*, X2 = X2 \ X1 ;
15         else
16             X1** = L_inter(X1, X2, f_, f) ;
17             nonseps = nonseps ∪ X1 ;
18             X1 = X1* \ X1 ∪ X1** ;
19             X2 = X2 \ X1* ;

20 if |X1| = 1 then seps = seps ∪ X1 ;
21 else nonseps = nonseps ∪ X1 ;
22 return seps and nonseps;
```

- If no interaction has been identified, i.e. if $|X_1^\star| = |X_1|$, the X_1 set is recognized as a nonseparable subset if it contains several variables, otherwise the only variable in X_1 is identified as a separable one (lines 8-9). The process moves on to the next variable that is yet to be classified (lines 10-11).

- Otherwise, some interaction has been identified between X_1 and X_2. The variables in X_2 responsible of the interaction have been identified during the recursive detection at line 5 but at this stage, the variables in X_1 responsible of the interaction have not yet been determined (and they should be to perform the overlapped decomposition). If X_1 contains only one variable, this is the one responsible of the interaction. In this case, the algorithm moves on to the next iteration while making the same update that for the RDG3 strategy (lines 13-14). Otherwise (i.e. if X_1 contains several variables), those interacting with X_2 are identified at line 16 using a recursive mechanism again (see Algorithm 3) and the update described in lines 17-19 produces the desired overlapped decomposition.

Algorithm 2: R_Inter($X_1, X_2, \underline{f}, f$)

```
1 if Interact(X1, X2, f_, f) then
2     if |X2| = 1 then
3         X1 = X1 ∪ X2 ;
4     else
5         Split X2 into equally-sized
          groups G1, G2 ;
6         X1^1 = R_Inter(X1, G1, f_, f);
7         X1^2 = R_Inter(X1, G2, f_, f);
8         X1 = X1^1 ∪ X1^2 ;
9 return X1 ;
```

Algorithm 3: L_Inter($X_1, X_2, \underline{f}, f$)

```
1 if Interact(X1, X2, f_, f) then
2     if |X1| = 1 then
3         return X1 ;
4     else
5         Split X1 into equally-sized
          groups G1, G2 ;
6         X1^1 = L_Inter(G1, X2, f_, f);
7         X1^2 = L_Inter(G2, X2, f_, f);
8         X1 = X1^1 ∪ X1^2 ;
9 return X1 ;
```

Note that the main difference between the `R_inter` and the `L_inter` functions lies in the fact that the former focuses on the set X_2 while the latter works on X_1. Furthermore, the two functions also differ in line 3 (see Algorithms 2 and 3) since the `R_inter` function adds variables from X_2 that interact with X_1 to X_1 while the `L_inter` function only returns variables from X_1 that interact with X_2. For both algorithms, the `Interact` function at line 1 relies on Theorem 1.

3.2 Overlapped CC Framework

The main layout of the overlapped CC framework is similar to the standard CC presented in Sect. 2.1 with the major exception that it is designed to detect and manage overlapping variables efficiently. For this purpose, the decomposition step performs the ORDG algorithm presented in the previous section.

The optimization still consists in iteratively evolving each subcomponent in a round-robin strategy. However, in this step, the cooperation between subproblems through the sharing of best solutions in the context vector needs to be revised. Since subcomponents overlap, a variable x_i belonging to one component S_k may also appear in another component S_l. This introduces the issue of which value of x_i has to be shared in the context vector. In a standard framework (i.e. without any overlapping), this value is the one of the variable x_i of the best individual in the (only) subpopulation containing x_i (see Fig. 2a for an illustrative example). For the overlapped framework, this idea is extended: the value of x_i in the context vector is the one of the best individual among all the individuals in the two subpopulations focusing on x_i (or in the only subpopulation if x_i is not overlapped). Such an arrangement is illustrated in Fig. 2b.

Note that, in order to choose the best individual within two different subpopulations, the function value of each individual used for comparison is the one that has been computed during the optimization of the corresponding subcomponents in the round-robin fashion loop. In this process, individuals in each subcomponent are completed with the variables of the context vector in order to be evaluated. The latter is updated each time a better solution is reached.

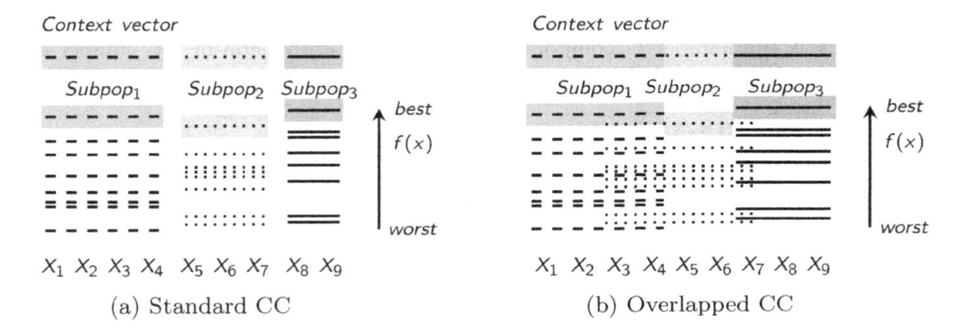

(a) Standard CC

(b) Overlapped CC

Fig. 2. Management of the context vector within a standard and an overlapped CC framework. The illustrative example relies on the interaction structure presented in Fig. 1. Dashed, dotted and solid lines represent individuals from subpopulations 1, 2 and 3 respectively. The context vector is built with the variables values of the best individual in each subpopulation.

4 Experimental Settings and Results

The performance of the new overlapped framework is assessed on large-scale overlapping benchmark problems derived from the CEC'2013 suite [17] and compared with the standard CC framework based on RDG3 decomposition [15]. The benchmark set contains 6 functions. Two of them, F_5 and F_6, are directly taken from [17]: F_6 is the 1000-d shifted Rosenbrock function and F_5 is the 905-d shifted Schwefel's function with conflicting[1] overlapping subcomponents. The four other functions, F_1 to F_4, are obtained by replacing the Schwefel basis function in F_5 by Ackley, Elliptic, Rastrigin and Rosenbrock functions respectively. Therefore, functions F_1 to F_5 contain 20 overlapping subcomponents that share 5 variables

[1] Note that the function f_{13} in [17] also contains overlapping subcomponents but it has not been included in the benchmark set because their overlapping subcomponents are conforming. It means that they have the same optimum value with respect to both subcomponent functions. It can be simply optimized in a standard CC framework.

with adjacent subcomponents. The F_6 function (Rosenbrock) can be seen as containing 999 subcomponents sharing one variable with adjacent ones.

In order to evaluate the decomposition effects of the newly proposed framework on overlapping problems, the RDG3 and ORDG strategies are used to decompose the benchmark problems presented above. For the RDG3, two different threshold values $\epsilon_n = 50$ and $\epsilon_n = 0$ are tested. The first value is the one used to study optimization results in [15] while the second value aims to identify as many components as possible and systematically cut the overlapping at shared variables. The number of components generated (k), the sum of the number of variables in each group (r) and the number of function evaluations computed (FEs) are reported in Table 1.

Table 1. Decomposition results of RDG3 (with $\epsilon_n = 50$ and $\epsilon_n = 0$) and ORDG strategies. k is the number of components generated, r is the sum of the number of variables in each group and FEs is the number of function evaluations.

Fun	RDG3 ($\epsilon_n = 50$)			RDG3 ($\epsilon_n = 0$)			ORDG		
	k	r	FEs	k	r	FEs	k	r	FEs
F_1	12	905	16273	20	905	16597	12	1011	16702
F_2	12	905	16252	19	905	16666	17	1000	18214
F_3	12	905	16249	20	905	16615	17	1000	18214
F_4	12	905	16252	20	905	16666	17	1000	18214
F_5	13	905	16288	21	905	16669	17	1003	18202
F_6	20	1000	49891	500	1000	25435	999	1998	59848

For the RDG3 decompositions, r is simply equal to the number of variables of the function because there is no overlap. For F_1 to F_5, the ORDG should capture the overlapping subcomponents introduced in [17] and therefore retrieve the 1000 variables involved in the benchmark construction. This is the case for F_2 to F_4. For F_1 and F_5, some additional interactions between independent variables have been identified due to computational round-off errors and lead to a slightly larger value of r. Still according to [17], the number of components k for functions F_1 to F_5 is equal to 20 in the benchmark construction. The RDG3 with $\epsilon_n = 50$ produces only 12 (or 13) components since components that contain less than 50 variables are merged with other ones. The RDG3 with $\epsilon_n = 0$ retrieves the 20 subcomponents (except for F_2 and F_5 for which the small difference is again due to computational round-off errors). The ORDG detects 17 components for functions F_2 to F_5[2]. They correspond to the ones formed in the benchmark construction except that some of them have been merged. Indeed, if the ORDG procedure starts the detection with a variable belonging to a component that share some overlapping variables with two adjacent components,

[2] Theoretically, 17 components should also be detected for F_1 but round-off erros affect the results for that particular function.

the latter are merged to form only one component. Thereafter, adjacent components to these components are also merged and so on. Although this prevents the detection of the 20 subcomponents, the obtained decomposition still agrees with the desired objective. In this particular case, the fact that some overlapping components contain two subsets of variables that do not directly interact will not affect the optimization efficiency. For the F_6 function, the obtained decomposition corresponds to the expected one, 20 (500) components of 50 (2) variables are formed for the RDG3 with $\epsilon_n = 50$ ($= 0$ resp.) and the ORDG produces 999 components of 2 variables. Finally, since the ORDG analyses additional interactions with respect to the RDG3, the cost in terms of FEs is higher. However, the additional cost remains reasonable and will be negligible with respect to the budget in terms of FEs allowed for the optimization.

The influence of the decomposition on the optimization results is analyzed by embedding each kind of decomposition in the overlapped CC framework presented above. In particular, when the latter is coupled with the RDG3 decompositions, it behaves like the standard CC. The evolutionary algorithm used to optimize the subcomponents is a genetic algorithm. In this study, the one implemented in the Minamo software is considered [10]. Here there is an overview of its main features: real-value representation of the individuals; tournament selection to pick up pairs of parents; arithmetic crossovers for recombination; mutation rate of 1 %; elitism of two individuals. Within the CC framework, the population size is set to 10 times the number of variables of the considered component. The round-robin fashion optimization loop is repeated until the maximum number of FEs is reached. It is set to 3×10^6 in total (for the decomposition and the optimization).

The median of the best solution over 51 independent runs and the standard deviation are reported in Table 2. The CC-ORDG produces better solution quality than the CC-RDG3 for 4 of the 6 functions. The CC-RDG3 with $\epsilon_n = 0$ generates the best results for the 2 other functions. Convergence graphs depicting the convergence behavior along the optimization process are also provided in Fig. 3. It can be seen that the three algorithms follow the same trend for functions F_1 to F_5[3]. Between the two variants of the CC-RDG3, the slightly different number of components does not have too much influence on the optimization quality. However, for F_6, the CC-RDG3 with $\epsilon_n = 0$ that produces many more subcomponents (each of them focusing on 2 variables) has a better handle of the optimization. Furthermore, the closed results between the CC-RDG3 with $\epsilon_n = 0$ and the CC-ORDG may be surprising. By analyzing the convergence behavior of the overlapping variables in the CC-ORDG in details, it can be seen that most of the time, the variables shared among two subcomponents converge to the same value at the same rate in the two subcomponents. In this context, the overlapped decomposition does not significantly contribute to a better cooperation between subcomponents in comparison with the cooperation through the sharing of the context vector performed in a standard CC framework. Therefore,

[3] Note that for F_2, the CC-ORDG is stuck in a pseudo-optima for a few runs. It causes the large green-colored area in Fig. 3b.

Table 2. Optimization results of the CC-RDG3 (with $\epsilon_n = 50$ and $\epsilon_n = 0$) and the CC-ORDG. The median of the best solution over 51 independent runs and the standard deviation are presented. Best median values are in bold.

Fun	RDG3 ($\epsilon_n = 50$)		RDG3 ($\epsilon_n = 0$)		ORDG	
	Median	Std	Median	Std	Median	Std
F_1	7.03e+07	1.77e+05	7.04e+07	1.48e+05	**7.01e+07**	2.84e+05
F_2	3.95e+13	3.56e+12	**3.53e+13**	5.24e+12	3.85e+13	1.67e+14
F_3	4.83e+08	2.81e+07	5.22e+08	4.45e+07	**4.17e+08**	8.11e+07
F_4	6.01e+11	1.82e+10	**4.27e+11**	1.40e+10	7.80e+11	3.58e+10
F_5	1.04e+11	2.15e+10	1.15e+11	2.66e+10	**9.64e+10**	1.75e+10
F_6	8.68e+05	9.80e+04	1.50e+03	1.21e+02	**1.34e+03**	1.01e+02

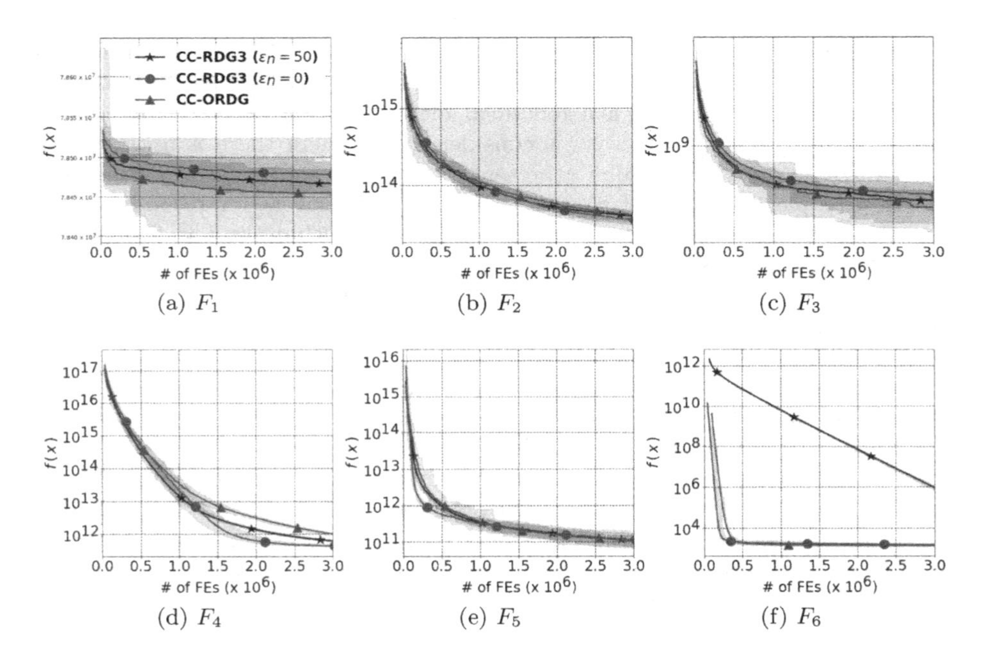

(a) F_1 (b) F_2 (c) F_3

(d) F_4 (e) F_5 (f) F_6

Fig. 3. Convergence graphs representing the evolution of $f(x)$ (in log-scale) with respect to the number of FEs. CC-RDG3 with $\epsilon_n = 50$ (blue stars), CC-RDG3 with $\epsilon_n = 0$ (red circles), CC-ORDG (green triangles). The solid line depicts the median value while the light-colored area represents the interval between the best and the worst value over the 51 runs.

although results in Table 2 might indicate that the CC-ORDG provides slightly better results, we can not definitely claim that a strategy is better than the other.

5 Discussion

The new CC framework introduced in this paper is designed to optimize overlapping LSGO problems with an overlapped decomposition strategy. In this context, an overlapped variant of the RDG has been developed to efficiently detect overlapping variables and share them among several subcomponents. The optimization step of the standard CC framework has also been extended in order to efficiently sha-re information between overlapped subcomponents through the context vector.

Numerical experiments were conducted on 6 benchmark functions. The extension of the method to a larger set of test functions is straightforward. However we believe the latter goes beyond the scope of this introductory paper and thus it will be considered in a further work. Similarly, the benchmark set is limited to 905-d and 1000-d functions, which is common practice in LSGO studies. Further research on the scalability may also be carried out to determine how the algorithm performs on more complex problems with larger dimensions.

The experiments presented in this paper show that the new approach produces the desired overlapped decomposition. However, although the optimization results might indicate that the new decomposition helps to get slightly better solutions, we can not definitely claim that the new framework outperforms the standard ones. This may be partly explained by the fact that the exchange of information between subcomponents in a standard CC framework through the context vector is stronger than we could expect. In any way, there is scope for even better progress to further develop the CC concept to deal with overlapping problems. We think that the new strategy that introduces overlapped subcomponents may be a promising way to achieve such an improvement.

References

1. Bellman, R.: Adaptive Control Processes : A Guided Tour. Princeton University Press, Princeton, New Jersey, USA (1961)
2. van den Bergh, F., Engelbrecht, A.P.: A cooperative approach to particle swarm optimization. IEEE Trans. Evol. Comput. **8**, 225–239 (2004)
3. Eiben, A., Smith, J.: Introduction to Evolutionary Computing. Natural Computing Series. Springer, Heidelberg, Germany (2015)
4. Garcia, D.J., You, F.: Supply chain design and optimization: challenges and opportunities. Comput. Chem. Eng. **81**, 153–170 (2015)
5. Li, L., Fang, W., Wang, Q., Sun, J.: Differential grouping with spectral clustering for large scale global optimization. In: 2019 IEEE Congress on Evolutionary Computation, pp. 334–341. IEEE, Piscataway, New Jersey, USA (2019)
6. Liu, Y., Yao, X., Zhao, Q., Higuchi, T.: Scaling up fast evolutionary programming with cooperative coevolution. In: Proceedings of the 2001 Congress on Evolutionary Computation, Vol. 2, pp. 1101–1108. IEEE, Piscataway, New Jersey, USA (2001)
7. Mahdavi, S., Shiri, M.E., Rahnamayan, S.: Metaheuristics in large-scale global continues optimization: a survey. Inf. Sci. **295**, 407–428 (2015)
8. Omidvar, M.N., Li, X., Mei, Y., Yao, X.: Cooperative co-evolution with differential grouping for large scale optimization. IEEE Trans. Evol. Comput. **10**(10), 1–17 (2013)

9. Potter, M.A., De Jong, K.A.: A cooperative coevolutionary approach to function optimization. In: Davidor, Y., Schwefel, H.P., Männer, R. (eds) Parallel Problem Solving from Nature - PPSN III, **866**, 249–257. Lecture Notes in Computer Science. LNCS. Springer, Heidelberg, Germany (1994). https://doi.org/10.1007/11539117_147
10. Sainvitu, C., Iliopoulou, V., Lepot, I.: Global optimization with expensive functions - sample turbomachinery design application. In: Diehl, M., et al. (eds.) Recent Advances in Optimization and its Applications in Engineering, 499–509. Springer-Verlag, Berlin, Heidelberg (2010). https://doi.org/10.1007/978-3-642-12598-0_44
11. Shi, Y., Teng, H., Li, Z.: Cooperative co-evolutionary differential evolution for function optimization. In: wang, L., Chen, K., Ong, Y.S. (eds) Advances in Natural Computation. ICNC 2005, **3611**, 1080–1088. Lecture Notes in Computer Science. LNCS, Springer, Heidelberg, Germany (2005). https://doi.org/10.1007/11539117_147
12. Song, A., Chen, W.N., Luo, P.T., Gong, Y.J., Zhang, J.: Overlapped cooperative co-evolution for large scale optimization. In: 2017 IEEE International Conference on Systems, Man, and Cybernetics, pp. 3689–3694. IEEE, Piscataway, New Jersey, USA (2017)
13. Strasser, S., Sheppard, J., Fortier, N., Goodman, R.: Factored evolutionary algorithms. IEEE Trans. Evol. Comput. **2**(21), 281–293 (2017)
14. Sun, L.S.Y., Cheng, X., Liang, Y.: A cooperative particle swarm optimizer with statistical variable interdependence learning. Inf. Sci. **1**(186), 20–39 (2012)
15. Sun, Y., Li, X., Ernst, A., Omidvar, M.N.: Decomposition for large-scale optimization problems with overlapping components. In: 2019 IEEE Congress on Evolutionary Computation, pp. 326–333. IEEE, Piscataway, New Jersey, USA (2019)
16. Sun, Y., Omidvar, M.N., Kirley, M., Li, X.: Adaptive threshold parameter estimation with recursive differential grouping for problem decomposition. In: Proceedings of the Genetic and Evolutionary Computation Conference, GECCO 2018, pp. 889–896. ACM, New York, NY, USA (2018)
17. Xi, L., Tang, K., Omidvar, M.N., Yang, Z., Qin, K.: Benchmark functions for the CEC 2013 special session and competition on large-scale global optimization. Technical report. RMIT University, Melbourne (2013)
18. Yang, Z., Tang, K., Yao, X.: Large scale evolutionary optimization using cooperative coevolution. Inf. Sci. **178**(15), 2985–2999 (2008)

Improved SAT Models for NFA Learning

Frédéric Lardeux[(✉)] and Eric Monfroy

LERIA, University of Angers, Angers, France
{frederic.lardeux,eric.monfroy}@univ-angers.fr

Abstract. Grammatical inference is concerned with the study of algorithms for learning automata and grammars from words. We focus on learning Nondeterministic Finite Automaton of size k from samples of words. To this end, we formulate the problem as a SAT model. The generated SAT instances being enormous, we propose some model improvements, both in terms of the number of variables, the number of clauses, and clauses size. These improvements significantly reduce the instances, but at the cost of longer generation time. We thus try to balance instance size vs. generation and solving time. We also achieved some experimental comparisons and we analyzed our various model improvements.

Keywords: Constraint problem modeling · SAT · Model reformulation

1 Introduction

Grammatical inference [7] is concerned with the study of algorithms for learning automata and grammars from words. It plays a significant role in numerous applications, such as compiler design, bioinformatics, speech recognition, pattern recognition, machine learning, and others. The problem we address in this paper is learning a finite automaton from samples of words $S = S^+ \cup S^-$, which consist of positive words (S^+) that are in the language and must be accepted by the automaton, and negative words (S^-) that must be rejected by the automaton. A non deterministic automaton (NFA) being generally a smaller description for a language than an equivalent deterministic automaton (DFA), we focus here on NFA inference. An NFA is represented by a 5-tuple $(Q, \Sigma, \Delta, q_1, F)$ where Q is a finite set of states, the vocabulary Σ is a finite set of symbols, the transition function $\Delta : Q \times \Sigma \rightarrow \mathcal{P}(Q)$ associates a set of states to a given state and a given symbol, $q_1 \in Q$ is the initial state, and $F \subseteq Q$ is the set of final states.

Not to mention DFA (e.g., [6]), the problem for NFA has been tackled from a variety of angles. In [15] a wide panel of techniques for NFA inference is given. Some works focus on the design of ad-hoc algorithms, such as *DeLeTe2* [3] that is based on state merging methods. More recently, a new family of algorithms for regular languages inference was given in [14]. Some approaches are based on metaheuristic, such as in [12] where hill-climbing is applied in the context of regular language, or [4] which is based on genetic algorithm. In contrast to metaheuristics, complete solvers are always able to find a solution if there exists

B. Dorronsoro et al. (Eds.): OLA 2021, CCIS 1443, pp. 267–279, 2021.
https://doi.org/10.1007/978-3-030-85672-4_20

one, to prove the unsatisfiability of the problem, and to find the optimal solution in case of optimization problems. In this case, generally, the problem is modeled as a Constraint Satisfaction Problem (CSP [11]). For example, in [15], an Integer Non-Linear Programming (INLP) formulation of the problem is given. Parallel solvers for minimizing the inferred NFA size are presented in [8,9]. The author of [10] proposes two strategies, based on variable ordering, for solving the CSP formulation of the problem.

In this paper, we are not interesting in designing or improving a solver, but we focus in improving models of the problem in order to obtain faster solving times using a standard SAT solver. Modeling is the process of translating a problem into a CSP consisting in decision variables and constraints linking these variables. The INLP model for NFA inference of [15] cannot be easily modified to reduce the instances: to our knowledge, only Property 1 of our paper could be useful for the INLP model, and we do not see any other possible improvement. We thus start with a rather straightforward conversion of the INLP model into the propositional satisfiability problem (SAT [5]). This is our base SAT model to evaluate our improvements. The model, together with a training sample, lead to a SAT instance that we solve with a standard SAT solver. The generated SAT instances are very huge: the order of magnitude is $|S|.(|\omega|+1).k^{|\omega|}$ clauses, where k is the number of states of the NFA, ω is the longest word of S, and $|S|$ is the number of words of the training sample. We propose three main improvements to reduce the generated SAT instances. The first one prevents generating subsumed constraints. Based on a multiset representation of words, the second one avoid generating some useless constraints. The last one is a weaker version of the first one, based on prefixes of words. The first improvement returns smaller instances than the second one, which in turn returns smaller instances than the third one. However, the first improvement is very long and costly, whereas the third one is rather fast. We are thus interested in balancing generation and solving times against instance sizes. We achieved some experiments with the Glucose solver [1] to compare the generated SAT instances. The results show that our improvements are worth: larger instances could be solved, and faster. Generating the smallest instances can be too costly, and the best results are obtained with a good balance between instance sizes and generation/solving time.

This paper is organized as follows. In Sect. 2, we describe the problem and we give the basic SAT model. We also evaluate the size of the generated instances. Section 3 presents 3 model improvements, together with sketches of algorithms to generate them. Section 4 exposes our experimental results and some analysis. We finally conclude in Sect. 5.

2 Modeling the Problem in SAT

The non-linear integer programming (INLP) model of [9,15] cannot be easily improved or simplified. Indeed, the only improvement proposed in [15] and [9] corresponds to Property 1 (given in the next section). In this section, we thus present a SAT formulation of the NFA inference problem. This SAT model permits many improvements to reduce the size of the generated SAT instances.

The NFA Inference Problem. Consider an alphabet $\Sigma = \{s_1, \ldots, s_n\}$ of n symbols; a training sample $S = S^+ \cup S^-$, where S^+ (respectively S^-) is a set of *positive words* (respectively *negative words*) from Σ^*; and an integer k. The problem consists in building a NFA of size k which validates words of S^+, and rejects words of S^-. The problem can be extended to an optimization problem: it consists in inferring a minimal NFA for S, i.e., an NFA minimizing k. However, we do not consider optimization in this paper.

Notations. Let $A = (Q, \Sigma, q, F)$ be a NFA with: $Q = \{q_1, \ldots, q_k\}$ a set of states, Σ a finite alphabet (a set of symbols), q the initial state, and F the set of final states. The symbol λ represents the empty word. We denote by K the set $\{1, \ldots, k\}$. A transition from q_j to q_k with the symbol s_i is denoted by $\tau_{s_i, q_j \to q_k}$. Consider the word $w = w_1 \ldots w_n$ with $w_1, \ldots, w_n$ in Σ. Then, the notion of transition is extended to w by $T_{w, q_{i_1} \to q_{i_{n+1}}}$ which is a sequence of transitions $\tau_{w_1, q_{i_1} \to q_{i_2}}, \ldots, \tau_{w_n, q_{i_n} \to q_{i_{n+1}}}$. The set of candidate transitions for w between the states q_{i_1} and q_{i_l} in a NFA of size k is $\mathcal{T}_{w, q_{i_1} \to q_{i_l}} = \{T_{w, q_{i_1} \to q_{i_l}} \mid \exists i_2, \ldots i_{i_l - 1} \in K,\ T_{w, q_{i_1} \to q_{i_l}} = \tau_{w_1, q_{i_1} \to q_{i_2}}, \ldots, \tau_{w_l, q_{i_{l}-1} \to q_{i_l}}\}$.

A SAT Model. Our base model is a conversion into SAT of the nonlinear integer programming problem given in [15] or [9]. Consider the following variables:

- k the size of the NFA we want to build,
- $F = \{f_1, \ldots, f_k\}$ a set of k Boolean variables determining whether states q_1 to q_k are final or not,
- and $\Delta = \{\delta_{s, q_i \to q_j} | s \in \Sigma \text{ and } i, j \in K\}$ a set of $n.k^2$ variables determining whether there is or not a transition $\delta_{s, q_i \to q_j}$, i.e., a transition from state q_i to state q_j with the symbol s, for each q_i, q_j, and s.

A transition $T_{w_1 \ldots w_n, q_{i_1} \to q_{i_{n+1}}} = \tau_{w_1, q_{i_1} \to q_{i_2}}, \ldots \tau_{w_n, q_{i_n} \to q_{i_{n+1}}}$ exists if and only if the conjunction $d = \delta_{w_1, q_{i_1} \to q_{i_2}} \wedge \ldots \wedge \delta_{w_n, q_{i_n} \to q_{i_{n+1}}}$ is true. We call d a c_transition, and we say that d models $T_{w_1 \ldots w_n, q_{i_1} \to q_{i_{n+1}}}$. We denote by D_{w, q_i, q_j} the set of all c_transitions for the word w between states q_i and q_j.

The problem can be modeled with 3 sets of equations:

1. If the empty word λ is in S^+ or in S^-, we can determine whether the first state is final or not:

$$\text{if } \lambda \in S^+, \quad f_1 \tag{1}$$

$$\text{if } \lambda \in S^-, \quad \neg f_1 \tag{2}$$

2. For each word $w \in S^+$, there is at least a transition starting in q_1 and ending in a final state q_j:

$$\bigvee_{j \in K} \bigvee_{d \in D_{w, q_1, q_j}} (d \wedge f_j) \tag{3}$$

With the Tseitin transformations [13], we create one auxiliary variable for each combination of a word w, a state $j \in K$, and a transition $d \in D_{w, q_1, q_j}$:

$$aux_{w,j,d} \leftrightarrow d \wedge f_j$$

For each w, we obtain a formula in CNF:

$$\bigwedge_{j \in K} \bigwedge_{d \in D_{w,q_1,q_j}} [(\neg aux_{w,j,d} \vee (d \wedge f_j))] \tag{4}$$

$$\bigwedge_{j \in K} \bigwedge_{d \in D_{w,q_1,q_j}} (aux_{w,j,d} \vee \neg d \vee \neg f_j) \tag{5}$$

$$\bigvee_{j \in K} \bigvee_{d \in D_{w,q_1,q_j}} aux_{w,j,d} \tag{6}$$

d is a conjunction, and thus $\neg aux_{w,j,d} \vee d$ is a conjunction of $|w|$ binary clauses: $(\neg aux_{w,j,d} \vee \delta_{w_1,q_1 \rightarrow q_{i_2}}) \wedge \ldots \wedge (\neg aux_{w,j,d} \vee \delta_{w_{|w|},q_{i_{|w|}} \rightarrow q_{i_{|w|+1}}})$.
$|D_{w,q_1,q_j}| = k^{|w|-1}$ since for each symbol of w there is k possible moves in the NFA, except for the last symbol which leads to q_j. Thus, we have $(|w|+1).k^{|w|}$ binary clauses for Constraints (4), $k^{|w|}$ $(|w|+2)$-ary clauses for Constraints (5), and one $k^{|w|}$-ary clause for Constraints (6). We have added $k^{|w|}$ auxiliary variables.

3. For each $w \in S^-$ and each state q_j, either there is no complete transition from state q_1 to q_j, or q_j is not final:

$$\neg \left[\bigvee_{j \in K} \bigvee_{d \in D_{w,q_1,q_j}} (d \wedge f_j) \right] \tag{7}$$

Constraints (7) are already in CNF, and we have $k^{|w|}$ $(|w+1|)$-ary clauses.

Thus, the constraint model M_k for building a NFA of size k is:

$$M_k = \bigwedge_{w \in S^+} \Big((4) \wedge (5) \wedge (6)\Big) \wedge \bigwedge_{w \in S^-} (7)$$

and is possibly completed by (1) or (2) if $\lambda \in S^+$ or $\lambda \in S^-$.

Size of the Models. Considering ω_+, the longest word of S^+, and ω_-, the longest word of S^-, the number of constraints in model M_k is bounded by:

- $|S^+|.(|\omega_+|+1).k^{|\omega_+|}$ binary clauses;
- $|S^+|.k^{|\omega_+|}$ $(|\omega_+|+2)$-ary clauses;
- $|S^+|$ $k^{|\omega_+|}$-ary clauses;
- $|S^-|.k^{|\omega_-|}$ $(|\omega_-|+1)$-ary clauses.

The number of Boolean variables is bounded by:

- k variables in F determining final states;
- $n.k^2$ variables determining existence of transitions;
- $|S^+|.k.^{|\omega_+|}$ auxiliary variables $aux_{w,j,d}$.

It is thus obvious that it is important to improve the model M_k.

3 Improving the SAT Model

We now give some properties that can be used for improving the SAT model. By abuse of language, we will say that a model M_1 is smaller than a model M_2 whereas we should say that the SAT instance generated with M_1 and data D is smaller than the instance generated with M_2 and D. A first and simple improvement is based on the following property.

Property 1 (Empty word λ). If $\lambda \in S^-$, then each c_transition ending in q_1 does not have to be considered when generating the constraints related to the word $w \in S$.

Indeed, if w is positive, it cannot be accepted by a transition ending in q_1; similarly, if w is negative, $\neg d \vee \neg f_1$ is always true. When $\lambda \in S^+$, the gain is not very interesting: f_1 can be omitted in Constraints (7), (4), and (5). This does not really reduce the instance, and a standard solver would simplify it immediately.

Whereas a transition is an ordered sequence, the order of conjuncts in a c_transition is not relevant, and equal conjuncts can be deleted. Thus, a c_transition may model several transitions, and may correspond to several words. By abuse of language, we say that a c_transition ends in a state q_j if it corresponds to at least a transition ending in q_j. Thus, a c_transition may end in several states. We consider an order on c_transitions. Let d and d'' be two c_transitions. Then, $d \preceq d''$ if and only if there exists a c_transition d' such that $d \wedge d' = d''$. In other words, each transition variable $\delta_{s,q_i \to q_j}$ appearing in d also appears in d''. This order is used in the two first model improvements which are based on c_transitions. The third model improvement is based on transitions. We now consider some redundant constraints.

Property 2 (Redundant constraints). When a state q_i cannot be reached, each outgoing transition becomes free (it can be assigned true or false), and q_i can be final or not. In order to help the solver, all the corresponding variables can be assigned an arbitrary value. For each state q_j, $j \neq 1$:

$$\Big(\bigwedge_{i \in K, i \neq j} \bigwedge_{s \in \Sigma} \neg \delta_{s,q_i \to q_j} \Big) \to \neg f_j \wedge \Big(\bigwedge_{i \in K} \bigwedge_{s \in \Sigma} \neg \delta_{s,q_j \to q_i} \Big)$$

In CNF, these constraints generate (for all q_j), $(k-1).(k.n+1)$ redundant clauses of size $n.(k-1)+1$.

These constraints are useful when looking for a NFA of size k when k is not the minimal size of the NFA. Compared to SAT instance size, these redundant constraints can be very helpful without being too heavy.

Note that in our implementation, for all the models, we always simplify instances using Property 1 and removing duplicate transition variables in c_transitions (i.e., $\delta_{s,q_i \to q_j} \wedge \ldots \wedge \delta_{s,q_i \to q_j}$ is simplified into $\delta_{s,q_i \to q_j} \wedge \ldots$). Moreover, we also generate the redundant constraints as defined in Property 2.

Improvement Based on c_transitions Subsumption. This first improvement consists in removing tautologies for negative words, and some constraints and unsatisfiable disjuncts for positive words.

Property 3 (c_transition subsumption). Let v be a negative word from S^-, and $\neg d_v \vee \neg q_j$ be a Constraint (7) generated for the c_transition d_v for v ending in state q_j. We denote this constraint c_{v,d_v,q_j}. Consider a positive word w from S^+, and d_w a c_transition for w ending in q_j such that $d_v \preceq d_w$. Then, each $d_w \wedge f_j$ will be false due to c_{v,d_v,q_j}. Thus, Constraints (4) and (5) corresponding to w, d_w, and q_j will force to satisfy $\neg aux_{w,j,d_w}$; hence, they can be omitted and aux_{w,j,d_w} can be removed from Constraints (7). Similarly, consider ω from S^-, and d_ω a c_transition for w ending in q_j such that $d_v \preceq d_w$. Then, Constraint (7), $\neg d_v \vee \neg q_j$, will always be true (due to the constraint c_{v,d_v,q_j}), and can be omitted.

We can compute the size of the reduced SAT instance when the smaller word is a prefix. Let $v \in S^-$ and $w \in S$ be words such that $w = v.v'$, i.e., $v \subseteq w$ and v is a prefix of w. Then, using Property 3: if $w \in S^-$, the number of clauses generated for w is reduced to $(k-1).k^{|w|-1}$ clauses of size $|w+1|$; if $w \in S^+$, the number of clauses generated for w is reduced to $(|w|+1).(k-1).k^{|w|-1}$ binary clauses for Constraints (4), $(k-1).k^{|w|-1}$ $(|w|+2)$-ary clauses for Constraints (5), and one clause of size $(k-1).k^{|w|-1}$ for Constraint (6). The number of auxiliary variables is reduced to $(k-1).k^{|w|-1}$.

Operationally, we have a two step mechanism. First, for each negative word, each c_transition together with its ending state is generated and stored in a database of couples (c_transition, ending state) that we call c_couple. Then, for generating constraints for a word w, each of its c_couple is compared to the database. If a c_transition for w ending in q_j is smaller than a c_transition from the database also ending in q_j, then the corresponding constraints are not generated, as shown above. We call $M_{k,all}$ this reduced model.

Improvement Based on Multisets. Although efficient in terms of generated instance sizes, the previous improvement is very costly in memory and time. It becomes rapidly intractable. This second improvement also uses Property 3. It is a weakening of the above operational mechanism that does not omit every subsumed c_transition. This mechanism is less costly. Hence, generated instances will be a bit larger, but the balance generation time against instance size is very good. The idea is to order words in order to search in a very smaller database of c_couples (c_transition, ending state) when generating constraints for a word w. Moreover, this order will also imply the order for generating constraints.

We associate each word to a multiset which support is the vocabulary Σ. The word w, is thus associated with the multiset $ms(w) = \{s_1^{|w|_{s_1}}, \ldots, s_n^{|w|_{s_n}}\}$ where $|w|_{s_i}$ is the number of occurrences of the symbol s_i in w. Note that several words can have the same multiset representation. Based on multiset inclusion $(\{s_1^{a'_1}, \ldots, s_n^{a'_n}\} \subseteq_{\mathcal{M}} \{s_1^{a_1}, \ldots, s_n^{a_n}\} \Leftrightarrow \forall i, a'_i \leq a_i)$, we can now define the notion of word inclusion, noted $\subseteq_\omega$. Consider w and w', two words of Σ^*, then:

$$w' \subseteq_\omega w \Leftrightarrow ms(w') \subseteq_{\mathcal{M}} ms(w)$$

Consider a sample $S = S^+ \cup S^-$. Let $\top(S)$ be the multiset defined as

$$\top(S) = \{s_1^{1+max_{w \in (S)}\{|w|_{s_1}\}}, \ldots, s_n^{1+max_{w \in (S)}\{|w|_{s_n}\}}\}$$

and $\bot = \{s_1^0, \ldots, s_n^0\}$. Then, $\top(S)$ represents words which are not in the sample S, and $\bot$ represents the empty word λ which may be in S.

Consider the sample $S = S^+ \cup S^-$. Let $MS(S) = \{ms(w) | w \in S^+ \cup S^-\}$ be the set of the representations of words of S. Then, $(MS(S) \cup \{\bot, \top(S)\}, \subseteq_{\mathcal{M}})$ is a lattice. Let m be a multiset of $MS(S)$. Then, $inf(m)$ is the set of multisets $\{m' \in MS(S) \mid m' \subseteq_{\mathcal{M}} m\}$. This lattice of multisets defines the data structure used for constraint generation. For generating constraint of a word w of a multiset m, we now only compare its c_couples with the database of c_couples of words $w' \in S^-$ with $w' \subseteq_{\omega} w$, i.e., words represented by multisets smaller than m.

The negative words that allow to reduce the most, are the ones represented by the smallest multiset. We thus also propose a mechanism to reduce the database (c_transition, ending state) with the most useful c_couples, i.e., the ones from smallest words. Let $level(m)$ be the "level" of the multiset defined by: $level(m) = 0$ if $m = \bot$, $1 + max_{m' \in inf(m)}(level(m'))$ otherwise. Given a multiset m, and a threshold l, the *base* function returns all the multisets m' of level smaller than l, and such that $m' \subseteq_{\mathcal{M}} m$: $base(p, l) = \{n \in inf(p) \mid level(n) \leq l\} \bigcup \left(\bigcup_{p' \in inf(p)} base(p', l)\right)$ if $p \neq \bot$, $\emptyset$ otherwise.

Based on Property 3, c_couples of the negative words of these multisets will be used to reduce constraint generation of the words of m. We call this model $M_{k,mset,l}$, with l a given threshold. If *base* is called with the threshold 0, the database will be empty and the complete instance will be generated: $M_{k,mset,0} = M_k$. If *base* is called with the maximum level of the lattice, then, the database will be the largest one built with all the smaller words, and we will thus obtain the smallest instances with this notion of lattice. However, the larger the threshold, the longer the generation time, and the smaller the SAT instance. With the maximal threshold, the generated instances will be a bit larger than with the previous improvement ($M_{k,all} \subseteq M_{k,mset,max}$), but the generation is significantly faster. For lack of space, we cannot give here the complete algorithms for generating this improved model.

Improvements Based on Prefixes. Although faster to generate, the second model is still costly. We now propose a kind of weakening of Property 3, restricting its use to prefix.

Property 4 (Prefix). Let $w \in S$ be a word from the sample. Consider D^*_{w,q_i,q_j} the set of c_transitions defined by:

$$D^*_{w,q_i,q_j} = \bigvee_{l \in K, l \neq j} \left(\left(\bigvee_{d_u \in D^*_{u,q_i,q_l}} d_u \wedge \left(\bigvee_{d_v \in D^*_{v,q_l,q_i}} d_v \right) \right) \right)$$

if $w = u.v$, and $u \in S^-$; otherwise, $D^*_{w,q_i,q_j} = D_{w,q_i,q_j}$. Then,

$$\forall d \in D_{w,q_i,q_j} \setminus D^*_{w,q_i,q_j}, \neg d \vee \neg f_j$$

Hence, this property allows us to directly generate the reduced constraints, for negative or positive words, without comparing c_couples with a database.

Let $w = u_1 \dots u_n$ be a word from S such that $u_1 \in S^-$, $u_1.u_2 \in S^-$, and $u_1 \dots u_{n-1} \in S^-$ and for each $i < n$, there does not exist a decomposition $u_i = u_i'.u_i''$ such that $u_1 \dots u_{i-1}.u_i' \in S^-$. Then, if $w \in S^+$, using several times Property 4, Constraints (4), (5), and (6) can be replaced by Constraints (8), (9), and (10) where $l_0 = q_1$ and $N = [1, \dots, n]$:

$$\bigwedge_{i \in N, l_i \in K \setminus \{l_j | 1 \le j < i\}} \bigwedge_{i \in N, d_i \in D_{u_i, q_{l_{i-1}}, q_l}} [(\neg aux_{w,l_1,\dots,l_n} \vee (d_1 \wedge \dots \wedge d_n \wedge f_j))] \quad (8)$$

$$\bigwedge_{i \in N, l_i \in K \setminus \{l_j | 1 \le j < i\}} \bigwedge_{i \in N, d_i \in D_{u_i, q_{l_{i-1}}, q_l}} (aux_{w,l_1,\dots,l_n} \vee \neg d_1 \vee \dots \vee \neg d_n \vee \neg f_j) \quad (9)$$

$$\bigvee_{i \in N, l_i \in K \setminus \{l_j | 1 \le j < i\}} \bigvee_{i \in N, d_i \in D_{u_i, q_{l_{i-1}}, q_l}} aux_{w,l_1,\dots,l_n} \quad (10)$$

Similarly, if $w \in S^-$, using several times Property 4, Constraints (7) can be replaced by Constraints (11):

$$\bigwedge_{i \in N, l_i \in K \setminus \{l_j | 1 \le j < i\}} \bigwedge_{i \in N, d_i \in D_{u_i, q_{l_{i-1}}, q_l}} (\neg d_1 \vee \dots \vee \neg d_n \vee \neg f_j) \quad (11)$$

The number of clauses and variables generated for $w \in S^+$ is reduced to:

- $(|w| + 1).\left(\prod_{i=1}^{n}(k - i + 1)\right).k^{|w|-n}$ binary clauses for Constraints (8),
- $\left(\prod_{i=1}^{n}(k - i + 1)\right).k^{|w|-n}$ $(|w| + 2)$-ary clauses for Constraints (9),
- one clause of size $\left(\prod_{i=1}^{n}(k - i + 1)\right)$ for Constraint (10),
- and the number of auxiliary variables is reduced to $\left(\prod_{i=1}^{n}(k - i + 1)\right)$.

For $w \in S^-$, Constraints (11) are already in CNF and they correspond to $\left(\prod_{i=1}^{n}(k - i + 1)\right).k^{|w|-n}$ $(|w + 1|)$-ary clauses. Interestingly, these new counts of clauses (and more especially the factor $k - i + 1$ with $i = n$) also give us a lower bound for k: k must be greater than or equal to n, the number of nested prefixes in a word. This new improved model, that we call $M_{k,pref}$, is not much larger than $M_{k,mset}$, but it is significantly faster to generate.

Improvement Order. We have defined various models for inference of NFA of size k that can be ordered by their sizes: $M_{k,all} \subseteq M_{k,mset,l\_max} \subseteq m_{k,pref} \subseteq M_k$. Note that $M_{k,mset,l}$ with $l \neq l\_max$, and $M_{k,pref}$ cannot be compared in the general case; their sizes depend on the instance, the number and size of prefixes, and on the given level l. In the next section, we compare these models not only in terms of instance size, but also in terms of generation and resolution time.

4 Experimental Results

We suspect that, with respect to their generation time, the models are in reverse order of the order given above. Thus, we are interested in finding the best balance between three parameters: model size v.s. generation time + SAT solving time.

The experiments were carried out on a computing cluster with Intel-E5-2695 CPUs and 128 GB of memory. Running times were limited to 2 h for the generation of SAT instances, and 3 h to solve them. We used the Glucose [1] SAT solver with the default options. The benchmarks are based on the training set of the StaMinA Competition (http://stamina.chefbe.net). We selected 12 instances[1] with a sparsity $s \in \{12.5\%, 25\%, 50\%, 100\%\}$ and an alphabet size $|\Sigma| \in \{2, 5, 10\}$. For each of them, we limited the number of words to $|S^+| = |S^-| = 10$ and 20 for a maximal size of words equal to 7 and to $|S^+| = |S^-| = 20$ for a maximal size of words equal to 10. We generate CNF instances for different NFA sizes ($k \in \{3, 4, 5\}$). Consequently, we obtained 96 instances.

Table 1 presents a synthetic view of our experiments. The 4 first columns detail the instances: size of the NFA (k), size of the longest word ($|\omega|$), number of positive (and negative) words ($|S^+|$), and the model. The next columns provide average values over the 12 instances for the modeling time (T_{Model}), the number of variables ($\#Var$), the number of clauses ($\#Cl$), the solving time (T_{solve}), and the total modeling+solving time (T_{total}). We do not indicate the standard deviations but they are very close to zero. "-" indicates that no result was obtained before the time-out. From Table 1, we can draw some general conclusions about model improvements. As expected, $M_{k,all}$ always returns the smallest instances, and also the instances that Glucose solve the fastest. However, the generation time of these instances is very long. Thus, the total CPU time, i.e., generation + solving, is not the best. We can also see that when we increase the maximum length of words, this model does not permit to generate the instances in less than 2 h (e.g., Table 1, for $k = 4$, $\omega = 10$, and $|S^+| = 20$). This model is thus tractable, but only for small instances, with short words and small samples.

$M_{k,mset,l_{max}}$ generates instances a bit larger than $M_{k,all}$. Consider the negative word $v = aaab$, and the positive word $w = ba$. $M_{k,all}$ uses some c_transitions of v to ignore some clauses of w that $M_{k,mset,l_{max}}$ will not detect. For example, a loop on aaa from v with the same transition in v is used in $M_{k,all}$ but not in $M_{k,mset,l_{max}}$. However, with the multiset data structure, we obtain a much faster generation of instances. The total time is thus more interesting with $M_{k,mset,l_{max}}$ than with $M_{k,all}$. The generation time of $M_{k,mset,l_{max}}$ is still very high, and its interest is not always significant. For large instances, not presented in the table, $M_{k,mset,l_{max}}$ could not be generated in less than 2 h.

For $M_{k,pref}$, we can see that the generation time becomes reasonable, and much smaller than with the two previous improvements. Although smaller than with M_k, the instances are larger than with $M_{k,mset,l_{max}}$. In various experiments, this improvement was the best for the total time. Note also that our training samples are not so big, and that the number of prefixes is not so important. With larger $|S^+|$, for a fixed k, we should obtain better performances of $M_{k,pref}$.

We also tried two more improvements of $M_{k,mset,l}$ with $l \in \{1, 3\}$. The generation time of these models is logically faster than the ones of $M_{k,mset,l_{max}}$; as planned, the SAT instances are also larger. However, we were pleasantly surprised by the total time which is much better than for $M_{k,mset,l_{max}}$. The three models

[1] We conserved the "official" name used during the Stamina Competition.

Table 1. Comparison on 96 generated instances between the models $m_{k,all}$, $m_{k,mset,l_{max}}$, $m_{k,mset,1}$, $m_{k,mset,3}$, and $m_{k,pref}$. Instances are grouped by size of the NFA (k), size of the longest word ($|\omega|$), and number of positive (and negative) words ($|S^+|$). For each line, obtained values are average on 12 instances.

k	$\lvert\omega\rvert$	$\lvert S^+\rvert$	Model	T_{model}	#Var.	#Cl.	T_{solve}	T_{total}
3	7	10	m_k	0.19	6742	61366	0.22	0.41
			$m_{k,all}$	0.68	4310	37789	0.14	0.82
			$m_{k,mset,l_{max}}$	0.17	4742	42020	0.14	0.31
			$m_{k,mset,1}$	0.18	5517	49484	0.16	0.34
			$m_{k,mset,3}$	0.17	4822	42850	0.14	0.31
			$m_{k,pref}$	0.18	6466	58645	0.2	0.38
		20	m_k	0.48	14830	134302	1.58	2.06
			$m_{k,all}$	2.62	8274	72569	1.64	4.26
			$m_{k,mset,l_{max}}$	0.42	8929	79030	1.22	1.64
			$m_{k,mset,1}$	0.45	11179	99811	1.39	1.84
			$m_{k,mset,3}$	0.46	9148	81188	1.27	1.73
			$m_{k,pref}$	0.43	13689	123390	1.71	2.14
	10	20	m_k	11	303519	3276974	397.68	408.68
			$m_{k,all}$	746.08	108417	1172093	79.98	826.06
			$m_{k,mset,l_{max}}$	9.87	122423	1313463	143.32	153.19
			$m_{k,mset,1}$	9.04	208610	2255307	233.97	243.01
			$m_{k,mset,3}$	9.06	134720	1443357	156.24	165.3
			$m_{k,pref}$	8.88	281408	3040802	270.04	278.92
4	7	10	m_k	1.46	45014	428775	10.3	11.76
			$m_{k,all}$	19.42	32956	302835	5.59	25.01
			$m_{k,mset,l_{max}}$	1.64	35362	328938	5.58	7.22
			$m_{k,mset,1}$	1.42	39242	369600	7.12	8.54
			$m_{k,mset,3}$	1.56	36048	336637	5.58	7.14
			$m_{k,pref}$	1.3	43655	414141	10.69	11.99
		20	m_k	3.93	100984	950473	83.55	87.48
			$m_{k,all}$	93.48	64428	588293	74.55	168.03
			$m_{k,mset,l_{max}}$	4.33	68041	628400	43.08	47.41
			$m_{k,mset,1}$	3.65	83463	777005	32.32	35.97
			$m_{k,mset,3}$	4.27	70720	653396	41.36	45.63
			$m_{k,pref}$	3.37	94829	887943	55.88	59.25
	10	20	m_k	187.59	4670833	53350566	2084.78	2272.37
			$m_{k,all}$	-	-	-	-	-
			$m_{k,mset,l_{max}}$	919.56	2304788	26010946	651	1570.56
			$m_{k,mset,1}$	173.82	3336332	38121787	658.7	832.52
			$m_{k,mset,3}$	375.34	2345238	26693196	107.13	482.47
			$m_{k,pref}$	162.45	4405201	50260648	1331.92	1494.37
5	7	10	m_k	6.61	201651	1962754	215.06	221.67
			$m_{k,all}$	232.47	161828	1526044	51.82	284.29
			$m_{k,mset,l_{max}}$	14.38	169816	1619550	171.92	186.3
			$m_{k,mset,1}$	7.24	182445	1759734	180.98	188.22
			$m_{k,mset,3}$	10.76	172660	1653301	210.1	220.86
			$m_{k,pref}$	6.26	196894	1908623	176.12	182.38
		20	m_k	19.37	456976	4382919	1268.18	1287.55
			$m_{k,all}$	1158.5	320689	2995308	631.14	1789.64
			$m_{k,mset,l_{max}}$	44.01	333799	3148787	1115.9	1159.91
			$m_{k,mset,1}$	20.24	398074	3784691	1192.49	1212.73
			$m_{k,mset,3}$	32.82	348339	3288509	1309.17	1341.99
			$m_{k,pref}$	16.54	434008	4141453	1203.36	1219.9

Table 2. Focus on 2 specific instances.

k	$\|\omega\|$	$\|S^+\|$	Model	T_{model}	#Var.	#Cl.	T_{solve}	T_{total}
25_training								
5	7	20	m_k	16.72	378030	3748314	934.92	951.64
			$m_{k,all}$	854.47	271338	2626880	841.22	1695.69
			$m_{k,mset,l_{max}}$	48.71	275331	2678349	1538.06	1586.77
			$m_{k,mset,1}$	14.25	280899	2733709	895.92	910.17
			$m_{k,mset,3}$	23.67	277359	2696089	1147.41	1171.08
			$m_{k,pref}$	11.76	338880	3377124	687.79	699.55
35_training								
4	10	20	m_k	163.10	5253332	59504339	-	-
			$m_{k,all}$	-	-	-	-	-
			$m_{k,mset,l_{max}}$	676.22	4234500	47661301	2322.42	2998.64
			$m_{k,mset,1}$	209.86	4969772	56092438	-	-
			$m_{k,pref}$	184.56	5253332	59504339	7145.62	7330.18

$M_{k,pref}$, $M_{k,mset,1}$, and $M_{k,mset,3}$ are very difficult to compare. Depending on the instance, on the number and size of prefixes, on multiset inclusion, one can be better than the other. But for all the instances we tried, one of this 3 models was always the best of the 6 models, and they were better than M_k. Table 2 presents a focus on 2 specific instances (25_training and 35_training, both with $|\Sigma| = 5$) with a fixed value for k, $|\omega|$, and $|S^+|$. The columns correspond exactly to those of Table 1. For the first instance, we clearly see the order presented in Sect. 3 for instance sizes of improved models. We can also see the reverse order in terms of generation time. When $|\Sigma|$ is small, the probability of having prefixes is higher than with larger vocabularies, and for this instance, $M_{k,pref}$ returns the best instance in terms of generation+solving time. For the second instance, $M_{k,all}$ could not be generated in less than 2 h. M_k and $M_{k,mset,3}$ could be generated rather quickly, but could not be solved. $M_{k,pref}$ was even faster for generating the SAT instance. However, we see that there was not prefix in the training set (the size of instances of M_k and $M_{k,pref}$ are the same). The overhead for taking prefixes into account is rather insignificant (12% of generation time). Since the solving time was close to the timeout, the M_k instance did not succeed to be solved while the $M_{k,pref}$ instance succeeded (the small difference of 55 s., i.e., less than 0,8 %, is certainly due to clause order in the SAT instance). This instance shows that $M_{k,mset,l_{max}}$ can be the best model in terms of total time. This is due to the fact that there is no negative word being prefix of another word from S, and that the lattice is rather "wide", with a long branch. Hence, $M_{k,mset,l}$ is interesting when l is large for this training sample.

5 Conclusion

In the context of grammatical inference, we proposed various model improvements for learning Nondeterministic Finite Automaton of size k from samples of words. Our base model, M_k, is a conversion from an INLP model [15]. The first improvement, $M_{k,all}$, leads to the smallest SAT instances, which are also solved quickly. However, generating this model is too costly. Thus, when problems grow (in terms of k, $|S|$, or length of words), $M_{k,all}$ cannot be generated anymore. We proposed a set of improvements based on multiset representation of words, $M_{k,mset,l}$. The generated SAT instances are a bit larger with the maximal level than with $M_{k,all}$, but generation is still costly. We thus defined a third improvement based on prefix. On average, the best balance between generation and solving time is obtained with $M_{k,pref}$, $M_{k,mset,1}$, or $M_{k,mset,3}$: the generation is rather light and the reductions are significant. The interest of our work is that, to our knowledge, we are the only ones working on CSP model improvements. It is very complicated to compare our results with previous works. Many works on this topics are only formal and experimental results are also difficult to compare. For examples, the authors of [8,9] focus on a parallel solver for optimizing k. In [10], experiments are based on samples issued from the Waltz-DB database [2] of amino acid sequences, i.e., all the words are of size 6, and there cannot be any prefix word: in the tests we performed, only anagrams could be used in multisets. Moreover, for all the 50 instances we tried issued from this database, the M_k model could be generated and solved in a reasonable time, without need of any model improvement.

In the future, we plan to hybridize $M_{k,mset,l}$ for small values of l with $M_{k,pref}$. The second idea is to simplify the work of the SAT solver and of the instance generation with simplified and incomplete training samples. We would then evaluate our SAT models with respect to the accurateness of the generated NFA on test set of words.

References

1. Audemard, G., Simon, L.: Predicting learnt clauses quality in modern SAT solvers. In: Proceedings of IJCAI 2009, pp. 399–404 (2009)
2. Beerten, J., et al.: WALTZ-DB: a benchmark database of amyloidogenic hexapeptides. Bioinform. **31**(10), 1698–1700 (2015)
3. Denis, F., Lemay, A., Terlutte, A.: Learning regular languages using RFSAs. Theoret. Comput. Sci. **313**(2), 267–294 (2004)
4. Dupont, P.: Regular grammatical inference from positive and negative samples by genetic search: the GIG method. In: Carrasco, R.C., Oncina, J. (eds.) Grammatical Inference and Applications, ICGI 1994. LNCS (LNAI), vol. 862, pp. 236–245. Springer, Heidelberg (1994). https://doi.org/10.1007/3-540-58473-0_152
5. Garey, M.R., Johnson, D.S.: Computers and Intractability, A Guide to the Theory of NP-Completeness. W.H. Freeman & Company, San Francisco (1979)
6. Heule, M., Verwer, S.: Software model synthesis using satisfiability solvers. Empir. Softw. Eng. **18**(4), 825–856 (2013)

7. de la Higuera, C.: Grammatical Inference: Learning Automata and Grammars. Cambridge University Press, Cambridge (2010)
8. Jastrzab, T.: On parallel induction of nondeterministic finite automata. In: Proceedings of ICCS 2016. Procedia Computer Science, vol. 80, pp. 257–268. Elsevier (2016)
9. Jastrzab, T.: Two parallelization schemes for the induction of nondeterministic finite automata on PCs. In: Wyrzykowski, R., Dongarra, J., Deelman, E., Karczewski, K. (eds.) Parallel Processing and Applied Mathematics, PPAM 2017. LNCS, vol. 10777, pp. 279–289. Springer, Cham (2017). https://doi.org/10.1007/978-3-319-78024-5_25
10. Jastrzab, T.: A comparison of selected variable ordering methods for NFA induction. In: Rodrigues, J., et al. (eds.) Computational Science, ICCS 2019. LNCS, vol. 11540, pp. 741–748. Springer, Cham (2019). https://doi.org/10.1007/978-3-030-22750-0_73
11. Rossi, F., van Beek, P., Walsh, T. (eds.): Handbook of Constraint Programming, 1st edn. Elsevier Science, Amsterdam (2006)
12. Tomita, M.: Dynamic construction of finite-state automata from examples using hill-climbing. In: Proceedings of the Fourth Annual Conference of the Cognitive Science Society, pp. 105–108 (1982)
13. Tseitin, G.S.: On the complexity of derivation in propositional calculus. In: Siekmann, J.H., Wrightson, G. (eds.) Automation of Reasoning. SYMBOLIC, pp. 466–483. Springer, Heidelberg (1983). https://doi.org/10.1007/978-3-642-81955-1_28
14. de Parga, M.V., García, P., Ruiz, J.: A family of algorithms for non deterministic regular languages inference. In: Ibarra, O.H., Yen, H.C. (eds.) Implementation and Application of Automata, CIAA 2006. LNCS, vol. 4094, pp. 265–274. Springer, Heidelberg (2006). https://doi.org/10.1007/11812128_25
15. Wieczorek, W.: Grammatical Inference - Algorithms, Routines and Applications. Studies in Computational Intelligence, vol. 673. Springer, Heidelberg (2017). https://doi.org/10.1007/978-3-319-46801-3

Applications of Learning and Optimization Methods

Synthesis of Scheduling Heuristics by Composition and Recombination

Dominik Mäckel[1], Jan Winkels[1], and Christin Schumacher[2(✉)]

[1] Department of Computer Science 14 – Software Engineering, TU Dortmund University, Dortmund, Germany
{dominik.maeckel,jan.winkels}@tu-dortmund.de
[2] Department of Computer Science 4 – Modeling and Simulation, TU Dortmund University, Dortmund, Germany
christin.schumacher@tu-dortmund.de

Abstract. In many machine scheduling studies, individual algorithms for each problem have been developed to cope with the specifics of the problem. On the other hand, the same underlying fundamentals (e.g. Shortest Processing Time, Local Search) are often used in the algorithms and only slightly modified for the different problems. This paper deals with the synthesis of machine scheduling algorithms from components of a repository. Especially flow shop and job shop problems with makespan objective are considered to solve with Shortes/Longest Processing Time, NEH, Giffler & Thompson algorithms. For these components, the paper includes an exemplary implementation of an agile scheduling system that uses the Combinatory Logic Synthesizer to recombine components of scheduling algorithms to solve a given scheduling problem. Special attention is given to the composition heuristics and the process of recombination to executable programs. The advantages of this componentization are discussed and illustrated with examples. It will be shown that algorithms can be generalized to deal with scheduling problems of different machine environments and production constraints.

1 Introduction

In production, machine scheduling algorithms help to decide automatically when a certain job should be executed on which machine. Many manufacturers have not yet automated their machine scheduling. One reason is that for each machine scheduling problem with its numerous specific characteristics, suitable algorithms have to be selected, adapted, and implemented individually. Each practical scheduling problem can be categorized into a problem class, for which dedicated heuristics are applicable. If a class is a subset of another class, the heuristics of the superset class can often also be applied to the subset class. Also, relationships and overlapping between categories can be identified which simplifies the transfer of heuristics between problem classes.

The assignment problem which a combination of heuristics or metaheuristics should be chosen for which practical production environment concerning the

B. Dorronsoro et al. (Eds.): OLA 2021, CCIS 1443, pp. 283–293, 2021.
https://doi.org/10.1007/978-3-030-85672-4_21

applicability, solution quality, and computing time represents a combinatorial challenge. The synthesis framework Combinatory Logic Synthesizer ((CL)S) [1] is suitable for the automated solution of this task. The (CL)S can construct software from a collection of individual components and it is possible to specify components semantically, which enables the (CL)S to select the appropriate components. The framework then automatically generates all possible combinations in the form of executable software.

The objective of the paper is to use the (CL)S-Framework to automatically select and combine different algorithms to solve a given scheduling problem. Therefore, we build a (CL)S repository of algorithms for different machine environments, which takes the relationships of the classes into account and automatically composes selected algorithms for instances of these problems.

This paper is structured as follows: First, we present the general classification scheme of machine scheduling problems. In the related work, we discuss algorithms for scheduling of flow shop and job shop problems and present the framework on which our implementation is based on, the (CL)S. The handling of this framework, as well as the generation and composition of algorithms, is shown in the fourth chapter with example runs. In detail, we show the potential of the tool and the resulting possibilities using the Giffler & Thompson's algorithm.

2 Classification of Machine Scheduling Problems

Machine scheduling problems can be specified by a tuple $\alpha|\beta|\gamma$ [2, pp. 288–290] [3, pp. 13–21] [4, pp. 1–2]. In the following, parameter values are specified which are considered in this paper.
The parameter α defines the amount and arrangement of machines [4, pp. 14–15]:

- 1: Single Maschine, one machine is available for production.
- Fm: flow shop, m machines with one machine per processing stage. All jobs follow the same route through the machines.
- Jm: job shop, m machines with one machine per stage. Each job has a prescribed route through the stages. The route may differ between the jobs.
- Om: Open Shop, m machines, where each job can visit the machines one after the other in an order that is determined by the planner.

Parameter β can contain as many entries as required and describes characteristics and limitations of the production process:

- $prmu$: Permutation, the processing sequence of jobs from the first processing stage through all machines is to be kept consistent [3, p. 17].
- $skip$: skipping stages of jobs is possible (further example, but not applicated in the paper) [5, pp. 1151–1155] [4, p. 13].

γ specifies the objective function:

- C_{max}: Makespan, interval between production start of the first scheduled job and finish time of the last job.

3 Related Work

In the following, important scheduling algorithms for these machine environments and β-constraints in combination with makespan minimization are described, as well as related work according to the (CL)S.

3.1 Machine Scheduling Algorithms for Flow Shops and Job Shops

In the context of machine scheduling, an enormous number of papers and algorithms are available. Literature overviews for flow shops and job shops can be found in Komaki, Sheikh, and Malakooti [6] Framinan, Gupta, and Leisten [7] (permutation flow shop with makespan minimization) and Zhang et al. [8]. A comparison between commonly used algorithms for constructive flow shop scheduling can be found in Ruiz and Maroto [9]. Different dispatching rules have been studied in Arisha, Young, and El Baradie [10]. In the following, selectively a few algorithms of the overviews are analyzed that dealt with flow shops or job shops to minimize the makespan and are related to our problem classes (see Sect. 2).

Some of the most commonly used constructive heuristics for flow shops and job shops are Shortest Processing Time First (SPT), Longest Processing Time First (LPT), and the NEH-heuristic (flow shops) and have therefore been considered in this paper. The benefits of dispatching rules like SPT and LPT are low computational complexities and therefore fast calculations, and transparent behavior for production planners. The NEH-Heuristic, firstly published by Nawaz, Enscore, and Ham [11, pp. 92–94] for permutation flow shops and makespan minimization ($Fm|prmu|C_{max}$) produces good results in most cases. Giffer and Thompson [12] published a constructive algorithm that also applies rules like SPT and LPT to job shops.

3.2 Giffler & Thompson Algorithm

Using the algorithm by Giffler & Thompson, job shop as well as flow shop problems can be solved. It schedules exactly one job on a machine in each iteration, so the algorithm returns complete schedules after $m * n$ iterations, where m is the number of machines and n the amount of jobs. The heuristic is only parameterized by the applied dispatching rule. In the algorithm, this dispatching rule decides between several competing jobs on the same machine. The implementation of the complete Giffler & Thompson algorithm is shown in Algorithm 1.1. The algorithm consists of four phases where steps 2 to 4 iterate until all jobs are scheduled [13, S. 75–76]. The calculated schedule and the completion times of the scheduled jobs and for all machines are returned.

1 Let Z_i be the completion time of machine i. Initialize $Z_i = 0$ for $i = 0, ..., m$. Select a dispatching rule.
2 Select machine i^* that first can finish a job out of the set of jobs, which are waiting to be processed next on one of the machines and are not scheduled yet.
3 From the set of all jobs waiting to be processed on this machine i^* select one job by the dispatching rule which is initalized in step 1.
4 Schedule selected job on machine i^* and update Z_{i^*}. If there are jobs left to be scheduled, return to step 2.

Algorithm 1.1: Implementation of the Giffler & Thompson algorithm

Algorithm 1.1 works as follows. In each iteration (step 2–4), the machine is determined, which can first complete a job. For this purpose, each not yet fully scheduled job is iterated and the end time after scheduling on the next machine to be visited is compared. Up to this point, it is a greedy algorithm that selects a machine according to the earliest completion time on the next machine the job has to be processed on. Once the machine to be scheduled has been determined, in the second phase the job is varied to meet a prioritization on the machine. This is done by determining all jobs that are also to be scheduled next on the selected machine, including the job determined in the previous phase. If two or more jobs are waiting to be scheduled on the selected machine, the jobs get ranked according to the selected dispatching rule. After selecting a job on the determined machine, it gets scheduled and Z_i, as well as the current end time of the job, gets updated.

3.3 Combinatory Logic Synthesizer

Combinatory Logic Synthesizer, short: (CL)S, is a type-based framework for the synthesis of software from a set of components specified in a repository [1]. The framework was developed in the programming language Scala and is used in this paper. In addition to the synthesis, the framework also allows the immediate execution of the synthesis result. Due to the implementation in the Scala programming language, the synthesis results can also access existing Java and Scala libraries. The framework (CL)S was developed at the chair 14 of the faculty for Computer Science at the TU Dortmund University.

The Combinatory Logic Synthesizer ((CL)S) is particularly suitable for handling unpredictable variability, which makes it well suited for the synthesis of machine allocation algorithms in production planning. (CL)S enables the specification of components, their implementation, as well as the modeling of variability and the automatic composition of components under consideration of the modeled variability rules [14]. All this is uniformly done within the framework. Thus, the framework provides a solid basis for mapping and specifying individual heuristics and algorithms, and is also suitable as a technological basis for the automatic composition of components [15]. The (CL)S has been used in the past for numerous applications of a similar nature. As an example, we mention the automatic configuration of factory planning projects [16], the automatic generation of BPMN processes [14], and the automated configuration of plans in construction projects [17]. The basis for the use of the framework is that within the target domain, results can be composed of specifiable components. In the (CL)S the specification is done by so-called semantic intersection types. How components can be specified and implemented, and which solutions are then generated automatically, is shown in the following chapters using an example.

4 Implementation

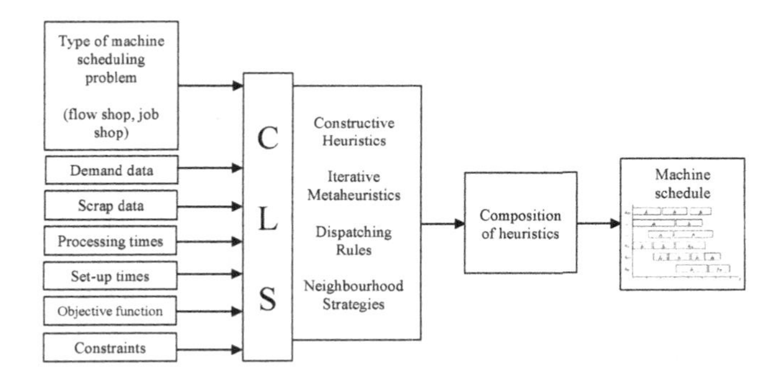

Fig. 1. Concept of schedule generation with (CL)S

The (CL)S-Repository contains all algorithm components as shown in Fig. 1, which can be combined into an executable scheduling system. Through a synthesis request to the (CL)S framework, production characteristics can be used to intersect with the defined types of the algorithm components. The (CL)S only selects those heuristics that are applicable to the given problem class. Available problem classes in this exemplary implementation are flow shop and job shop. After composing the algorithms, they can be utilized to solve the given scheduling problem and produce valid machine schedules. The synthesized algorithms work as transition functions and transfer the given data object into an applicable machine schedule. After scheduling, the makespan is calculated.

Further problem classes can be integrated by adding further possible parameter assignments and therefore extending the intersection types. By specifying additional parameters, further β constraints can be realized, which may exclude further heuristics because they are not applicable for the problem, or include others because they require certain assumptions or additional data such as deadlines.

$\Gamma = \{$

Scheduler: (String → String) ∩ (Algorithm ∩ shopClass → Scheduler(shopClass))
NEH: String ∩ (Algorithm ∩ FS)
FSDispatch: (String → String) ∩ (PriorityRule → Algorithm ∩ FS)
GifflerThompson: (String → String) ∩ (PriorityRule → Algorithm ∩ JS ∩ FS)
LPT: String ∩ PriorityRule
SPT: String ∩ PriorityRule

$\}$

Fig. 2. (CL)S repository

Our defined (CL)S repository is shown in Fig. 2 and the solution tree calculated by the (CL)S across all combinators of the repository is illustrated in Fig. 3. The repository's first combinator *Scheduler* of Fig. 2 is a wrapping base module, which serves as the common target type for all synthesis requests. Accordingly, it is found on the first level of the solution tree (left square in Fig. 3). As parameter *shopClass* (see Fig. 2) it receives information about the problems' machine environment (α-component). Starting from the base module, the different algorithms for flow shop and job shop problems of the type *Algorithm* are now available according to the parameter *shopClass*. By concretizing the parameter when calling the synthesis, the number of applicable combinators is reduced in such a way that only the algorithms for the corresponding problem class can be used. This is done by using the parameter also as an intersection type of the base module and thus an intersection with combinators of other problem classes is no longer possible.

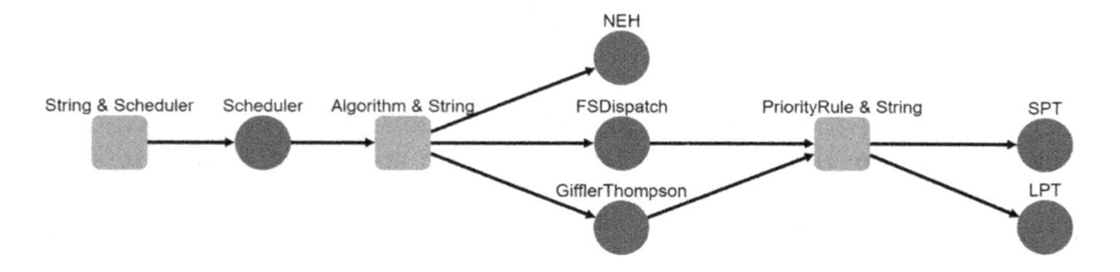

Fig. 3. (CL)S solution tree for flow shops

The first two algorithms *NEH* and *FSDispatch* in our implementation can only be applied to flow shops while the algorithm of Giffler & Thompson can be applied to job shops, which implies that it can also be used for flow shops because flow shops are a real subset of job shops as shown in Fig. 4.

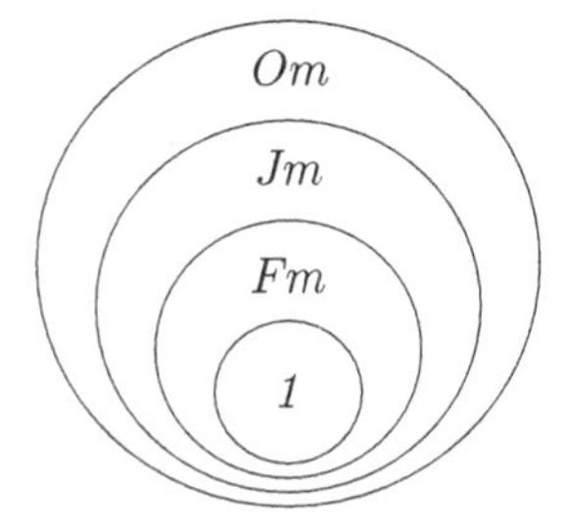

Fig. 4. Relationships between considered scheduling problem classes

The algorithms *FSDispatch* and *GifflerThompson* additionally require a dispatching rule. Figure 3 shows the reuse of these dispatching rules SPT and LPT for *FSDispatch* and *GifflerThompson*. This shows again one advantage of such a composing method. It is easily possible to integrate and combine new algorithms, heuristics, and dispatching rules into the tool by inserting them into the repository as combinators with corresponding intersection types. New components can reuse already existing ones. Individual components can also be replaced by other possibly better performing components without having to replace them individually at all points. Furthermore, the derivation graph in Fig. 3 shows similarities and differences between algorithms in the sense that the use of similar components is immediately recognizable. The procedure of disassembling an algorithm into reusable components and representing them as (CL)S-combinators is now explained in detail using the example of the Giffler & Thompson algorithm.

5 Results

To show that the same implementation of an algorithm can be effectively used for different machine environments, the Giffler & Thompson algorithm and its implementation is shown in Algorithm 1.1 has been applied to a flow shop and a job shop problem. The selection of the dispatching rule takes place inside the dispatching rule combinator that has been selected by CLS and parsed into the program code at this point. The dispatching rule is varied by replacing the code at this point.

To give a concrete example, processing times in Table 1 have been randomly generated from a triangular distribution with lower limit 5s, upper limit 15s, and mode 8s. For the job shop problem, also the processing order has been randomized across the stages as shown in Table 2. The entry "4" in row "S1" and column "job 1" indicates that job 1 has to be processed on the first stage (S1) in the fourth production step. Before, the job has to visit stage 3, then stage 2 and stage 2 in exactly this sequence. The calculated job shop schedule of the Giffler & Thompson algorithm with LPT-rule is shown in Fig. 5.

Since the algorithm was not particularly designed for flow shop problems, it is reasonable to compare its result with the NEH heuristic. The two schedules

Table 1. Generated processing times

Job	1	2	3	4	5	6	7	8	9	10
S1	6	12	8	9	10	8	9	9	11	7
S2	7	11	7	7	9	7	10	9	10	6
S3	8	12	11	8	9	8	12	13	7	9
S4	9	12	11	7	6	9	10	8	11	10

Table 2. Order for job shop production

Job	1	2	3	4	5	6	7	8	9	10
S1	4	1	4	3	4	4	1	3	2	3
S2	3	3	1	4	1	2	2	2	3	1
S3	1	2	2	1	3	3	4	4	4	4
S4	2	4	3	2	2	1	3	1	1	2

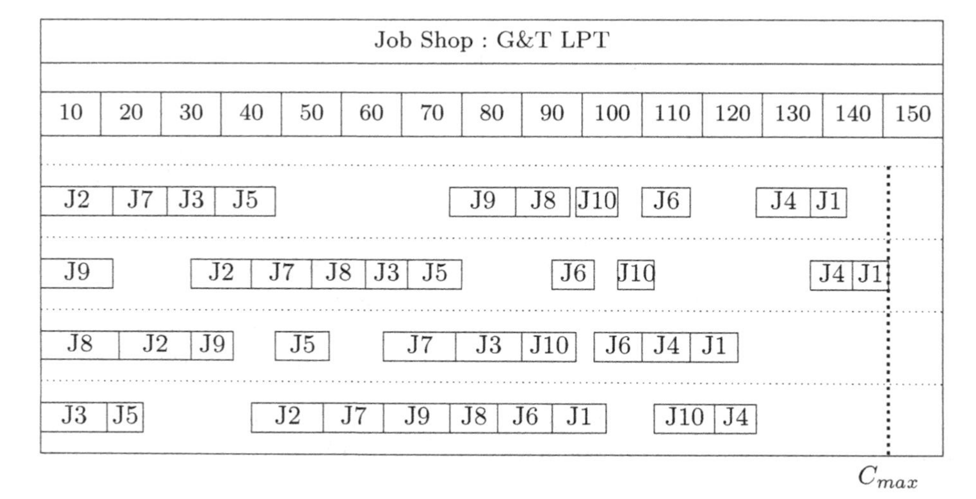

Fig. 5. Jop shop schedule with Giffler & Thompson

are shown in Fig. 6. As expected, the NEH heuristic creates a better schedule than the Giffler & Thompson algorithm. It is worth mentioning that Giffler & Thompson created a valid schedule that can keep up with algorithms specially designed for flow shop algorithms and can therefore be for example used as a starter solution for an iterative algorithm or it can be used if no better solution is available. In addition, Giffler & Thompson algorithm can be executed with different priority rules. To execution of the algorithm with different priority rules as input parameters lead to multiple solutions, the planner team can choose from. The benefit is not having to implement an algorithm for flow shop problems as the job shop algorithm can already handle it.

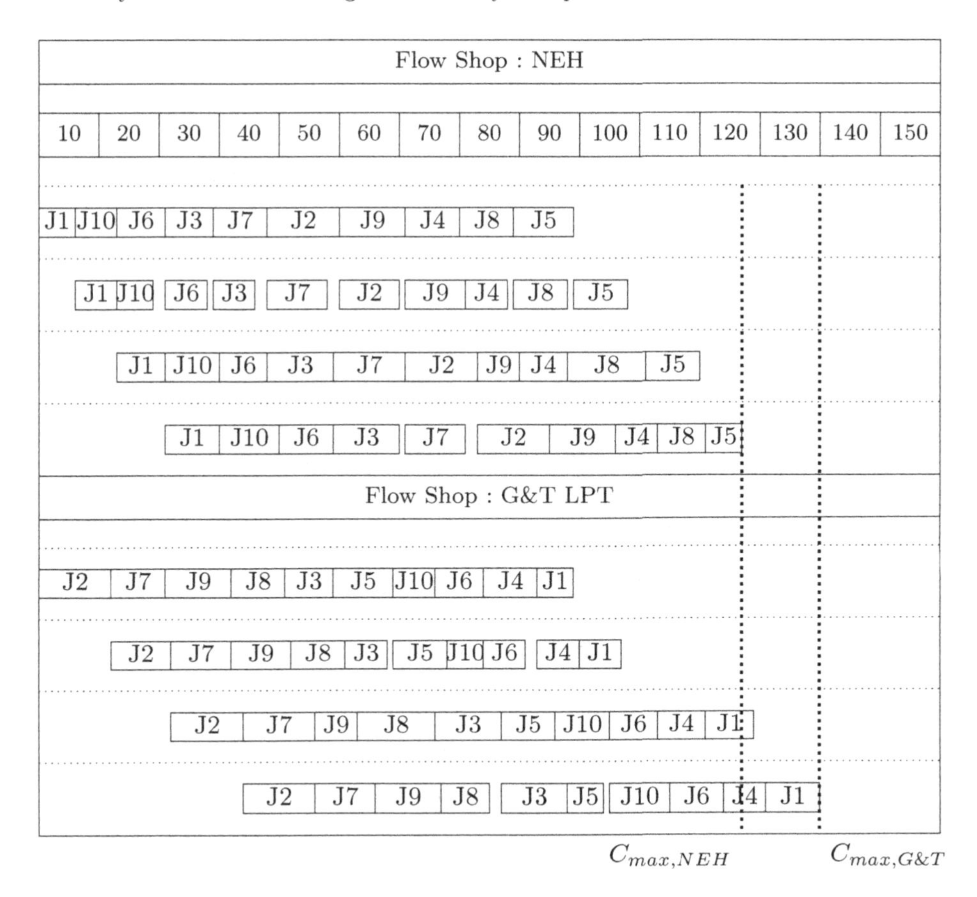

Fig. 6. Comparison of flow shop schedules with NEH and Giffler & Thompson

6 Conclusion

In this paper, we presented a repository for machine scheduling algorithms using the (CL)S, a framework that can generate algorithms automatically and to create solutions that are specially tailored to a previously specified problem. We used this framework for the problem area of machine scheduling in order to solve flow shop and job shop problems with SPT, LPT, NEH and Giffler & Thompson.

We have classified scheduling algorithms and mapped them as components in a (CL)S repository. Through componentization, different algorithms can be integrated into a framework via a uniform interface. This makes it easy to generate different algorithmen to scheduling problems. The recombined algorithms generate valid schedules according to their functionalities. Algorithms can be defined for various problem classes and constraints. According to the synthesis request, only those algorithms are recombined that apply to the current problem.

The shown concept is not limited to constructive algorithm as presented in this study and can also be applied to any iterative metaheuristic in further studies if the given data object already contains a constructive start solution. Concatenations of different constructive and iterative heuristics are conceivable as well. Also, extensions of other objective functions are possible.

References

1. Bessai, J., Dudenhefner, A., Düdder, B., Martens, M., Rehof, J.: Combinatory logic synthesizer. In: Margaria, T., Steffen, B. (eds.) ISoLA 2014. LNCS, vol. 8802, pp. 26–40. Springer, Heidelberg (2014). https://doi.org/10.1007/978-3-662-45234-9_3 ISBN 978-3-662-45233-2
2. Graham, R.L., et al.: Optimization and approximation in deterministic sequencing and scheduling: a survey. In: Hammer, P.L., Johnson, E.L., Korte, B.H. (ed.) Annals of Discrete Mathematics: Proceedings of the Advanced Research Institute on Discrete Optimization and Systems Applications of the Systems Science Panel of NATO and of the Discrete Optimization Symposium co-sponsored by IBM Canada and SIAM Banff, Aha and Vancouver, vol. 5, pp. 287–326. Elsevier (1979). https://doi.org/10.1016/S0167-5060(08)70356-X
3. Pinedo, M.: Scheduling: Theory, Algorithms, and Systems, 5th edn. Springer, Cham (2016). https://doi.org/10.1007/978-3-319-26580-3. ISBN 9783319265780
4. Ruiz, R., Vázquez-Rodríguez, J.A.: The hybrid ow shop scheduling problem. Eur. J. Oper. Res. **205**(1), 1–18 (2010). https://doi.org/10.1016/j.ejor.2009.09.024. ISSN 0377-2217
5. Ruiz, R., Şerifoğlu, F.S., Urlings, T.: Modeling realistic hybrid flexible flowshop scheduling problems. Comput. Oper. Res. **35**(4), 1151–1175 (2008). https://doi.org/10.1016/j.cor.2006.07.014. ISSN 03050548
6. Komaki, G.M., Sheikh, S., Malakooti, B.: Flow shop scheduling problems with assembly operations: a review and new trends. Int. J. Prod. Res. **57**(10), 2926–2955 (2019). https://doi.org/10.1080/00207543.2018.1550269. ISSN 0020-7543
7. Framinan, J.M., Gupta, J.N.D., Leisten, R.: A review and classification of heuristics for permutation flow-shop scheduling with makespan objective. J. Oper. Res. Soc. **55**(12), 1243–1255 (2004). https://doi.org/10.1057/palgrave.jors.2601784
8. Zhang, J., et al.: Review of job shop scheduling research and its new perspectives under Industry 4.0. J. Intell. Manuf. **30**(4), 1809–1830 (2019). https://doi.org/10.1007/S10845-017-1350-2. ISSN 0956-5515
9. Ruiz, R., Maroto, C.: A comprehensive review and evaluation of permutation flowshop heuristics to minimize flowtime. Eur. J. Oper. Res. **40**, 479–494 (2005). https://doi.org/10.1016/j.ejor.2004.04.017. ISSN 0956-5515
10. Arisha, A., Young, P., El Baradie, M.: Flow shop scheduling problem: a computational study. In: Sixth International Conference on Production Engineering and Design for Development (PEDD6). Dublin Institute of Technology, Cairo, Egypt, 1 Jan 2002, pp. 543–557 (2002)
11. Nawaz, M., Enscore, E.E., Ham, I.: A heuristic algorithm for the m-machine, n-job flow-shop sequencing problem. Omega **11**(1), 91–95 (1983). https://doi.org/10.1016/0305-0483(83)90088-9. http://www.sciencedirect.com/science/article/pii/0305048383900889. ISSN 03050483

12. Giffer, B., Thompson, G.L.: Algorithms for solving production scheduling problems. Oper. Res. **8**(4) 487–503 (1960). https://doi.org/10.1287/opre.8.4.487. http://search.ebscohost.com/login.aspx?direct=true&db=bsu&AN=7687426&site=ehost-live
13. Jaehn, F., Pesch, E.: Ablaufplanung: Einführung in Scheduling, 1st edn. Springer, Berlin (2014). https://doi.org/10.1007/978-3-642-54439-2. ISBN 978-3-642-54439-2
14. Bessai, J., Dudenhefner, A., Düdder, B., Martens, M., Rehof, J.: Combinatory process synthesis. In: Margaria, T., Steffen, B. (eds.) ISoLA 2016. LNCS, vol. 9952, pp. 266–281. Springer, Cham (2016). https://doi.org/10.1007/978-3-319-47166-2_19
15. Winkels, J.: Automatisierte Komposition und Konfiguration von Work-flows zur Planung mittels kombinatorischer Logik. Technische Universität Dortmund. https://doi.org/10.17877/DE290R-20469
16. Winkels, J., Graefenstein, J., Schäfer, T., Scholz, D., Rehof, J., Henke, M.: Automatic composition of rough solution possibilities in the target planning of factory planning projects by means of combinatory logic. In: Margaria, T., Steffen, B. (eds.) ISoLA 2018. LNCS, vol. 11247, pp. 487–503. Springer, Cham (2018). https://doi.org/10.1007/978-3-030-03427-6_36 ISBN 9783030034283
17. Lenz, L.T., et al.: Smart factory adaptation planning by means of BIM in combination of constraint solving techniques. In: Proceedings of the International Council for Research and Innovation in Building and Construction (CIB), World Building Congress 2019 – Constructing Smart Cities (2019)

Solving QAP with Auto-parameterization in Parallel Hybrid Metaheuristics

Jonathan Duque[1(✉)], Danny A. Múnera[1], Daniel Díaz[2], and Salvador Abreu[3]

[1] Facultad de Ingeniería, Universidad de Antioquia, Medellín, Colombia
{jonathan.duque,danny.munera}@udea.edu.co
[2] CRI, Université Paris 1, Paris, France
daniel.diaz@univ-paris1.fr
[3] NOVA-LINCS, Universidade Évora, Évora, Portugal
spa@uevora.pt

Abstract. The Quadratic Assignment Problem (QAP) is one of the most challenging combinatorial optimization problems with many real-life applications. Currently, the best solvers are based on hybrid and parallel metaheuristics, which are actually highly complex and parametric methods. Finding the best set of tuning parameters for such methods is a tedious and error-prone task. In this paper, we propose a strategy for auto-parameterization of QAP solvers. We show evidence that auto-parameterization can further improve the quality of computed solutions. Our auto- parameterization scheme relieves the user from having to find the right parameters while providing a high quality solution.

Keywords: QAP · Auto-parametrization · Heuristics · Parallelism

1 Introduction

The Quadratic Assignment Problem (QAP) is a hard combinatorial optimization problem with many real-life applications such as scheduling, facility location, electronic chipset layout, production, process communications, among others [3]. QAP has been shown to be NP-Hard and finding effective algorithms to solve it is an active research topic in recent years.

Medium size problems can be solved using exact methods (e.g., size ≤30), which can find an optimal solution or prove that a problem has no solution [1]. Exact methods consider the entire search space: either explicitly by exhaustive search or implicitly, by pruning some portions of the search space that have been detected as irrelevant for the search.

To tackle harder problems, one must resort to incomplete methods which provide good, albeit potentially sub-optimal solutions in a reasonable time. Such is the case for metaheuristics, which are high-level procedures that make choices to efficiently explore part of the search space, so as to make problems tractable.

B. Dorronsoro et al. (Eds.): OLA 2021, CCIS 1443, pp. 294–309, 2021.
https://doi.org/10.1007/978-3-030-85672-4_22

Metaheuristics usually have several parameters to adjust their behavior depending on the problem to solve [7]. Examples of metaheuristics include genetic algorithms, tabu search, local search and simulating annealing.

Metaheuristics operate on two main working principles: intensification and diversification. The former refers to the method's ability to explore more deeply a promising region of the search space, while the latter refers to the exploration of different regions of the search space. By design, some metaheuristics methods are better at intensifying the search while others are so at diversifying it. However, the behavior of most metaheuristics can be controlled via a set of parameters. A fine tuning of these parameters is therefore crucial to achieve an effective trade-off between intensification and diversification, and hence good performance in solving a given problem. Unfortunately, selecting the best set of parameters is a tedious and error-prone task. This process is even harder because the best parameters values vary with the problem structure and even just for different instances of the same problem, as stated by the Non-Free-Lunch theorem [41].

Each metaheuristic has its own strengths and weaknesses, which may vary according to the problem or even to the instance being solved. The trend is thus to design hybrid metaheuristics, which combine diverse methods in order to benefit from the individual advantages of each one [5]. However, this increases the number of parameters (parameters of individual metaheuristics and new parameters to control the hybridization). The design and implementation of a hybrid metaheuristic is a complex process; tuning the resulting parameters, to reach the best performance, is also very challenging.

Despite the good results obtained using hybrid metaheuristics, it is still necessary to reduce the processing times needed for the hardest instances [36]. One of the most plausible options entails parallelism [13]. In parallel metaheuristics one can have multiple instances of the same (or different) metaheuristics running in parallel, either independently or cooperatively through concurrent process communications [10,37]. Not only does parallelism help to decrease processing time, but it can also be a means to easily implement hybridization.

In previous work we proposed a Cooperative Parallel Local Search solver, called CPLS [30,32]. CPLS embeds various simple local search metaheuristics and then relies on cooperative parallelization to concurrently execute several metaheuristic instances, which cooperate during the search process. We later extended CPLS, by proposing PHYSH (Parallel HYbridization of simple Heuristics) [26,27]. PHYSH supports the combination of population-based and single-solution metaheuristics. CPLS and PHYSH also require the fine tuning of a larger number of parameters, since more metaheuristics (of different types) are involved. Moreover, the configuration of the parallel interaction itself (communication between the methods) involves yet another set of parameters which need to be adjusted. Tuning this increasing number of parameters makes it even more difficult to find the appropriate setting for the algorithm to behave optimally.

Automating the task of finding good parameters is thus desirable and has attracted significant attention from researchers. We may identify two kinds of strategies for automatic tuning: *parameter tuning* and *parameter control* [20].

In parameter tuning (off-line tuning) the set of parameters are defined before applying the algorithm to a specific problem (static definition of parameters). Several strategies for automatic parameter tuning of metaheuristics have been proposed [21,22]. In contrast, parameter control strategies (online-tuning) adapts the values of the controlled parameters during the algorithm execution (dynamic adaptation of parameters). The idea is to find the best parameters setting during the solving process, using some mechanism to alter the parameter values according to the algorithm performance.

Parameter tuning can be seen as a pre-process pass which is executed before the solving in order to determine the adequate values for parameters. This does not affect the implementation of the solver. On the other hand, *parameter control* has to be implemented in the kernel of the solver. The former may appear easier but when the number of parameters become large it is hard to use in practice. Indeed, it usually requires many runs to identify the best parameter settings, making this a time-consuming process. These methods are often limited by the number of parameters and the computational power available. In that case, parameter control strategies emerge as a viable solution to deal with the high complexity of current solvers (hybrid and/or parallel).

In this paper we propose a parallel hybrid method with a parameter control strategy for solving the QAP, called DPA-QAP. DPA-QAP embeds multiple metaheuristic methods in a parallel hybrid execution and self-adapts the parameters of the metaheuristics using an iterative process, adaptation is performed based on performance measures. We carried out an experimental evaluation which shows that the auto-parametrization strategy outperforms a simpler version of DPA-QAP with no auto-parametrization, i.e., a parallel hybrid method with static parametrization. We perform the evaluation using the classical QAPLIB instances and also a particular set of very hard QAP instances.

In the remaining of this paper we present the related work on Sect. 2. Section 3 presents the general structure of DPA-QAP and Sect. 4 introduces the auto-parametrization strategy. Section 5 contains the experimental evaluation performed which validates our strategy. A short conclusion ends the paper.

2 Related Work

The Quadratic Assignment Problem (QAP) was first proposed by Koopmans and Beckmann in 1957 [25] as a model for a facilities location problem. This problem consists in assigning a set of n facilities to a set of n locations, while minimizing the cost associated with the *flows* of items among facilities and the *distance* between them.

Metaheuristic methods have been successfully applied for solving QAP. From the 90s, several metaheuristic methods have emerged as a suitable option to solve this problem, e.g., Tabu Search [38], Genetic Algorithms [40], among several others. These methods perform well on a wide range of QAP instances, however, some hard instances still require very long runs to achieve quality solutions. Moreover, no method was able to get good performance on an extensive set of instances. The aforementioned problems spurred the emergence of new techniques based on hybridization and parallelization. For instance, one of the fundamental methods of hybrid metaheuristics is the memetic algorithm (MA) [29]. MA is an effective approach which combines an evolutionary algorithm with a local search procedure. Hybrid metaheuristics are intricate procedures, tricky to design, implement, debug and tune, therefore, it is unsurprising that hardly any of them only combine more that a couple of methods.

Parallelizing metaheuristics grants access to using powerful computational platforms with the aim of speeding up the search process [12]. A straightforward implementation of parallel metaheuristics is the *Independent multi-walks* approach which speeds up the search process by performing concurrent executions of multiple metaheuristic instances, therefore augmenting the probability to get quickly a good solution [11]. Another kind of parallel metaheuristics allows the concurrent instances to cooperate by exchanging information during the search process, aiming to improve the efficiency of the solver [28,39]. We identify these methods as *Cooperative multi-walk* approaches.

We proposed a way to create hybridization through cooperative parallelization in our CPLS framework [30,32]. CPLS allows the user to code the individual metaheuristics, and the framework manages parallelism and communications. In CPLS, different local search metaheuristics concurrently interact by exchanging relevant information about the search. This interaction provides a cooperative way to intensify the search. This framework has been successfully used to solve hard variants of Stable Matching Problems [33] and hard instances of QAP [30,31]. Since CPLS does not support population-based methods, we proposed an extension of the framework called PHYSH [26,27], which provides an efficient strategy to promote cooperation between population-based and single-solution methods, metaheuristics of a different nature. Both CPLS and PHYSH have proved able to efficiently solve several hard QAP instances.

Parallel hybrid metaheuristics often have many parameters which modify the algorithm behaviour. Setting these parameters has influence on the performance of the method, however, finding the optimal values for these parameters is usually a hard task [21]. Using hybridization and parallelism makes this task even more difficult for mainly two reasons: First, hybrid metaheuristics inherit the parameters of each "low level" metaheuristic, so one needs to find the setting of more parameters, since a parameter configuration for one algorithm usually is not suitable for another. Second, cooperative parallel strategies require parameters to define their behaviour, e.g., for determining how frequently metaheuristics should interact or how each metaheuristic has to use the received information,...

Tuning metaheuristic parameters (i.e., *offline-tuning*) has been carried out in different ways, in earlier times the tuning process was done by hand, another approach was to take parameters values from similar algorithms reported in the literature. More recently, the use of specialized tools for automatic parameter tuning has become prevalent, these techniques use advanced methodologies and tools from a theory of experiment design to machine learning approaches, among others [19]. Several methods have been proposed for parameter tuning, for instance F-Race [4], ParamILS [21], SMAC [22], HORA [2]. However, these methods have limitations when tuning a large number of parameters or when they require significant computational resources to perform the test runs [20].

Parameter control (*online-tuning*) emerges as a reasonable option. Some strategies have been proposed for specific metaheuristics such as [35] for swarm intelligence and [23] for evolutionary algorithms. Also, some specific strategies has been proposed for the QAP, such as [16] which proposes a strategy for self-control parameters on a Tabu Search method.

Hyper-heuristics present another way to face the problem of metaheuristic parameter control. These form a novel research approach in which a high level strategy selects or generates the best metaheuristics with their respective parameters and acceptance criteria. Aiming to have more general methods, not designed for a single problem or for a few instances of a problem [9]. To the best of our knowledge, only one hyperheuristic method solves the QAP and uses parallelism in its design [14]: the authors propose a parameter control method using a genetic algorithm (GA) acting as a high-level strategy in the hyper-heuristic approach. The GA, generation by generation, performs the adaptation of parameters through cross-over and mutation operators, ending up with the parameters at their best adjustment for each method.

We achieve a form of hyperheuristic using cooperative parallelism. The key idea is to use the parallel computational power to not only create a hybrid metaheuristic but also to automatically control the parameters of the metaheuristic involved in the parallel hybrid method.

3 DPA-QAP Method

This section presents the general structure of DPA-QAP, a Dynamic Parameter Adaptation method for solving the Quadratic Assignment Problem. DPA-QAP is build on the top of a parallel hybrid metaheuristic solver, similar to the one presented in [30]. Figure 1 presents the two main components of DPA-QAP, the *Worker* nodes and a *Master* node (workers and master to simplify). Workers run a set of metaheuristics, in parallel, carrying out the search process. We design each worker to run in a separate thread, ideally bound to its own dedicated core, each thread runs a specific metaheuristic instance.

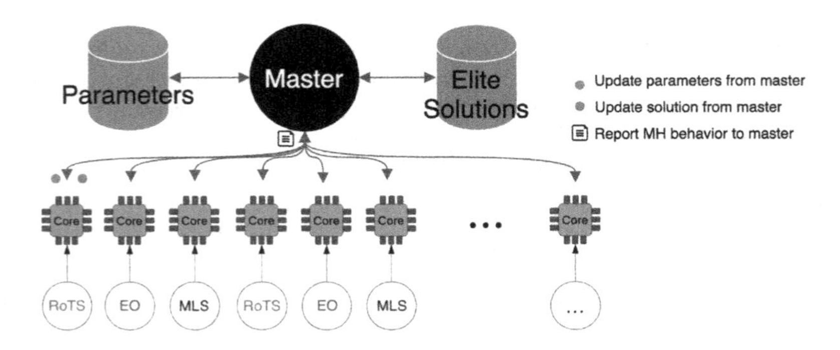

Fig. 1. DPA-QAP top-level view.

Each worker reports periodically its current candidate solution and some contextual information (e.g., solution cost, performance metrics, etc.) to the master, which stores best intermediate solutions into an elite pool. When the master receives a solution from a worker, it merges it into the elite pool. If the incoming solution is already present, it gets mutated by performing two random swaps. This mechanism promotes some diversity for the candidate solutions in the pool. When the elite pool is full, the master sends solutions to workers, ensuring the receiver implements a different metaheuristic from the one that inserted that solution into the pool. This process constitutes a flexible interaction feature which eases the hybridization of metaheuristics promoting cross-fertilization among different types. The size of this pool is equal to the number of workers, since the pool must have a solution for each method.

On the top of this cooperative parallel search, DPA-QAP implements a dynamic adaptation strategy which is tasked to automatically adjust the parameters of the metaheuristics during their execution, looking for the best setting and trying to ensure a balance between intensification and diversification.

3.1 Metaheuristics Used in the DPA-QAP Method

We select three different metaheuristic methods for the workers: Robust Tabu Search (RoTS), Extremal Optimization (EO), and Multistart Local Search (MLS). We select these metaheuristics because they are commonly used in combinatorial optimization problems particularly for the QAP. We now present a brief description of each of these methods.

Robust Tabu Search. The name Tabu Search (TS) refers to the use of an adaptive memory and special problem-solving strategies, to get a better local search method [17]. The idea is to memorize within a structure the elements that for the LS will be forbidden to use (*tabu*) and thus avoid getting trapped in local optima. TS looks for the best solution within the neighborhood but does not visit the solutions of previous neighbors if they have been visited before or have been marked as prohibited locations [14]. RoTS is an adaptation of TS to the QAP and has been one of the best performing methods for this problem [38].

Extremal Optimization. EO is a metaheuristic inspired by self-organizing processes as frequently found in nature: for EO this is *self-organized criticality* (SOC). The version proposed by [6] has only one adjustable parameter: τ, and uses of a Probability Distribution Function (PDF). EO proceeds like this: it inspects all candidate configurations assigning a fitness value, by means of the goal function. The configurations are then ranked from worst to best. EO resorts to the PDF to choose a solution from organized configurations. The role of the τ parameter is to provide different search strategies from pure random walk ($\tau = 0$) to deterministic (greedy) search ($\tau \Rightarrow \infty$). In previous work, we extended the basic EO metaheuristic to support not only a power-law PDF, but also an exponential and a gamma-law PDFs [31].

Multistart Local Search. Local Search (LS) is one of the oldest and most frequently used metaheuristics. LS starts from an initial solution and repeatedly improves it within a defined neighborhood. Neighbor solutions can be generated by applying minor changes to the initial solution. LS ends when no improved solutions are found in the neighborhood achieving a local optima [42]. Multistart Local Search (MLS) is a modification of LS that iteratively performs multiple different searches, executing each LS from a different starting point. When MLS reaches a local optimum, it tries to escape by restarting the search from scratch or performing some random moves in the current solution.

Table 1. Metaheuristic's parameter ranges (n stands for QAP instance's size).

Metaheuristic	Parameter name	Range
RoTS	Tabu duration factor	$[4n–20n]$
	Aspiration factor	$[n^2–10n^2]$
EO	PDF	Power - Exponential - Gamma
	τ	[0, 1]
MLS	Start type	Restart from scratch - Random swaps

Metaheuristics Parameters. Table 1 presents the parameters considered for each metaheuristic, together with the range of variation for each parameter. These ranges are picked from the best performances, as reported in the literature. For RoTS we use the parameters reported in [38], for EO we select the parameters reported in [31] and for MLS, the only parameter used is the restart process, then no range is needed.

4 Automatic Parameter Adaption in DPA-QAP

The DPA-QAP method operates within an iterative process. At the beginning, workers are initialized with random parameters. DPA-QAP dynamically adapts the best setting of parameters in every worker (which is executing a metaheuristic instance). Parameter control depends on the performance in the solving process for an individual worker at each iteration. Each worker periodically reports relevant information to the master. With this information, the master evaluates the worker's performance and tweaks its parameters, trying to strike a balance between intensification and diversification in the search. Figure 2 depicts the flow diagram of this process. Gray boxes represent the functionality executed by workers, in parallel. White boxes specify the iterative adaptation process by the master. The master waits while the workers perform the search. When it receives a metaheuristic report, it develops a *performance evaluation* for each worker and executes the *parameters' adaptation* procedure. The master then sends a new, evolved, set of parameters and a new configuration back to the workers. Workers resume the search with the settings they received: parameters and restarting from a new initial solution (from the master's elite pool). DPA-QAP repeats this process until an established number of iterations is accomplished or when the solution target is reached.

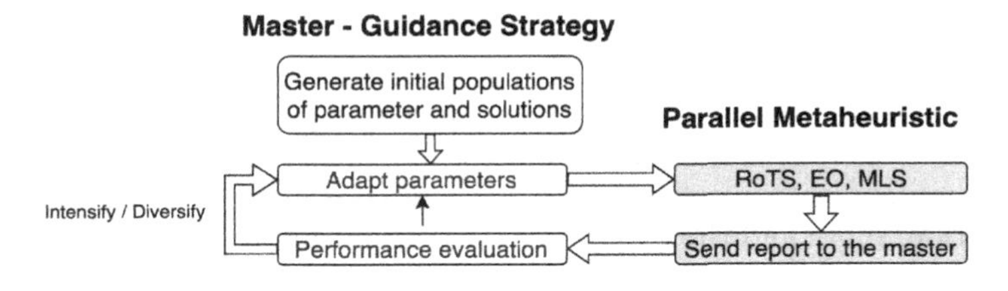

Fig. 2. DPA-QAP flow diagram.

4.1 Metaheuristics Performance Metrics

At each iteration of the parameters' adaptation process, each metaheuristic runs for a given time, `iteration_time`. When the `iteration_time` is running out, workers report to the master the initial solution and the best found solution in the interval with theirs associated costs. In order to assess the performance of a worker using a specific set of parameters, the master computes the distance between the initial and final solution (pair-wise difference) and the percentage gain for that iteration. The percentage gain is defined as: $\text{gain} = \frac{cost_{initial} - cost_{final}}{cost_{initial}}$.

Evaluating the performance of the metaheuristics is a critical process, and selecting the right set of metrics affects the overall performance of the parameters' adaptation process. In this work we consider two classical metrics, the percentage gain in the cost of the objective function and the distance between solutions. The gain acts as a direct indicator of the metaheuristic's performance, meanwhile the distance is assessing how diverse the search is. Other metrics can be also considered, for instance, the time spend on local optima, the number of iterations without improvements, among many others.

4.2 Performance Evaluation

The parameters' adaptation process evaluates the workers' performance by processing the percentage gain and the similarity between the initial and final solution. Through experimentation we verify that the gain is usually bigger at initial stages of the search than at the final stage. For this reason, DPA-QAP changes the value of the diversification gain limit during the search process, inspired by how the temperature decreases in simulated annealing [24]. Figure 3 shows how the diversification gain limit decreases in DPA-QAP during the search process. Using this dynamic limit, DPA-QAP diversifies the search more easily at the beginning than at the end of the search process. The similarity criterion is computed comparing the distance between the initial and final solutions. If this distance is lower than one-third of the QAP size (i.e., 66% of the variables are equal), we consider both solutions as "very similar".

Considering these two criteria, we defined the following rules to determine which action must be taken for adapting the worker's parameters: If the gain obtained by the method and its pair-wise difference is lower than the corresponding limits, the component adapts the metaheuristic parameters to *diversify* the search. If the gain is higher than the corresponding diversification gain limit or the pair-wise difference is higher than the distance solution limit, the component adapts the parameters to *intensify* the search. Both the dynamic diversification limits and the distance solution limit are hyper-parameters of the auto-parametrization strategy. We plan to test different limits in future work.

Adapting the Parameters. The evaluation of the worker's performance outputs a mandate which can be, intensify or diversify. This output is used as input for the parameters adaptation process. For each possible case we define a behavior depending of the metaheuristic type.

In EO the parameter τ is in the range 0 to 1 and, depending on its value and the PDF, this may lead the metaheuristic to intensify or diversify the search, by adding or subtracting a delta value belonging to the range (see Fig. 4). The parameters are then adjusted by adding to their values using deltas, so the master performs a search process that looks for the best parameters setting for a given metaheuristic.

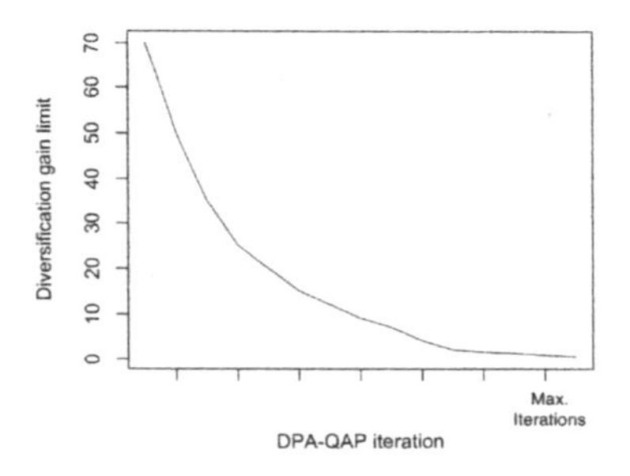

Fig. 3. Gain diversification limits.

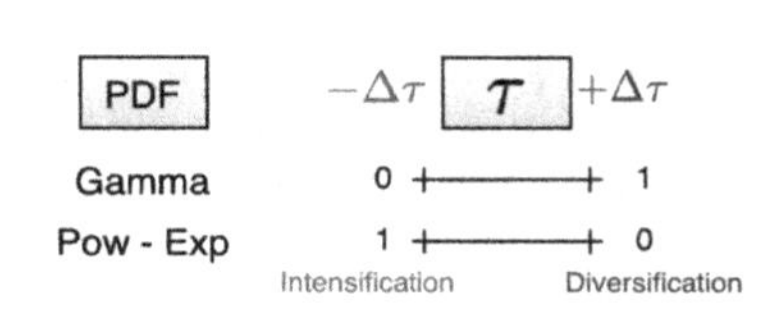

Fig. 4. EO parameters adaptation.

We define the parameter adaptation process for Robust Tabu search as follows: if the parameter adaptation component returns *diversify*, a delta of $n/2$ is added to the tabu duration and a delta of $n^2/2$ is added to the aspiration parameters. If the parameter adaptation component returns *intensify*, the tabu duration is subtracted by $n/2$ and the aspiration is decreased by $n/2$. For intensification, the delta for the adaptation of the aspiration parameter is different to diversification. This is done intending to slow down the intensification process, avoiding to stagnates on a local optimum. For the case of MLS, if there is any gain in cost, the type of restart is retained. If there is no gain, the algorithm changes to the other option.

5 Experimental Evaluation

In this section we present an experimental evaluation of our proposed method, DPA-QAP, comparing its performance against an independent parallel hybrid metaheuristic method. We consider three sets of very hard benchmarks: the 20 hardest instances of QAPLIB [8] and two sets of even harder instances: Drezner's [15] **dre***XX* and Palubeckis's [34] **Inst***XX* instances. Each instance is executed 30 times stopping as soon as the Best Known Solution (BKS) is found or when a time limit of 5 min is hit, in case the BKS is not reached. All experiments have been carried out on a quad-AMD Opteron 6380 system, totaling 64 cores running at 2.5 GHz and 128 GB of RAM.

At present, DPA-QAP is systematically configured with 30 worker nodes: 10 running RoTS, 10 running EO and 10 running MLS. Each worker node randomly initializes each parameter of its metaheuristic by randomly picking a value from the admissible values (see Table 1). These parameters are then periodically adapted as explained in the previous section. In this experiment, parameter control is triggered every 15 or 20 s, depending on the size of the problem. Each metaheuristic can thus adapt its parameters up to 20 times during the 5 minutes global execution cap. We plan to study the impact of varying this interval and determine if it is also possible and useful to dynamically adapt it.

We compare DPA-QAP to a *base solver* (BASE-QAP) which is statically parametrized (this solver is actually derived from DPA-QAP by disabling the parameter control mechanisms). Other than that, BASE-QAP is identical to DPA-QAP: it also creates 30 metaheuristic instances (10 of each type of metaheuristic); each metaheuristic instance also randomly initializes its parameters, which instead remain fixed during the execution. Our goal is to compare this pre-process parameterization (parameters fixed) with self-parameterization. Usually the parameter tuning pre-process is a time-consuming task, the idea is to avoid this offline tuning step by having an online method able to adapt its parameters meanwhile the problem solution is carry out.

Both methods are similarly implemented in Java 11 using the `ForkJoinPool` and `Atomic`*`Type`* classes to handle the parallelism in a shared memory model [1]. In all cases we made sure that each worker node is actually mapped by the JVM onto a different physical core, at runtime.

5.1 Evaluation on QAPLIB

We evaluated the performance of the DPA-QAP on QAPLIB, a well-known collection of 134 QAP problems of various sizes and difficulties [8]. The instances are named as `name`*XX* where `name` corresponds to the first letters of the author and *XX* is the size of the problem. For each instance, QAPLIB also includes the Best Known Solution (BKS), which is sometimes the optimum. Many QAPLIB instances are easy for a parallel solver, we therefore selected the 20 hardest

[1] Source code and instances can be found here.

ones (removing all systematically solved instances). We ran both DPA-QAP and BASE-QAP under the same conditions (30 repetitions, time limit of 5 min).

Table 2 presents the results. For each solver, the table lists the number of times the BKS is reached across the 30 executions (**#BKS**), the Average Percentage Deviation (**APD**), which is the average of the 30 relative deviation percentages computed as follows: $100 \times \frac{Avg-BKS}{BKS}$, where Avg is the average of the 30 found costs, and finally the average execution time (**Time**). Execution times are given in seconds (as a decimal number). This time is the elapsed (wall clock) time, and includes the time to install all solver instances, solve the problem, communications and the time to detect and propagate the termination. To compare the performance of both solvers, we first compare the number of BKS found, then (in case of tie), the APDs and finally the execution times. For each benchmark, the best-performing solver row is highlighted and the discriminant field is enhanced in bold font.

DPA-QAP outperforms BASE-QAP on 14 out of 20 of the hardest QAPLIB instances, while the reverse only occurs for 6 instances. 7 instances can never been solved by any solver. Clearly, a time limit of 5 min is too short for those hard problems: we plan to experiment with larger time limits. The "summary" row shows that DPA-QAP obtains a better *#BKS* than BASE-QAP (192 vs. 153, a 25% increase). The average APD is also better (0.174 vs. 0.180). Notice that solutions of better quality are obtained in a slightly shorter average execution time (269.5 s vs. 276.7 s).

Notice that BASE-QAP is indeed an efficient solver for this benchmark, it implements a parallel hybridization strategy and its parameters, despite being randomly initialized, are selected within a range taken from state-of-the-art solvers which report competitive results. Still, DPA-QAP managed to outperform BASE-QAP in most instances.

Table 2. Evaluation of dynamic adaptation on 20 hardest instances of QAPLIB.

		DPA-QAP			BASE-QAP		
	BKS	#BKS	APD	Time	#BKS	APD	Time
sko72	66256	**28**	0.010	130.9	24	0.012	161.2
sko81	90998	**20**	0.012	209.6	10	0.011	242.3
sko90	115534	**9**	0.022	262.2	8	0.016	274.9
sko100a	152002	**12**	0.027	245.0	4	0.029	279.3
sko100b	153890	**20**	0.012	223.1	14	0.014	242.9
sko100c	147862	**27**	0.010	268.7	20	0.010	287.2
sko100d	149576	6	0.024	287.7	**9**	0.021	285.9
sko100e	149150	**20**	0.012	266.1	16	0.015	271.2
sko100f	149036	8	0.018	267.8	**9**	0.017	265.9
tai40a	3139370	**4**	0.082	272.7	3	0.085	290.6
tai50a	4938796	0	**0.386**	300.0	0	0.401	300.0
tai60a	7205962	0	**0.479**	300.0	0	0.519	300.0
tai80a	13499184	0	**0.689**	300.0	0	0.780	300.0
tai100a	21044752	0	**0.647**	300.0	0	0.685	300.0
tai80b	818415043	**14**	0.031	282.1	13	0.028	254.5
tai100b	185996137	5	0.084	282.9	**10**	0.077	285.3
tai150b	498896643	0	0.654	300.0	0	**0.601**	300.0
tai256c	44759294	0	0.183	300.0	0	**0.179**	300.0
tho150	8133398	0	0.095	300.0	0	**0.086**	300.0
wil100	273038	**19**	0.011	292.0	13	0.013	293.0
Summary		**192**	**0.174**	**269.5**	153	0.180	276.7

5.2 Evaluation on Harder Instances

We evaluated DPA-QAP on two more sets of instances, artificially crafted to be very difficult for metaheuristics: the `dre`*XX* instances proposed by Drezner [15] and the `Inst`*XX* instances by Palubeckis [34]. These problems are generated with a known optimum. For this test we used the same machine and configuration as for QAPLIB (30 cores and a time limit of 5 min with 30 repetitions).

Table 3 presents the results for Drezner's instances. We have omitted small instances which are systematically solved by both solvers in less than 15 s. We start with `dre42` which is solved by both solvers at each replication; even on this case DPA-QAP is much faster than BASE-QAP: 34 s vs. 61 s. In all instances, DPA-QAP outperforms BASE-QAP. As a whole, DPA-QAP reaches more BKS (60 vs. 38) and, when the optimum is not reached, solutions provided by DPA-QAP are of much better quality than BASE-QAP as shown by the APDs (23.558 vs. 32.408), and it does so in a shorter period of time.

Table 3. Evaluation on Drezner instances.

	OPT	DPA-QAP #BKS	DPA-QAP APD	DPA-QAP Time	BASE-QAP #BKS	BASE-QAP APD	BASE-QAP Time
dre42	764	30	0.0	**34**	30	0.0	61
dre56	1086	**21**	14.1	213	8	21.0	259
dre72	1452	**9**	27.4	265	0	34.9	300
dre90	1838	0	**22.1**	300	0	28.0	300
dre110	2264	0	**36.1**	300	0	52.1	300
dre132	2744	0	**41.7**	300	0	58.3	300
SMRY		**60**	**23.6**	**235**	38	32.4	254

Table 4. Evaluation on Palubeckis' instances.

	OPT	DPA-QAP #BKS	DPA-QAP APD	DPA-QAP Time	BASE-QAP #BKS	BASE-QAP APD	BASE-QAP Time
Inst40	837900	**29**	0.15	108	26	0.17	151
Inst50	1840356	**23**	0.10	199	18	0.12	238
Inst60	2967464	**20**	0.16	188	11	0.15	249
Inst70	5815290	**9**	0.12	267	3	0.16	293
Inst80	6597966	2	**0.18**	292	2	0.19	292
Inst100	15008994	0	**0.18**	300	0	0.18	300
Inst150	58352664	0	**0.14**	300	0	0.14	300
Inst200	75405684	0	**0.14**	300	0	0.14	300
SMRY		**83**	**0.15**	**244**	60	0.16	265

Table 4 presents the results for Palubeckis' instances. As in the previous case, we did not include small instances which are systematically solved by both solvers in less than 15 s. Here again, DPA-QAP performs better than BASE-QAP on all instances of the benchmark. As for Drezner's instances, the time limit of 5 min appears too short to solve large instances. However, DPA-QAP does find more BKS (83 vs. 60) and dynamic parameter adaptation makes it possible to improve the quality of solutions wrt. BASE-QAP as shown by the APDs (0.147 vs. 0.157).

6 Conclusions and Future Work

We have proposed a dynamic parameter adaptation scheme for parallel and hybrid solvers based on metaheuristics to solve the QAP. The basic principle of this approach is to have a master node which periodically collects the progress of each metaheuristic. This node has a global view of the overall search progress, therefore it can provide each metaheuristic with new parameter values in order to increase its effectiveness. We proposed DPA-QAP: an implementation of this architecture in Java, embedding three well-known metaheuristics: Robust Tabu Search, Extremal Optimization and Multistart Local Search. The first experiments performed on very difficult instances of QAP validate our approach by significantly improving solution quality.

We plan to extend this work in several directions. First, we will experiment on machines with more cores and with time limits greater than the 5 minutes cap which was allowed in this work. We will also try to determine the best settings for parameter reporting and adjustment: in this experiment we used a constant interval which needs to be refined. Another line of potential experiments consists in including efficient metaheuristics, such as Ant Colony Optimization [18]; or embedding population-based methods, e.g. genetic algorithms. Finally, we plan to address larger instances of the QAP as well as other difficult problems. As an outcome, we aim to design and propose a general framework for self-adaptation able to address a wide variety of combinatorial search and optimization problems.

Acknowledgements. This research was supported by the CODI project PRV19-1-01 funded by the University of Antioquia in Medellín, Colombia.

References

1. Abdel-Basset, M., Manogaran, G., Rashad, H., Zaied, A.N.H.: A comprehensive review of quadratic assignment problem: variants, hybrids and applications. J. Ambient Intell. Human. Comput. 1–24 (2018). https://doi.org/10.1007/s12652-018-0917-x
2. Barbosa, E.B., Senne, E.L.: A heuristic for optimization of metaheuristics by means of statistical methods. In: ICORES 2017 - Proceedings of the 6th International Conference on Operations Research and Enterprise Systems 2017-January (Icores), pp. 203–210 (2017). https://doi.org/10.5220/0006106402030210
3. Bhati, R.K., Rasool, A.: Quadratic assignment problem and its relevance to the real world: a survey. Int. J. Comput. Appl. **96**(9), 42–47 (2014)
4. Birattari, M., Kacprzyk, J.: Tuning Metaheuristics: A Machine Learning Perspective, vol. 197. Springer, Heidelberg (2009). https://doi.org/10.1007/978-3-642-00483-4
5. Blum, C.: Hybrid metaheuristics in combinatorial optimization: a tutorial. In: Dediu, A.-H., Martín-Vide, C., Truthe, B. (eds.) TPNC 2012. LNCS, vol. 7505, pp. 1–10. Springer, Heidelberg (2012). https://doi.org/10.1007/978-3-642-33860-1_1
6. Boettcher, S., Percus, A.: Nature's way of optimizing. Artif. Intell. **119**(1), 275–286 (2000). https://doi.org/10.1016/S0004-3702(00)00007-2
7. Boussaïd, I., Lepagnot, J., Siarry, P.: A survey on optimization metaheuristics. Inf. Sci. **237**, 82–117 (2013). https://doi.org/10.1016/j.ins.2013.02.041
8. Burkard, R.E., Karisch, S., Rendl, F.: QAPLIB - a quadratic assignment problem library. Eur. J. Oper. Res. **55**(1), 115–119 (1991). https://doi.org/10.1016/0377-2217(91)90197-4
9. Burke, E.K., Hyde, M.R., Kendall, G., Ochoa, G., Özcan, E., Woodward, J.R.: A classification of hyper-heuristic approaches: revisited. In: Gendreau, M., Potvin, J.-Y. (eds.) Handbook of Metaheuristics. ISORMS, vol. 272, pp. 453–477. Springer, Cham (2019). https://doi.org/10.1007/978-3-319-91086-4_14
10. Caniou, Y., Codognet, P., Richoux, F., Diaz, D., Abreu, S.: Large-scale parallelism for constraint-based local search: the costas array case study. Constraints **20**(1), 1–27 (2014). https://doi.org/10.1007/s10601-014-9168-4
11. Codognet, P., Munera, D., Diaz, D., Abreu, S.: Parallel local search (2018). https://doi.org/10.1007/978-3-319-63516-3_10
12. Cotta, C., Talbi, E.G., Alba, E.: Parallel hybrid metaheuristics. In: Parallel Metaheuristics, pp. 347–370. Wiley, Hoboken (2005). https://doi.org/10.1002/0471739383.ch15
13. Crainic, T.G., Toulouse, M.: Parallel meta-heuristics. In: Gendreau, M., Potvin, J.Y. (eds.) Handbook of Metaheuristics. ISOR, vol. 146, pp. 497–541. Springer, US (2010). https://doi.org/10.1007/978-1-4419-1665-5_17
14. Dokeroglu, T., Cosar, A.: A novel multistart hyper-heuristic algorithm on the grid for the quadratic assignment problem. Eng. Appl. Artif. Intell. **52**, 10–25 (2016). https://doi.org/10.1016/j.engappai.2016.02.004
15. Drezner, Z.: The extended concentric tabu for the quadratic assignment problem. Eur. J. Oper. Res. **160**(2), 416–422 (2005)

16. Fescioglu-Unver, N., Kokar, M.M.: Self controlling tabu search algorithm for the quadratic assignment problem. Comput. Ind. Eng. **60**(2), 310–319 (2011). https://doi.org/10.1016/j.cie.2010.11.014
17. Glover, F.: Tabu search—part II. ORSA J. Comput. **2**, 4–32 (1990). https://doi.org/10.1287/ijoc.2.1.4
18. Hani, Y., Amodeo, L., Yalaoui, F., Chen, H.: Ant colony optimization for solving an industrial layout problem. Eur. J. Oper. Res. **183**(2), 633–642 (2007). https://doi.org/10.1016/j.ejor.2006.10.032
19. Hoos, H.H.: Automated algorithm configuration and parameter tuning. In: Hamadi, Y., Monfroy, E., Saubion, F. (eds.) Autonomous Search, pp. 37–71. Springer, Heidelberg (2011). https://doi.org/10.1007/978-3-642-21434-9_3
20. Huang, C., Li, Y., Yao, X.: A survey of automatic parameter tuning methods for metaheuristics. IEEE Trans. Evol. Comput. **24**(2), 201–216 (2020). https://doi.org/10.1109/TEVC.2019.2921598
21. Hutter, F., Hoos, H., Leyton-Brown, K., Stützle, T.: ParamILS: an automatic algorithm configuration framework. J. Artif. Intell. Res. **36**, 267–306 (2009). https://doi.org/10.1613/jair.2808
22. Hutter, F., Hoos, H.H., Leyton-Brown, K.: Sequential model-based optimization for general algorithm configuration. In: Coello, C.A.C. (ed.) LION 2011. LNCS, vol. 6683, pp. 507–523. Springer, Heidelberg (2011). https://doi.org/10.1007/978-3-642-25566-3_40
23. Karafotias, G., Hoogendoorn, M., Eiben, Á.E.: Parameter control in evolutionary algorithms: trends and challenges. IEEE Trans. Evol. Comput. **19**(2), 167–187 (2014)
24. Kirkpatrick, S., Gelatt, C.D., Vecchi, M.P.: Optimization by simulated annealing. Science **220**(4598), 671–80 (1983)
25. Koopmans, T.C., Beckmann, M.: Assignment problems and the location of economic activities. Econometrica **25**(1), 53–76 (1957). https://doi.org/10.2307/1907742
26. Lopez, J., Munera, D., Diaz, D., Abreu, S.: On integrating population-based metaheuristics with cooperative parallelism. In: Proceedings - 2018 IEEE 32nd International Parallel and Distributed Processing Symposium Workshops, IPDPSW 2018 (2018). https://doi.org/10.1109/IPDPSW.2018.00100
27. López, J., Múnera, D., Diaz, D., Abreu, S.: Weaving of metaheuristics with cooperative parallelism. In: Auger, A., Fonseca, C.M., Lourenço, N., Machado, P., Paquete, L., Whitley, D. (eds.) PPSN 2018. LNCS, vol. 11101, pp. 436–448. Springer, Cham (2018). https://doi.org/10.1007/978-3-319-99253-2_35
28. Loukil, L., Mehdi, M., Melab, N., Talbi, E.G., Bouvry, P.: A parallel hybrid genetic algorithm-simulated annealing for solving Q3AP on computational grid. In: IPDPS 2009 - Proceedings of the 2009 IEEE International Parallel and Distributed Processing Symposium (2014). https://doi.org/10.1109/IPDPS.2009.5161126
29. Moscato, P., Cotta, C., Mendes, A.: Memetic algorithms. In: Handbook of Heuristics, vol. 1–2, pp. 53–85. Springer, Heidelberg (2004). https://doi.org/10.1007/978-3-540-39930-8_3
30. Munera, D., Diaz, D., Abreu, S.: Hybridization as cooperative parallelism for the quadratic assignment problem. In: Blesa, M.J., et al. (eds.) HM 2016. LNCS, vol. 9668, pp. 47–61. Springer, Cham (2016). https://doi.org/10.1007/978-3-319-39636-1_4

31. Munera, D., Diaz, D., Abreu, S.: Solving the quadratic assignment problem with cooperative parallel extremal optimization. In: Chicano, F., Hu, B., García-Sánchez, P. (eds.) EvoCOP 2016. LNCS, vol. 9595, pp. 251–266. Springer, Cham (2016). https://doi.org/10.1007/978-3-319-30698-8_17
32. Munera, D., Diaz, D., Abreu, S., Codognet, P.: A parametric framework for cooperative parallel local search. In: Blum, C., Ochoa, G. (eds.) EvoCOP 2014. LNCS, vol. 8600, pp. 13–24. Springer, Heidelberg (2014). https://doi.org/10.1007/978-3-662-44320-0_2
33. Munera, D., Diaz, D., Abreu, S., Rossi, F., Saraswat, V., Codognet, P.: Solving hard stable matching problems via local search and cooperative parallelization. In: AAAI, Austin, TX, USA (2015)
34. Palubeckis, G.: An algorithm for construction of test cases for the quadratic assignment problem. Informatica, Lith. Acad. Sci. **11**(3), 281–296 (2000)
35. Parpinelli, R.S., Plichoski, G.F., Silva, R.S.D., Narloch, P.H.: A review of techniques for online control of parameters in swarm intelligence and evolutionary computation algorithms. Int. J. Bio-Inspired Comput. **13**(1), 1–20 (2019)
36. Saifullah Hussin, M.: Stochastic local search algorithms for single and bi-objective quadratic assignment problems. Ph.D. thesis, Université de Bruxelles (2016)
37. Silva, A., Coelho, L.C., Darvish, M.: Quadratic assignment problem variants: a survey and an effective parallel memetic iterated tabu search. Eur. J. Oper. Res. (xxxx) (2020). https://doi.org/10.1016/j.ejor.2020.11.035
38. Taillard, E.: Robust taboo search for the quadratic assignment problem. Parallel Comput. **17**(4–5), 443–455 (1991). https://doi.org/10.1016/S0167-8191(05)80147-4
39. Talbi, E.G., Bachelet, V.: COSEARCH: a parallel cooperative metaheuristic. J. Math. Modell. Algorithms **5**(1), 5–22 (2006). https://doi.org/10.1007/s10852-005-9029-7
40. Tate, D.M., Smith, A.E.: A genetic approach to the quadratic assignment problem. Comput. Oper. Res. **22**(1), 73–83 (1995). https://doi.org/10.1016/0305-0548(93)E0020-T
41. Wolpert, D.H., Macready, W.G.: No free lunch theorems for optimization. IEEE Trans. Evol. Comput. **1**(1), 67–82 (1997). https://doi.org/10.1109/4235.585893
42. Yagiura, M., Ibaraki, T.: Local search (2002). https://doi.org/10.1201/9781420010749

Theoretical Analysis of a Dynamic Pricing Problem with Linear and Isoelastic Demand Functions

Mourad Terzi[1,2(✉)], Yassine Ouazene[1,2], Alice Yalaoui[1,2], and Farouk Yalaoui[1,2]

[1] Chaire Connected Innovation, Université de Technologie de Troyes, 12 rue Marie Curie, CS 42060, 10004 Troyes Cedex, France
{mourad.terzi,yassine.ouazene,alice.yalaoui,farouk.yalaoui}@utt.fr
[2] LIST3N, Université de Technologie de Troyes, 12 rue Marie Curie, CS 42060, 10004 Troyes Cedex, France

Abstract. Dynamic pricing strategies are usually adopted to dynamically adjust the products' prices taking into account demand function characteristics to maximize the revenue. This paper addresses the problem in which a firm has to make decisions about its selling prices in each period to maximize the total profit over the whole horizon. We propose a theoretical analysis of this problem from which we show that: first, when the demand function is linear, the problem can be formulated as a quadratic programming problem. We also present the Karush-Kuhn-Tucker system, which can be used to find the optimal pricing policy when the objective function is concave. Then, when the demand is isoelastic, we also show that the problem can be reduced to the maximization of N independent functions in bounded intervals. Some numerical examples are provided to illustrate the results obtained for both the linear and isoelastic cases.

Keywords: Revenue maximization · Dynamic pricing · Linear and isoelastic demand · Quadratic programming · KKT conditions

1 Introduction

Dynamic pricing is a pricing strategy where the firms adjust dynamically the prices of the products and services according to the perceived demand at different times Narahari et al. [12]. One of the key elements when dealing with a dynamic pricing problem is the demand function which characterizes the relation between different factors like (selling price, advertising, seasonality,...) and the demand. In the paper of Huang et al. [11], a survey on the demand functions was presented. The factors considered are price, rebate, lead time, space,

Supported by the European Regional Development Fund (FEDER) and the Industrial Chair Connected-Innovation (https://chaire-connected-innovation.fr/).

B. Dorronsoro et al. (Eds.): OLA 2021, CCIS 1443, pp. 310–323, 2021.
https://doi.org/10.1007/978-3-030-85672-4_23

quality, and advertising. The authors observed that: 1) the linear and isoelastic demand functions are the two widely used in the literature, and 2) the majority of publications consider the price and quality factors.

Initially, the dynamic pricing has been applied to the service industries such as airline [16] and hotels [4]. According to Elmaghraby et al. [10], factors like 1) the availability of demand data and decision-support system to track the changes in prices and, 2) the simplicity of prices adjusting due to the recent developments in technologies lead to several works on dynamic pricing on a wide range of industries like retails [6].

Several studies dealing with the coordination of dynamic pricing and production decisions with the discrete-time horizon and multiple products are conducted. The work of Bajwa and Sox [2] presented a joint pricing, production, and advertising decisions model for a firm that produces and sells multiple products as different brands. The authors assumed that the demand is a function of the price and advertising money and demonstrated that coordinating the marketing and operational decisions leads the firm to increase its profitability. The paper of Bajwa, Fontem, and Sox [1] considered a manufacturer with a limited production capacity. They proposed a model that allows lost sales under a price-dependent demand function. Ouazene et al. [13] studied the problem in which the products can be sold through multiple channels and the demand is a price-dependent function. The authors compared the dynamic and constant pricing strategies. The paper of Couzon et al. [8] presented an extension of the classical capacitated lot-sizing problem by considering a production system with variable capacity under a price-dependent demand function. In [9], the same authors as in [8] improved the model studied in Bajwa, Fontem, and Sox [1] by introducing new lower and upper bounds that reduced the search space. They also proposed new constructive efficient heuristics to solve the model. All the papers cited above assumed that the demand in a given period is a function of the price of the product in the same period and can take the linear or isoelastic form.

Several surveys on dynamic pricing have been published, Elmaghraby et al. [10], and Chen et al. [7] reviewed the literature on dynamic pricing with the presence of inventory considerations. A survey on dynamic pricing and learning was conducted in den Boer et al. [5]. The authors reviewed the literature on dynamic pricing with demand uncertainty.

The presented work investigates the problem in which a firm has to make decisions about its selling prices in each period to maximize the total profit over the whole horizon. This problem has been initially tackled by Shakya et al. [15] and solved by combining neural networks and evolutionary algorithms. Their study is based on linear, exponential, and multinomial logit demand functions. In the presented paper, a theoretical analysis in which we consider the linear and isoelastic demand functions will be conducted. The mathematical properties of the problem will be studied and some theoretical results that lead to finding the optimum pricing policy will be provided.

The remainder of this paper is organized as follows. Section 2 presents the dynamic pricing problem assumptions and mathematical formulation. Section 3 describes the resolution approach under the linear and isoelastic demand. A numerical experiments are presented in Sect. 4. A conclusion is to be found in Sect. 5.

2 Problem Description

The dynamic pricing problem addressed in this study is the same as the model presented in [15]. The model is denoted as (P_0). The problem considers a firm that produces and sells its product. The goal is to find the product's price in each period to maximize the firm's total profit over a given horizon. Following the notations used to describe the model:

N	Number of periods in the horizon
t	Time index, $t = 1, ..., N$
Q_t	Number of production (sales) at period t
P_t	Price of a product at period t (decision variable)
C_t	Cost of one unit production in period t
Π	Total profit during the entire planning horizon
$\overline{P_t}$	Upper bound for the price at period t
$\underline{P_t}$	Lower bound for price at period t
K_t	Upper bound for the capacity at period t
M_t	Lower bound for the capacity at period t

The initial mathematical model P_0 is detailed below:

$$\max \Pi = \sum_{t=1}^{N}(P_tQ_t - C_tQ_t) \tag{1}$$

$$st : M_t \leq Q_t \leq K_t, \quad t = 1, 2, ..., N \tag{2}$$

$$\underline{P_t} \leq P_t \leq \overline{P_t}, \quad t = 1, 2, ..., N \tag{3}$$

$$P_t > 0, \quad t = 1, 2, ..., N \tag{4}$$

The objective function represents the total profit over all the horizon to maximize. P_tQ_t is the total revenues in period t and C_tQ_t is the cost per production in period t. Constraints 2 consider the production capacity, the objective is to regulate the use of the available capacity in each period (machines, labor, etc. ...) by considering production values that are at least equal to the minimum available capacity and don't exceed the maximum production capacity. Constraints (3) bound the selling price of each period by $\underline{P_t}$ and $\overline{P_t}$ to avoid a lower profit value, and a lower demand. Finally, constraints (4) are the non-negativity constraints. Note that the decisions variables P_t $(t = 1, ..., N)$ are a strict positive real numbers.

3 Resolution Approach

In the presented work, the linear and isoelastic demand functions are considered. Both are price-dependent demand. The linear demand is adopted from [15] and the isoelastic function is the same as the demand studied in [9]. The following notations will be considered in the presented work. Some new notations will be introduced throughout the analytical study.

$P^T = (P_1, P_2, ..., P_N)$	The pricing policy's vector
M	The total constraints' number

3.1 Case with Linear Demand Function

Following the same assumption as in [15], the demand in period t (Eq. 5) is linear and depends on the price of the product in the same period and on the price of the product in other periods. $a_t(> 0)$ is defined as the intercept parameter, it represents the number of customers willing to buy the product at period t. $b_{t't}$ are the slope parameters. They represent the impact of price in period t' on the demand in period t. b_{tt} is generally assumed to be negative because when the product's price in a period t increases, the corresponding demand in the same period decreases.

$$Q_t = \psi(P_1, P_2, ..., P_N) = a_t + \sum_{t'=1}^{N} b_{t't} P_{t'} \tag{5}$$

Replacing Q_t by its value from (5), P_0 can be written as:

$$\max_{P_1, P_2, .., P_N} \Pi = \sum_{t=1}^{N} \left(a_t + \sum_{t'=1}^{N} b_{t't} P_{t'} \right) (P_t - C_t) \tag{6}$$

$$st : M_t \leq a_t + \sum_{t'=1}^{N} b_{t't} P_{t'} \leq K_t, \quad t = 1, 2, ..., N \tag{7}$$

$$\underline{P_t} \leq P_t \leq \overline{P_t}, \quad t = 1, 2, ..., N \tag{8}$$

$$P_t > 0, \quad t = 1, 2, ..., N \tag{9}$$

Proposition 1. *The total profit function Π is quadratic and its expression is given in Eq. (10). W is a $(N \times N)$ symmetric matrix and V^T is a $(1 \times N)$ vector of real numbers. D is a real constant number and it is independent from the selling price vector.*

$$\Pi = \frac{1}{2} P^T W P + V^T P + D \tag{10}$$

Proof. Π can be rewritten as follows:

$$\Pi = \sum_{t=1}^{N} a_t P_t - \sum_{t=1}^{N} a_t C_t + \sum_{t=1}^{N} P_t \sum_{t'=1}^{N} b_{t't} P_{t'} - \sum_{t=1}^{N} C_t \sum_{t'=1}^{N} b_{t't} P_{t'}$$

$$\Pi = S_1 + S_2 + S_3$$

$$With \quad S_1 = -\sum_{t=1}^{N} a_t C_t = D$$

$$S_2 = \sum_{t=1}^{N} a_t P_t - \sum_{t=1}^{N} C_t \sum_{t'=1}^{N} b_{t't} P_{t'}$$

$$S_2 = (a_1 P_1 - C_1(b_{11} P_1 + b_{21} P_2 + ... + b_{N1} P_N)) + (a_2 P_2 - C_2(b_{12} P_1 + b_{22} P_2 + ... + b_{N2} P_N)) + \ldots + (a_N P_N - C_N(b_{1N} P_1 + b_{2N} P_2 + ... + b_{NN} P_N))$$

$$S_2 = P_1(a_1 - C_1 b_{11} - C_2 b_{12} - \ldots - C_N b_{1N}) + P_2(a_2 - C_1 b_{21} - C_2 b_{22} - \ldots - C_N b_{2N}) + \ldots + P_N(a_N - C_1 b_{N1} - C_2 b_{N2} - \ldots - C_N b_{NN})$$

$$S_2 = P_1(a_1 - \sum_{t=1}^{N} C_t b_{1t}) + P_2(a_2 - \sum_{t=1}^{N} C_t b_{2t}) + \ldots + P_N(a_N - \sum_{t=1}^{N} C_t b_{Nt})$$

$$S_2 = V^T P$$

With

$$V_{1,N}^T = \left(a_1 - \textstyle\sum_{t=1}^{N} b_{1t} C_t, \;\; a_2 - \sum_{t=1}^{N} b_{2t} C_t, \;\; \ldots, \;\; a_N - \sum_{t=1}^{N} b_{Nt} C_t\right)$$

and

$$P^T = \left(P_1, P_2, \ldots P_N\right)$$

$$S_3 = \sum_{t=1}^{N} P_t \sum_{t'=1}^{N} b_{t't} P_{t'} = P_1(b_{11} P_1 + b_{21} P_2 + \ldots + b_{N1} P_N) + P_2(b_{12} P_1 + b_{22} P_2 + \ldots + b_{N2} P_N) + \ldots P_N(b_{1N} P_1 + b_{2N} P_2 + \ldots + b_{NN} P_N)$$

Let consider :

$$S_4 = \frac{1}{2} P^T W P$$

with P is the same vector as defined for S_2 and :

$$W_{N,N} = \begin{pmatrix} 2b_{11}, & b_{12} + b_{21}, & b_{13} + b_{31}, & \ldots & , b_{1N} + b_{N1} \\ b_{12} + b_{21}, & 2b_{22}, & b_{23} + b_{32}, & \ldots & , b_{2N} + b_{N2} \\ \vdots & \vdots & \vdots & \vdots & \vdots \\ b_{1N} + b_{N1}, & b_{2N} + b_{N2}, & b_{3N} + b_{N3}, & \ldots, & 2b_{NN} \end{pmatrix}$$

$$S_4 = \frac{1}{2}(P_1, P_2, .., P_N) \begin{pmatrix} 2b_{11}, & b_{12} + b_{21}, & b_{13} + b_{31}, & \ldots & , b_{1N} + b_{N1} \\ b_{12} + b_{21}, & 2b_{22}, & b_{23} + b_{32}, & \ldots & , b_{2N} + b_{N2} \\ \vdots & \vdots & \vdots & \vdots & \vdots \\ b_{1N} + b_{N1}, & b_{2N} + b_{N2}, & b_{3N} + b_{N3}, & \ldots, & 2b_{NN} \end{pmatrix} \begin{pmatrix} P_1 \\ P_2 \\ \vdots \\ P_N \end{pmatrix}$$

$$S_4 = \frac{1}{2}\Big(2b_{11} P_1 + (b_{12} + b_{21}) P_2 + ... + (b_{1N} + b_{N1}) P_N), (b_{12} + b_{21}) P_1 + 2b_{22} P_2 +$$

$$\ldots + (b_{2N} + b_{N2})P_N, \ldots, (b_{1N} + b_{N1})P_1 + (b_{2N} + b_{N2})P_2 + \ldots + 2b_{NN}P_N)\Big) \begin{pmatrix} P_1 \\ P_2 \\ \vdots \\ P_N \end{pmatrix}$$

$$S_4 = \frac{1}{2}\Big(P_1\Big(2b_{11}P_1 + (b_{12} + b_{21})P_2 + \ldots + (b_{1N} + b_{N1})P_N\Big) + P_2\Big((b_{12} + b_{21})P_1 +$$

$$2b_{22}P_2 + \ldots + (b_{2N} + b_{N2})P_N\Big) + \ldots + P_N\Big((b_{1N} + b_{N1})P_1 + (b_{2N} + b_{N2})P_2 + \ldots$$

$$+ 2b_{NN}P_N\Big)\Big)$$

$$S_4 = \frac{1}{2}\Big(P_1\Big(2b_{11}P_1 + 2b_{21}P_2 + \ldots + 2b_{N1}P_N\Big) + P_2\Big(2b_{12}P_1 + 2b_{22}P_2 + \ldots +$$

$$2b_{N2}P_N\Big) + \ldots P_N\Big(2b_{1N}P_1 + 2b_{2N}P_2 + \ldots + 2b_{NN}P_N\Big)\Big)$$

$$S_4 = \Big(P_1\Big(b_{11}P_1 + b_{21}P_2 + \ldots + b_{N1}P_N\Big) + P_2\Big(b_{12}P_1 + b_{22}P_2 + \ldots + b_{N2}P_N\Big) +$$

$$\ldots + P_N\Big(b_{1N}P_1 + b_{2N}P_2 + \ldots + b_{NN}P_N\Big)\Big)$$

$$S_4 = S_3$$

Then, the following relation is tune :

$$\Pi = \frac{1}{2}P^T W P + V^T P + D$$

Considering the constraints of P_0, they can be rewritten as:

$$\sum_{t'=1}^{N} b_{t't}P_{t'} \leq K_t - a_t \qquad t = 1, 2, .., N \tag{11}$$

$$-\sum_{t'=1}^{N} b_{t't}P_{t'} \leq a_t - M_t \qquad t = 1, 2, .., N \tag{12}$$

$$P_t \leq \overline{P_t} \qquad t = 1, 2, .., N \tag{13}$$

$$-P_t \leq -\underline{P_t} \qquad t = 1, 2, .., N \tag{14}$$

From Eqs. (10), (11), (12), (13) and, (14) we have a quadratic objective function and linear constraints, as a result, P_0 is a quadratic programming problem and it can be represented as:

$$\begin{aligned} \max_{P_1, P_2, .., P_N} \quad & \Pi = \frac{1}{2}P^T W P + V^T P \\ st: \quad & AP \leq E \end{aligned} \tag{15}$$

The last term D is omitted from the objective function because it's a constant and it doesn't have any influence on the optimal pricing policy. The matrix $A_{M \times N}$ is defined from the M constraints and the vector E contains the right side of each constraint. The values of A and E are given in the following two equations.

$$A_{M,N} = \begin{pmatrix} b_{11}, & b_{21}, & \dots, & b_{N1} \\ -b_{11}, & -b_{21}, & \dots, & -b_{N1} \\ 1, & 0, & \dots, & 0 \\ -1, & 0, & \dots, & 0 \\ \vdots & \vdots & \vdots & \vdots \\ b_{1N}, & b_{2N}, & \dots, & b_{NN} \\ -b_{1N}, & -b_{2N}, & \dots, & -b_{NN} \\ 0, & 0, & \dots, & 1 \\ 0, & 0, & \dots, & -1 \end{pmatrix} \quad (16)$$

$$E_{M,1} = \begin{pmatrix} K_1 - a_1 \\ a_1 - M_1 \\ \overline{P_1} \\ -\underline{P_1} \\ \vdots \\ K_N - a_N \\ a_N - M_N \\ \overline{P_N} \\ -\underline{P_N} \end{pmatrix} \quad (17)$$

Since the problem P_0 is quadratic and all the constraints are linear, two cases are distinguished regarding the convexity of the objective function Π. When Π is not concave i.e. the matrix W is not definite or semi-definite negative, the problem is not convex and it can be solved using nonlinear programming algorithms such as interior-point method, gradient methods, etc. However, all these methods reach in generally a local optimum. When Π is concave i.e. the matrix W is definite or semi-definite negative, P_0 is a convex programming problem and it can be solved optimally.

First, let consider the case when Π is concave, one way to find the optimal solution of P_0, is the resolution of Karush-Kuhn-Tucker or KKT system related to P_0. The KKT conditions generally aren't sufficient i.e. if a point P^* is a solution for the KKT system, then P^* can be a local optimum, a global optimum, or saddle point. However, when dealing with a convex programming problem, the KKT conditions became sufficient and any solution of the KKT system is a global optimum of the considering problem. In the rest of the section, the KKT system for the problem P_0 when this later is a convex programming problem is presented.

Let $\Pi' = -\Pi$ and P_0' the problem presented as follows:

$$\begin{aligned} \min_{P_1, P_2, .., P_N} \quad & \Pi' = \frac{1}{2} P^T W' P + V'^T P \\ st: \quad & AP \leq E \end{aligned} \quad (18)$$

Note that the matrix $W' = -W$ and the vector $V'^T = -V^T$. The resolution of P_0 to optimality is equivalent to the resolution of P_0' to optimality. Furthermore,

P_0' is considered to define the KKT system. Before the presentation of the KKT system, some new notations are introduced:

$A_jP - E_j = g_j(P) \qquad j = 1, 2, .., M$
$\lambda^T = (\lambda_1, \lambda_2, .., \lambda_M) \qquad \lambda_j$ is the jth KKT multiplier with $j = 1, 2, .., M$

The KKT system related to P_0' is detailed below:

$$\nabla \Pi'(P) + \sum_{j=1}^{M} \lambda_j \nabla g_j(P) = 0 \qquad N\ equations \tag{19}$$

$$g_j(P) \leq 0 \qquad j = 1, 2, .., M \qquad M\ equations \tag{20}$$

$$\lambda_j g_j(P) = 0 \qquad j = 1, 2, ..., M \qquad M\ equations \tag{21}$$

$$\lambda_j \geq 0, \qquad j = 1, 2, ..., M \tag{22}$$

After computing the gradients related to the first N equations, they are represented as $P^T W' + V'^T + \lambda^T A = 0$. Regarding the Eqs. (20), they can be replaced by $(AP - E) \leq 0$. The value of $\lambda_j g_j(P) \leq 0\ \forall j$, since from the Eqs. (20) and (22), we have $g_j(P) \leq 0\ \forall j$ and $\lambda_j \geq 0\ \forall j$, respectively. As a result, the M equations related to $\lambda_j g_j(P) = 0\ \forall j$ are replaced by the constraint $\lambda_1 g_1(P) + \lambda_2 g_2(P) + ... + \lambda_M g_M(P) = 0$ which corresponds to $\lambda^T(AP - E) = 0$. The last sum is equal to 0 if and only if each term $\lambda_j g_j(P) = 0\ \forall j$. The KKT system for P_0' can be represented as:

$$P^T W' + V'^T + \lambda^T A = 0 \qquad N\ equations \tag{23}$$

$$AP - E \leq 0 \qquad M\ equations \tag{24}$$

$$\lambda^T(AP - E) = 0 \qquad 1\ equation \tag{25}$$

$$\lambda_j \geq 0, \qquad j = 1, 2, ..., M \tag{26}$$

Regarding the first N equations, they are represented as a one line vector $(1 \times N)$. The constraints remain the same if we consider the transpose of $P^T W' + V'^T + \lambda^T A = 0$ which is equal to $W'P + V' + A^T\lambda$. Now, considering the equations $AP - E \leq 0$ and $\lambda^T(AP - E) = 0$, the vector $S = (s_1, s_2, ..., s_M)^T$ is added, with $s_i \geq 0$ such that $AP - E + S = 0$ and $\lambda^T(-S) = 0$. The last term $\lambda^T(-S) = 0$ is equal to $-\lambda_1^T s_1 - \lambda_2^T s_2 - ... - \lambda_M^T s_M = 0$. As each term $-\lambda_j^T s_j \leq 0\ \forall j$, the term $\lambda^T(-S) = 0$ can be replaced by $\lambda^T(S) = 0$. The KKT system is represented as follows:

$$W'P + V' + A^T\lambda = 0 \qquad N\ equations \tag{27}$$

$$AP - E + S = 0 \qquad M\ equations \tag{28}$$

$$\lambda^T(S) = 0 \qquad 1\ equation \tag{29}$$

$$\lambda_j \geq 0, \qquad j = 1, 2, ..., M \tag{30}$$

Finally, the matrix representation of the KKT system of P_0' is:

$$\begin{pmatrix} W', & A^T & 0 \\ A, & 0 & I_M \end{pmatrix} \begin{pmatrix} P \\ \lambda \\ S \end{pmatrix} = \begin{pmatrix} -V' \\ E \end{pmatrix}$$

$$\lambda_j s_j = 0, \qquad j = 1, 2, ..., M$$
$$\lambda_j \geq 0, \qquad j = 1, 2, ..., M$$
$$s_j \geq 0, \qquad j = 1, 2, ..., M$$

3.2 Case with Isoelastic Demand Function

The isoelastic demand function also called the constant elasticity function is the simplest nonlinear demand function. One of its advantages is that it does not require a finite upper limit of the price [11]. The same demand as the one studied in [9] is considered. The demand in a period t depends only on the price of the product in the same period (Eq. 31). ($\gamma \geq 0$) is the seasonality factor. β is the price elasticity of demand, it measures the percentage change in the quantity demanded for a product in relation to percentage change in its price. According to Phillips [14], the price elasticity is defined as $\beta = \frac{-pD(p)}{d'(p)}$. Since $d'(p) \leq 0$ (downward-slopping of the demand), the value of $\beta \geq 0$.

$$Q_t = \alpha \gamma_t P_t^{-\beta} \tag{31}$$

Replacing Q_t its value from (31), the total profit function Π is equal to:

$$\Pi = \sum_{t=1}^{N} \alpha \gamma_t P_t^{-\beta} (P_t - C_t) \tag{32}$$

$$\Pi = \alpha \Big[\gamma_1 P_1^{-\beta} (P_1 - C_1) + \gamma_2 P_2^{-\beta} (P_2 - C_2) + \ldots + \gamma_N P_N^{-\beta} (P_N - C_N) \Big] \tag{33}$$

$$\Pi = \alpha \sum_{t=1}^{N} f_t(P_t) \tag{34}$$

$$with\ f_t(P_t) = \gamma_t P_t^{-\beta} (P_t - C_t) \tag{35}$$

Regarding the capacity constraints $M_t \leq Q_t \leq K_t\ \forall t$, they are represented as:

$$M_t \leq \alpha \gamma_t P_t^{-\beta} \leq K_t \tag{36}$$
$$\frac{M_t}{\alpha \gamma_t} \leq P_t^{-\beta} \leq \frac{K_t}{\alpha \gamma_t} \tag{37}$$
$$\left(\frac{K_t}{\alpha \gamma_t} \right)^{-\frac{1}{\beta}} \leq P_t \leq \left(\frac{M_t}{\alpha \gamma_t} \right)^{-\frac{1}{\beta}} \tag{38}$$

We define I_t as : $I_t = [\underline{P_t}, \overline{P_t}] \cap \left[\left(\frac{K_t}{\alpha \gamma_t} \right)^{-\frac{1}{\beta}}, \left(\frac{M_t}{\alpha \gamma_t} \right)^{-\frac{1}{\beta}} \right] = [a_t, b_t]$.

P_0 can be written as:

$$\max_{P_1, P_2, .., P_N} \Pi = \alpha \sum_{t=1}^{N} f_t(P_t)$$

$$s.t\ P_t \in [a_t, b_t], \quad t = 1, 2, ..., N$$

Finding the optimal pricing policy is equivalent to find the value P_{t_max} which maximizes f_t, i.e. $P^* = (\max f_1(P_1), \max f_2(P_2)..., \max f_N(P_N))$. The optimal selling price for each f_t is obtained analytically through the study of f_t's variation. The derivative of f_t, and the value P_{0t} for which $f_t'(P_t) = 0$ are presented in Eqs. (39) and (40) respectively.

When $\beta \leq 1$, $f_t'(P_t) > 0\ \forall\ P_t \geq 0$, which implies that f_t is increasing in $[0, +\infty[$ specially in $[a_t, b_t]$, then $P_{t\_max} = b_t$. When $\beta > 1$, $P_{0t} > 0$ and $0 < C_t < P_{0t}$. As a result, $f_t' \geq 0$ for P_t in $]0, P_{0t}]$ and $f_t' \leq 0$ for $P_t \geq P_{0t}$. This means that, f_t is increasing in $]0, P_{0t}]$ and decreasing in $[P_{0t}, +\infty[$. Regarding the order between P_{0t}, a_t and b_t. The following cases are considered:

1. If $a_t \leq P_{0t} \leq b_t$ then $P_{t\_max} = P_{0t}$
2. If $b_t \leq P_{0t}$ then $P_{t\_max} = b_t$
3. If $a_t \geq P_{0t}$ then $P_{t\_max} = a_t$

$$f_t' = \alpha\gamma_t\left(P_t^{-(\beta+1)}\left(P_t(1-\beta) + \beta.C_t\right)\right) \tag{39}$$

$$P_{0t} = \frac{\beta}{\beta - 1}C_t \tag{40}$$

4 Numerical Experiments

In this section, two numerical examples are presented to illustrate the proposed approach.

Example 1. The linear demand is considered, the instance's parameters are generated randomly (Table 1). The number of periods is fixed to $N = 2$. The total profit function Π is concave since ($2b_{11} = -2 < 0$, $2b_{22} = -6 < 0$ and $2b_{11}.2b_{22} - (b_{12} + b_{21})^2 = 3 > 0$), therefore the optimization problem P_0 is a convex programming problem. The feasible region X and the function Π are shown in Figs. (1a) and (1b) respectively.

Table 1. Instance 1 Parameters values

Parameters	$t = 1$	$t = 2$
C_t	2	3
a_t	3	8
M_t	3	2
K_t	4	9
$\overline{P_t}$	8	8
$\underline{P_t}$	1	1
b_{tj}	$b_{11} = -1, b_{12} = 1$	$b_{21} = 2, b_{22} = -3$

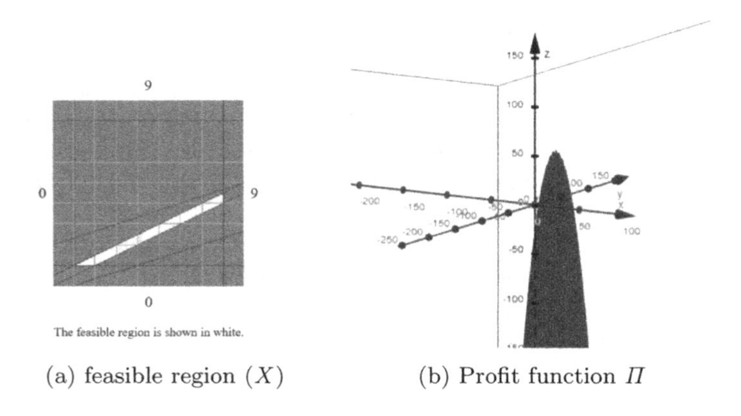

(a) feasible region (X) (b) Profit function Π

Fig. 1. Total profit function and feasible area

Before the resolution, the nature of the optimal pricing policy P^* is considered. Since the total profit function is concave and the global maximum of Π in $\mathbb{R}^2$ is $P_{max} = (17, \frac{32}{3}) \notin X$, any interior point of X is not an optimal solution for the problem P_0. As a result, P^* belongs to the boundary of X.

We apply the KKT system as defined in the Sect. 3.1 (see Appendix A for details). The KKT system is implemented and resolved using **Gekko** Beal et al. [3] on Python3. The optimal pricing policy is $P^* = (P_1^* = 8, P_2^* = \frac{9}{2})$ and the optimal total profit value is $\Pi^* = 27.75$. One can remark that $P_1^* = \overline{P_1} = 8$ and $-P_1^* + 2P_2^* = 1$, which confirm that P^* belongs to the boundary of X.

Example 2. The isoelastic demand is considered with $\beta = 2$ and, $\alpha = 100$. The values of N, C_t, M_t, K_t, $\underline{P_t}$ and $\overline{P_t}$ are the same as for Example 1. The seasonality parameters are fixed to $\gamma_1 = \gamma_2 = 0.5$. The red and blue curves in the following figure represent $f_1(P_1)$ and $f_2(P_2)$ respectively (Fig. 2).

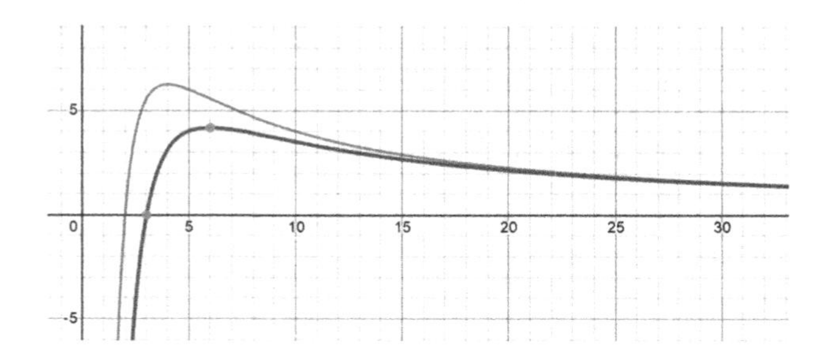

Fig. 2. Curves of f_1 and f_2 (Color figure online)

Table 2 shows the values of a_t, b_t, P_{0t}, and P_t^* which are computed by following the steps described in the Sect. 3.2. For f_1, the value of P_{01} belongs to $[a_1, b_1]$, as a result $P_1^* = P_{01} = 4$. Regarding the function f_2, the value of $P_{02} \geq b_2$, thus, $P_2^* = b_2 = 5$. The total profit function $\Pi = f_1(P_1^*) + f_2(P_2^*) = 6.25 + 4 = 10,25$.

Table 2. Prices' intervals and optimal pricing policy

Parameters	$t = 1$	$t = 2$
a_t	3.53	2.35
b_t	4.08	5
P_{0t}	4	6
P_t^*	4	5

5 Conclusion

This paper investigated the dynamic pricing problem adopted from Shakya et al. [15], in which a firm produces and sells its product over a finite horizon. The problem considers constraints such as limited production capacity and production costs. The firm has to set its selling prices such that the total profit is maximized.

As a first contribution, the case when the demand at a period t is a linear function of the price in the same period and the prices of the other periods is studied. It has been shown that under these assumptions the problem can be formulated as a quadratic programming problem. The Karush-Kuhn-Tucker system to obtain the optimal pricing policy when the total profit function is concave is presented.

The second contribution consists of the consideration of the isoelastic demand function which is commonly used in the literature. It has been proven that when dealing with this demand function, the objective function is the sum of N univariate functions over N bounded intervals. As a result, the optimal pricing policy is resumed to find the maximum of each function.

The presented work assumes that the selling price is the only factor that influences demand. However, consumers are generally sensitive to other parameters like the lead time, rebate, and competitor prices. One extension of this work is the incorporation of these parameters to the demand function to achieve a more accurate representation of the real market behavior's.

Appendix A KKT system for Example 1

From the parameters values presented in the Table 1, the problem optimization problem is formulated as:

$$\min_{P_1, P_2} \Pi' = \frac{1}{2}(P_1, P_2) \begin{pmatrix} 2 & -3 \\ -3 & 6 \end{pmatrix} \begin{pmatrix} P_1 \\ P_2 \end{pmatrix} + (-2, -13) \begin{pmatrix} P_1 \\ P_2 \end{pmatrix}$$

$$s.t: \begin{pmatrix} -1 & 2 \\ 1 & -2 \\ 1 & 0 \\ -1 & 0 \\ 1 & -3 \\ -1 & 3 \\ 0 & 1 \\ 0 & -1 \end{pmatrix} \begin{pmatrix} P_1 \\ P_2 \end{pmatrix} \leq E_{8,1} = \begin{pmatrix} 1 \\ 0 \\ 8 \\ -1 \\ 1 \\ 6 \\ 8 \\ -1 \end{pmatrix}$$

The KKT system is defined as:

$$\begin{pmatrix} 2 & -3 \\ -3 & 6 \end{pmatrix} \begin{pmatrix} P_1 \\ P_2 \end{pmatrix} + \begin{pmatrix} -2 \\ -13 \end{pmatrix} + \begin{pmatrix} -1 & 1 & 1 & -1 & 1 & -1 & 0 & 0 \\ 2 & -2 & 0 & 0 & -3 & 3 & 1 & -1 \end{pmatrix} \begin{pmatrix} \lambda_1 \\ \lambda_2 \\ \lambda_3 \\ \lambda_4 \\ \lambda_5 \\ \lambda_6 \\ \lambda_7 \\ \lambda_8 \end{pmatrix} = \begin{pmatrix} 0 \\ 0 \end{pmatrix}$$

$$\begin{pmatrix} -1 & 2 \\ 1 & -2 \\ 1 & 0 \\ -1 & 0 \\ 1 & -3 \\ -1 & 3 \\ 0 & 1 \\ 0 & -1 \end{pmatrix} \begin{pmatrix} P_1 \\ P_2 \end{pmatrix} - \begin{pmatrix} 1 \\ 0 \\ 8 \\ -1 \\ 1 \\ 6 \\ 8 \\ -1 \end{pmatrix} + \begin{pmatrix} s_1 \\ s_2 \\ s_3 \\ s_4 \\ s_5 \\ s_6 \\ s_7 \\ s_8 \end{pmatrix} = \begin{pmatrix} 0 \\ 0 \\ 0 \\ 0 \\ 0 \\ 0 \\ 0 \\ 0 \end{pmatrix}$$

$$(\lambda_1, \lambda_2, \lambda_3, \lambda_4, \lambda_5, \lambda_6, \lambda_7, \lambda_8) \begin{pmatrix} s_1 \\ s_2 \\ s_3 \\ s_4 \\ s_5 \\ s_6 \\ s_7 \\ s_8 \end{pmatrix} = 0$$

$$\lambda_j, s_j \geq 0, \qquad j = 1, 2, ..., 8$$

References

1. Bajwa, N., Fontem, B., Sox, C.R.: Optimal product pricing and lot sizing decisions for multiple products with nonlinear demands. J. Manag. Anal. **3**(1), 43–58 (2016)
2. Bajwa, N., Sox, C.R.: Coordination of pricing, advertising, and production decisions for multiple products. Int. J. Serv. Oper. Manag. **22**(4), 495–521 (2015)

3. Beal, L.D., Hill, D.C., Martin, R.A., Hedengren, J.D.: Gekko optimization suite. Processes **6**(8), 106 (2018)
4. Bitran, G.R., Mondschein, S.V.: An application of yield management to the hotel industry considering multiple day stays. Oper. Res. **43**(3), 427–443 (1995)
5. den Boer, A.V.: Dynamic pricing and learning: historical origins, current research, and new directions. Surv. Oper. Res. Manag. Sci. **20**(1), 1–18 (2015)
6. Chen, L., Mislove, A., Wilson, C.: An empirical analysis of algorithmic pricing on Amazon marketplace. In: Proceedings of the 25th International Conference on World Wide Web, pp. 1339–1349 (2016)
7. Chen, X., Simchi-Levi, D.: Pricing and inventory management. Oxford Handb. Pricing Manag. **1**, 784–824 (2012)
8. Couzon, P., Ouazene, Y., Yalaoui, F.: Joint pricing and lot-sizing problem with variable capacity. IFAC-PapersOnLine **52**(13), 106–111 (2019)
9. Couzon, P., Ouazene, Y., Yalaoui, F.: Joint optimization of dynamic pricing and lot-sizing decisions with nonlinear demands: theoretical and computational analysis. Comput. Oper. Res. **115**, 104862 (2020)
10. Elmaghraby, W., Keskinocak, P.: Dynamic pricing in the presence of inventory considerations: research overview, current practices, and future directions. Manag. Sci. **49**(10), 1287–1309 (2003)
11. Huang, J., Leng, M., Parlar, M.: Demand functions in decision modeling: a comprehensive survey and research directions. Decis. Sci. **44**(3), 557–609 (2013)
12. Narahari, Y., Raju, C., Ravikumar, K., Shah, S.: Dynamic pricing models for electronic business. Sadhana **30**(2–3), 231–256 (2005)
13. Ouazene, Y., Yalaoui, F., Kelly, R., Idjeraoui, T.: Coordination and optimization of dynamic pricing and production decisions. In: 2017 IEEE Symposium Series on Computational Intelligence (SSCI), pp. 1–6. IEEE (2017)
14. Phillips, R.L.: Pricing and Revenue Optimization. Stanford University Press (2005)
15. Shakya, S., Kern, M., Owusu, G., Chin, C.M.: Neural network demand models and evolutionary optimisers for dynamic pricing. Knowl.-Based Syst. **29**, 44–53 (2012)
16. Smith, B.C., Leimkuhler, J.F., Darrow, R.M.: Yield management at American airlines. Interfaces **22**(1), 8–31 (1992)

A Hybrid FLP-AHP Approach for Optimal Product Mix in Pulp and Paper Industry

Meenu Singh(✉) and Millie Pant

Department of Applied Science and Engineering, Indian Institute of Technology (IIT) Roorkee, Roorkee, India
{msingh1,pant.milli}@as.iitr.ac.in

Abstract. Pulp and Paper Industries (PPI) manufactures a wide range of papers based on three different GSM (Grams/sq. meter). i.e., lower GSM, middle GSM and higher GSM. In order to maximize the profit, the PPI must efficiently utilize its available resources thereby producing optimal units of three different GSMs. Such problems lie under the category of product mix problems and forms an important part of production planning for every process industry like paper mill. In the present study, this problem is represented as a Fuzzy Linear Programming (FLP) model, to include the inherent vagueness and uncertainties. The solutions obtained through FLP are further refined with the help of AHP (Analytical Hierarchy Process) to determine the most profitable solution. Results indicate that ranking results obtained by integrating AHP into FLP may help in providing a better guidance to the Decision Maker (DM) for determining an optimal product mix.

Keywords: Indian Pulp and Paper Industry (IPPI) · Product mix optimization · Fuzzy linear programming (FLP) · Multi-criteria decision making (MCDM)

1 Introduction

The Pulp and Paper Industry (PPI) plays an important role in Indian economy due to several reasons [1]. Different types of paper produced by a paper mill can be broadly classified into Cultural and Industrial papers [2]. The major production of mills deals with the cultural paper involving all types of writing and printing papers with three different levels of GSM (Grams/sq. meter). i.e., lower GSM, medium GSM, and higher GSM. The wrapping, packing, photographic and other functional papers are called industrial papers. The various stages of pulping and papermaking process are presented in Fig. 1.

Production planning is an important decision making for any process industry including PPI, where the main objective is to maintain a tradeoff between production and consumption.

The focus of the present study is to suggest an optimal production plan for Indian Pulp and Paper Industries (IPPI) producing a variety of papers. The objective here is to maximize the profit by suggesting an optimal product mix on the basis of different levels of GSM. The problem is formulated as a Fuzzy Linear Programming (FLP) model due to the inherent uncertainties in the model parameters. Further, sensitivity analysis

B. Dorronsoro et al. (Eds.): OLA 2021, CCIS 1443, pp. 324–336, 2021.
https://doi.org/10.1007/978-3-030-85672-4_24

of the proposed model is done by examining the effect of alpha-cuts on profit. Finally, Analytical Hierarchical Process (AHP) is introduced into the model for evaluating the feasible solutions as alternatives.

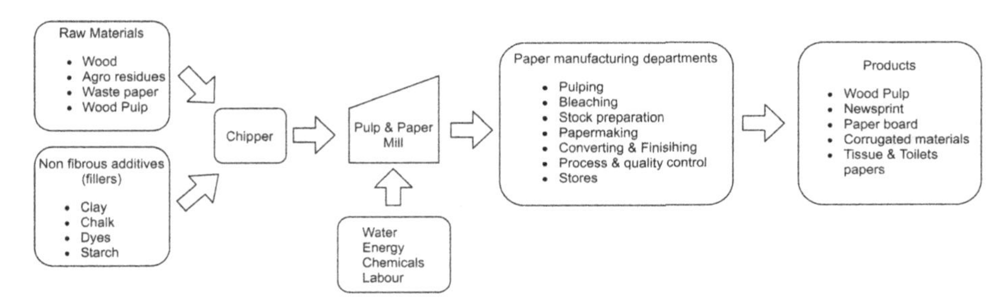

Fig. 1. Process flow diagram of PPI.

Rest of the paper is organized as follows. In Sect. 2, literature review on FLP for production planning is given. In Sect. 3, brief description of the methodology is provided. Section 4 and 5, presents a hypothetical but realistic case study illustrating the applicability of the proposed approach through a mathematical model. Section 6 provide the results and discussion of the model and the decision making process respectively. Finally, the paper concludes with Sect. 7, summarizing the present study and providing future research directions.

2 Literature Review

Linear Programming (LP) has gained its reputation as one of the best decision making tools for maximizing the goal achievements or minimizing the costs while satisfying all the constraints and restrictions. Literature is full of instances advocating the effectiveness of LP in different areas [3] and [4].

Introduction of fuzzy logic in LP was suggested long back in 1970's after the concept of fuzzy theory was established by Zadeh in 1965 [5]. In [6–8], the authors suggested that fuzzy set theory can be integrated with other mathematical programming approaches like non-linear programming, quadratic programing, dynamic programming and goal programming for a more realistic representation of the problem. The fuzzy theory is used to optimize the solutions for which constraints have fuzzy coefficients, fuzzy inequalities or fuzzy variables.

Several instance are available in literature where the researchers have successfully implemented FLP by substituting the crisp coefficient with fuzzy numbers in an LP problem [3, 4, 9, 10]. Researchers have also shown that most of real life problems with intensive decision making like product mix, manpower allocation, flow shop scheduling, transportation, production planning [9–13] can be dealt efficiently through an FLP approach. However, there are work in the area of production planning activities [14–16] but the authors were not able to find any relevant literature relating to production planning in a fuzzy environment for an IPPI.

The proposed methodology integrates AHP, a well-known Multi-Criteria Decision Making (MCDM) technique with FLP to select the most profitable solution from the set of solutions achieved in the optimization process.

3 Methodology

An integrated FLP-AHP model is proposed for determining the optimal product mix for an Indian Pulp and Paper Industry (IPPI). It is a two phase methodology, as illustrated in Fig. 2.

In Phase I, fuzzification of the problem is done, while in Phase 2, AHP is invoked to select the best possible alternative, out of the solutions obtained in Phase I. These phases are further divided into a number of steps as described below and as illustrated in Fig. 2.

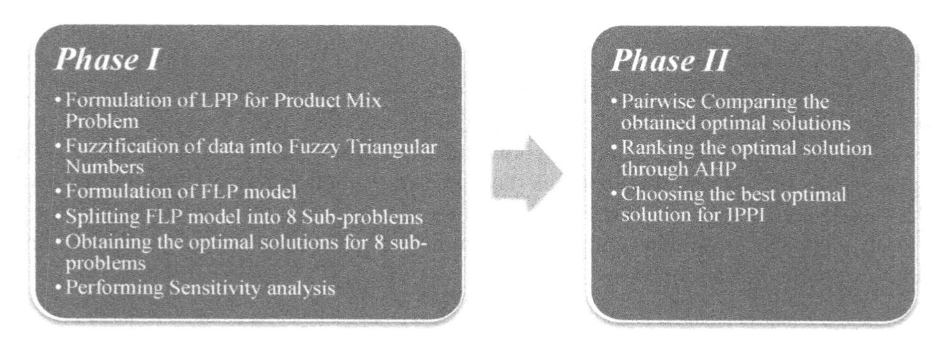

Fig. 2. Schematic presentation of FLP-AHP model

Phase I

Step 1. Formulation of LPP: develop the LPP model with the help of the available crisp data.
Step 2. Fuzzification of data: the available crisp data is changed into fuzzy triangular numbers (FTN).
Step 3. Formulation of FLP model: using the data generated in Step 2, the LPP developed in Step 1 is converted into an FLP model.
Step 4. Splitting: the model developed in Step 3, is split into a number of sub problems.
Step 5. Solution: Obtain the solution through different values of aspirations.

Phase II

Step 1. Decision Making through AHP: the solutions obtained for different values of aspirations are treated as alternatives and the best solution is ranked through AHP.

4 Case Study

4.1 Background

A hypothetical but realistic PPI in India is considered, that produces three types of writing/printing paper based on three different levels of GSM: lower GSM (58 to 64 GSM), medium GSM (68 to 80 GSM) and higher GSM (90 to 120 GSM) papers. Each kind of GSM papers are produced by a combination of raw material fibers such as eucalyptus and poplar with fillers such as china clay, GCC, PCC etc. The range of percentage composition of the raw material fibers with the fillers are presented in Table 1.

Table 1. The % of raw material and fillers used in each grade of paper.

Raw material	Lower GSM	Medium GSM	Higher GSM
Pulp	90%	86%	85%
Filler	10%	14%	15%

The data and details for the case study were constructed following a series of discussions with the experts from management and R&D department of an Indian paper mill.

During the papermaking process, the raw material goes through different processes including pulping, bleaching, stock preparation, papermaking, converting and finishing. During each process, the yield of pulp is lost to some extent due to the processing of a particular grade of paper while the remaining amount undergoes the next process.

The expected average losses from one ton of input for each grade of paper; average capacity of plant at each unit and the maximum output of the three kinds of papers produced in a day is summarized in Table 2. A blank entry represents "no loss" implying that the same amount is carried on to the next process.

The profit of each GSM for a ton is evaluated after estimating the selling price and other expenses of each GSM as per the current market trend and as per the opinion of experts. The profit for the three GSM with the details of plant capacity is given in Table 3.

Table 2. Processing of paper with net losses at each department.

Departments	Lower GSM			Medium GSM			Higher GSM		
	Input	*Losses*	*Output*	*Input*	*Losses*	*Output*	*Input*	*Losses*	*Output*
Input	1.000	–	–	1.000	–	–	1.000	–	–
Pulping process	1.000	0.01	0.990	1.000	0.02	0.983	1.000	0.02	0.981
Bleaching process	0.990	0.09	0.901	0.983	0.14	0.845	0.981	0.19	0.795
Stock preparation process	0.901	–	0.901	0.845	–	0.845	0.795	–	0.795
Papermaking process	0.901	0.05	0.856	0.845	0.07	0.786	0.795	0.08	0.734
Converting & finishing process	0.856	0.07	0.796	0.786	0.07	0.738	0.735	0.08	0.676
Output	–	–	0.796	–	–	0.738	–	–	0.676

Table 3. Details of plant capacity.

Papermaking process	Tons/day	Max production of each grade	Tons/day	Profit of each grade of paper produced	Rs./Ton
Pulping process	230	Lower GSM	34	Lower GSM	5600
Bleaching process	120	Medium GSM	158	Medium GSM	5000
Stock preparation process	220	Higher GSM	34	Higher GSM	5100
Converting & finishing process	220				

5 Mathematical Model

The above-mentioned data for the papermaking processes and objective function was analyzed to obtain estimates for LP problem model parameters. The decision variables of the model are notated as x_1, x_2 and x_3 representing the amount of: lower GSM, medium GSM and higher GSM to be produced. The objective function Z is to maximize the profit per ton for the three kinds of papers.

The LPP model of the problem can be formulated as below, by using the given data in Tables 1, 2 and 3:

$$
\begin{array}{ll}
Max\ Z = 5600x_1 + 5000x_2 + 4200x_3 & \\
s.t. & \\
0.90x_1 + 0.86x_2 + 0.85x_3 \leq 230 & \text{(Raw Material Constraint)} \\
0.12x_1 + 0.16x_2 + 0.17x_3 \leq 22.7 & \text{(Filler Constraint)} \\
x_1 + x_2 + x_3 \leq 230 & \text{(Pulping process constraint)} \\
0.9x_1 + 0.983x_2 + 0.988x_3 \leq 120 & \text{(Bleaching process constraint)} \\
0.901x_1 + 0.891x_2 + 0.869x_3 \leq 220 & \text{(Stock preparation process constraint)} \\
0.869x_1 + 0.860x_2 + 0.835x_3 \leq 220 & \text{(Converting \& Finishing Process)} \\
0.15x_1 + 0.7x_2 + 0.14x_3 \leq 24 & \text{(Time constraint)} \\
x_1 \leq 34 & \text{(Input constraint for Lower GSM)} \\
x_2 \leq 158 & \text{(Input constraint for Medium GSM)} \\
x_3 \leq 34 & \text{(Input constraint for Higher GSM)}
\end{array} \tag{1}
$$

However, there are several factors that affect the preciseness of data like: non-availability of raw materials or fillers, improper management of resources, inadequate power supply, and under utilization of capacity etc.

Presence of uncertainty in data justifies the application of fuzzy set theory for modeling such problems [11–13]. In this study, FLP is implemented to maximize the value of Z obtained through LP by using fuzzy triangular numbers. Fuzzy numbers generated for each coefficient are given in Table 4.

Table 4. Fuzzy numbers of each coefficient.

Process	Lower GSM			Medium GSM			Higher GSM		
	Lower	*Crisp*	*Higher*	*Lower*	*Crisp*	*Higher*	*Lower*	*Crisp*	*Higher*
Pulp	0.88	0.90	0.90	0.84	0.86	0.86	0.83	0.85	0.85
Filler	0.10	0.12	0.12	0.14	0.16	0.16	0.15	0.17	0.17
Bleaching process	0.989	0.990	0.990	0.983	0.985	0.985	0.980	0.981	0.981
Stock preparation process	0.890	0.901	0.901	0.837	0.845	0.845	0.790	0.795	0.795
Converting & finishing process	0.846	0.856	0.856	0.783	0.786	0.786	0.731	0.735	0.735
Time	0.14	0.15	0.15	0.696	0.7	0.7	0.14	0.15	0.15
Max production	34	34	35	158	158	161	34	34	35

5.1 Formulation of FLP

STEP 1. The crisp data mentioned in the Tables 1, 2 and 3 is utilized to calculate the fuzzy numbers for each coefficient as presented in Table 4.
STEP 2. Using the fuzzy data in Table 4, the Fuzzy Linear Programming model for the case company is formulated as (2):

$$\begin{aligned}
&Max\ Z = (5600, 5600, 6500)x_1 + (5000, 5000, 6200)x_2 + (4200, 4200, 5800)x_3 \\
&s.t. \\
&(0.88, 0.90, 0.90)x_1 + (0.84, 0.86, 0.86)x_2 + (0.83, 0.85, 0.85)x_3 \le (230, 280, 280) \\
&(0.10, 0.12, 0.12)x_1 + (0.14, 0.16, 0.16)x_2 + (0.15, 0.17, 0.17)x_3 \le (22.7, 34.1, 34.1) \\
&(0.989, 0.990, 0.990)x_1 + (0.983, 0.985, 0.985)x_2 + (0.980, 0.981, 0.981)x_3 \le (120, 120, 120) \\
&(0.890, 0.901, 0.901)x_1 + (0.837, 0.845, 0.845)x_2 + (0.790, 0.795, 0.795)x_3 \le (220, 220, 220) \\
&(0.846, 0.856, 0.856)x_1 + (0.783, 0786, 0.786)x_2 + (0.731, 0.735, 0.735)x_3 \le (220, 220, 220) \\
&(0.14, 0.15, 0.15)x_1 + (0.696, 0.7, 0.7)x_2 + (0.14, 0.15, 0.15)x_3 \le (24, 24, 24) \\
&x_1, x_3 \le (34, 34, 35) \\
&x_2 \le (158, 158, 161)
\end{aligned} \tag{2}$$

STEP 3. To obtain the better optimization result from the fuzzy numbers, the FLP has been split into eight sub-problems of linear crisp programming, as seen in Table 5, by taking in account, the total number of possible combinations of objective values with all the constraints values.
STEP 4. The optimal solutions of each sub-problem (3) to (9) are obtained through Lingo 18.0 software and are presented in Table 6.
STEP 5. The crisp fuzzy linear model for maximizing the value of aspiration (λ) is written as below:

$$\begin{aligned}
&Max\ Z = \lambda \\
&s.t. \\
&118300.0\lambda - 6500x_1 - 6200x_2 - 5800x_3 + 550071.4 \le 0 \\
&(0.90 - 0.02\lambda)x_1 + (0.86 - 0.02\lambda)x_2 + (0.85 - 0.02\lambda)x_3 - 50\lambda - 280 \le 0 \\
&(0.12 - 0.02\lambda)x_1 + (0.16 - 0.02\lambda)x_2 + (0.17 - 0.02\lambda)x_3 - 11.4\lambda - 34.1 \le 0 \\
&(0.990 - 0.001\lambda)x_1 + (0.985 - 0.002\lambda)x_2 + (0.981 - 0.001\lambda)x_3 - 120 \le 0 \\
&(0.901 - 0.011\lambda)x_1 + (0.845 - 0.008\lambda)x_2 + (0.795 - 0.005\lambda)x_3 - 220 \le 0 \\
&(0.856 - 0.010\lambda)x_1 + (0.786 - 0.003\lambda)x_2 + (0.735 - 0.004\lambda)x_3 - 220 \le 0 \\
&(0.15 - 0.01\lambda)x_1 + (0.7 - 0.004\lambda)x_2 + (0.15 - 0.001\lambda)x_3 - 24 \le 0 \\
&x_1 - \lambda - 34 \le 0 \\
&x_2 - (3.0\lambda) - 158 \le 0 \\
&x_3 - \lambda - 34 \le 0 \\
&0 \le \lambda \le 0 \\
&x_1, x_2, x_3 \ge 0
\end{aligned} \tag{11}$$

STEP 6. The solutions in Table 7 are obtained by varying the value of λ in Lingo 18.0 software. It is observed that the maximum profits of the case company varies depending on the value of λ and variables of product mix. Also, only 3 solutions are found to be feasible (highlighted in bold) while 4 solutions are infeasible.

Table 5. The sub-problems of FLP.

Sub-problem 1

$$\begin{array}{l} Max = 5600x_1 + 5000x_2 + 4200x_3 \\ s.t. \\ 0.90x_1 + 0.86x_2 + 0.85x_3 \le 280 \\ 0.10x_1 + 0.14x_2 + 0.15x_3 \le 34.1 \\ 0.990x_1 + 0.983x_2 + 0.981x_3 \le 120 \\ 0.901x_1 + 0.845x_2 + 0.795x_3 \le 220 \\ 0.856x_1 + 0.786x_2 + 0.735x_3 \le 220 \\ 0.15x_1 + 0.7x_2 + 0.14x_3 \le 24 \\ x_1, x_3 \le 34 \\ x_2 \le 158 \end{array} \tag{3}$$

Sub-problem 2

$$\begin{array}{l} Max = 6500x_1 + 6200x_2 + 5800x_3 \\ s.t. \\ 0.90x_1 + 0.86x_2 + 0.85x_3 \le 280 \\ 0.10x_1 + 0.14x_2 + 0.15x_3 \le 34.1 \\ 0.990x_1 + 0.983x_2 + 0.981x_3 \le 120 \\ 0.901x_1 + 0.845x_2 + 0.795x_3 \le 220 \\ 0.869x_1 + 0.786x_2 + 0.735x_3 \le 220 \\ 0.15x_1 + 0.7x_2 + 0.14x_3 \le 24 \\ x_1, x_3 \le 34 \\ x_2 \le 158 \end{array} \tag{4}$$

Sub-problem 3

$$\begin{array}{l} Max = 5600x_1 + 5000x_2 + 4200x_3 \\ s.t. \\ 0.90x_1 + 0.86x_2 + 0.85x_3 \le 230 \\ 0.10x_1 + 0.14x_2 + 0.15x_3 \le 22.7 \\ 0.990x_1 + 0.983x_2 + 0.981x_3 \le 120 \\ 0.901x_1 + 0.845x_2 + 0.795x_3 \le 220 \\ 0.856x_1 + 0.786x_2 + 0.735x_3 \le 220 \\ 0.15x_1 + 0.7x_2 + 0.14x_3 \le 24 \\ x_1, x_3 \le 35 \\ x_2 \le 161 \end{array} \tag{5}$$

Sub-problem 4

$$\begin{array}{l} Max = 6500x_1 + 6200x_2 + 5800x_3 \\ s.t. \\ 0.90x_1 + 0.86x_2 + 0.85x_3 \le 230 \\ 0.10x_1 + 0.14x_2 + 0.15x_3 \le 22.7 \\ 0.990x_1 + 0.983x_2 + 0.981x_3 \le 120 \\ 0.901x_1 + 0.845x_2 + 0.795x_3 \le 220 \\ 0.869x_1 + 0.786x_2 + 0.735x_3 \le 220 \\ 0.15x_1 + 0.7x_2 + 0.14x_3 \le 24 \\ x_1, x_3 \le 35 \\ x_2 \le 161 \end{array} \tag{6}$$

Sub-problem 5

$$\begin{array}{l} Max = 5600x_1 + 5000x_2 + 4200x_3 \\ s.t. \\ 0.88x_1 + 0.84x_2 + 0.83x_3 \le 280 \\ 0.12x_1 + 0.16x_2 + 0.17x_3 \le 34.1 \\ 0.989x_1 + 0.983x_2 + 0.980x_3 \le 120 \\ 0.890x_1 + 0.837x_2 + 0.790x_3 \le 220 \\ 0.846x_1 + 0.783x_2 + 0.731x_3 \le 220 \\ 0.15x_1 + 0.696x_2 + 0.15x_3 \le 24 \\ x_1, x_3 \le 34 \\ x_2 \le 158 \end{array} \tag{7}$$

Sub-problem 6

$$\begin{array}{l} Max = 6500x_1 + 6200x_2 + 5800x_3 \\ s.t. \\ 0.88x_1 + 0.84x_2 + 0.83x_3 \le 280 \\ 0.12x_1 + 0.16x_2 + 0.17x_3 \le 34.1 \\ 0.989x_1 + 0.983x_2 + 0.980x_3 \le 120 \\ 0.890x_1 + 0.837x_2 + 0.790x_3 \le 220 \\ 0.846x_1 + 0.783x_2 + 0.731x_3 \le 220 \\ 0.15x_1 + 0.696x_2 + 0.15x_3 \le 24 \\ x_1, x_3 \le 34 \\ x_2 \le 158 \end{array} \tag{8}$$

Sub-problem 7

$$\begin{array}{l} Max = 5600x_1 + 5000x_2 + 4200x_3 \\ s.t. \\ 0.88x_1 + 0.84x_2 + 0.83x_3 \le 230 \\ 0.12x_1 + 0.16x_2 + 0.17x_3 \le 22.7 \\ 0.989x_1 + 0.983x_2 + 0.980x_3 \le 120 \\ 0.890x_1 + 0.837x_2 + 0.790x_3 \le 220 \\ 0.846x_1 + 0.783x_2 + 0.731x_3 \le 220 \\ 0.15x_1 + 0.696x_2 + 0.15x_3 \le 24 \\ x_1, x_3 \le 35 \\ x_2 \le 161 \end{array} \tag{9}$$

Sub-problem 8

$$\begin{array}{l} Max = 6500x_1 + 6200x_2 + 5800x_3 \\ s.t. \\ 0.88x_1 + 0.84x_2 + 0.83x_3 \le 230 \\ 0.12x_1 + 0.16x_2 + 0.17x_3 \le 22.7 \\ 0.989x_1 + 0.983x_2 + 0.980x_3 \le 120 \\ 0.890x_1 + 0.837x_2 + 0.790x_3 \le 220 \\ 0.846x_1 + 0.783x_2 + 0.731x_3 \le 220 \\ 0.15x_1 + 0.696x_2 + 0.15x_3 \le 24 \\ x_1, x_3 \le 35 \\ x_2 \le 161 \end{array} \tag{10}$$

5.2 AHP

In order to select the best optimal solution from the Table 7, Analytical Hierarchy Process (AHP) is employed [17] as discussed in the next section.

The hierarchical structure in Fig. 3 presents the three feasible solutions (S1, S2, S3) for selection of the best solution. The priority conditions and alternatives were decided in the interview of three decision-makers (DMs), such as (DM 1, DM 2, DM 3). In order

Table 6. Optimal solutions of eight sub-problems.

	x_1	x_2	x_3	Z	
X_{11}	**34**	**19.71**	**34**	**431771.4**	***Lower bound***
X_{12}	34	19.71	34	540428.6	
X_{13}	35	19.29	35	439428.6	
X_{14}	35	19.29	35	550071.4	
X_{15}	34	20.80	34	437223.0	
X_{16}	34	19.71	34	540428.6	
X_{17}	35	19.29	35	439428.6	
X_{18}	**35**	**19.28**	**35**	**550071.4**	***Upper bound***

Table 7. Sensitivity analysis of value of aspirations.

λ	x_1	x_2	x_3	Z	Solution type
0	**34.5**	**20.603**	**34.5**	**₹ 5,52,091.39**	***Feasible***
0.3	**34.62**	**26.150**	**34.62**	**₹ 4,70,027.35**	***Feasible***
0.5	34.7	29.848	34.7	₹ 4,89,300.75	*Non-feasible*
0.7	34.78	33.546	34.78	₹ 5,08,574.15	*Non-feasible*
0.9	34.86	37.244	34.86	₹ 5,27,847.55	*Non-feasible*
1	34.9	39.093	34.9	₹ 5,37,484.25	*Non-feasible*
0.3309	**34.63237**	**26.722**	**34.63237**	**₹ 4,73,006.38**	***Feasible***

to prevent biases in their judgement, the DMs advocated for the equal. The same weights were therefore assigned to them as (1/3, 1/3, 1/3). Implementation of AHP consisted of the following steps:

1. Different comparison matrices were formed for the alternatives with respect to each criterion individually for DMs.
2. The final comparison matrix is aggregated as geometric mean of individual DMs matrix.

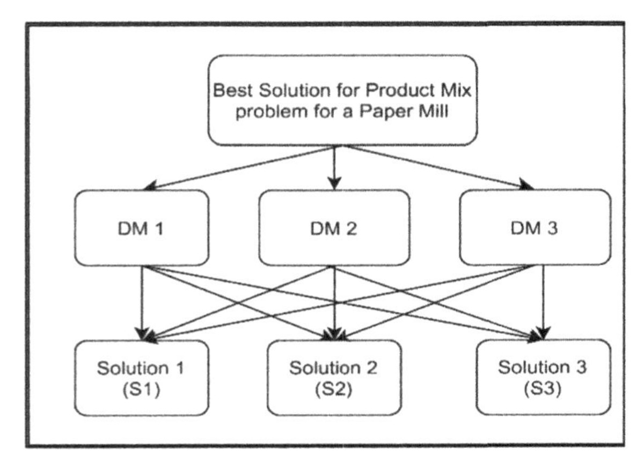

Fig. 3. A hierarchical structure of the problem.

6 Results and Discussions

6.1 Sensitivity Analysis

Membership of aspiration values contributed towards managing the profit of the mill is discussed in this section. The trends of profits and the trend of decision variables against the value of λ are presented in Fig. 4(a) & Fig. 4(b). The following observations were made:

1. The profit of product mix decreased in tandem with the decrease in the value of aspiration level. The maximum and minimum profits were obtained for $\lambda = 1$ and $\lambda = 0.3$ respectively. This is shown graphically in Fig. 4(a).
2. Decision variables x_1 and x_3 have the same values and the gap between them with x_2 is smaller as the value of λ increases. At $\lambda = 0.9$ or 1, the production amount of x_2 is greater than the x_1 & x_3. This trend can be visualized through Fig. 4(b).
3. Through the trends mentioned in 4(a) and 4(b), it can be said that the profit (in Rupees) in a closed interval of [470027.35, 552091.39] can be obtained if the lower GSM, medium GSM and higher GSM were produced (in Tons) in the interval of [34.5, 34.62], [20.6, 26.15] and [34.5, 34.62], respectively.
4. The optimal feasible solutions were obtained by λ value in the range of [0, 0.330].
5. The crisp profit earned through LPP was Rs. 434200.0, which improved up to 21.35%. i.e., Rs. 117891.40 by the use of FLP.

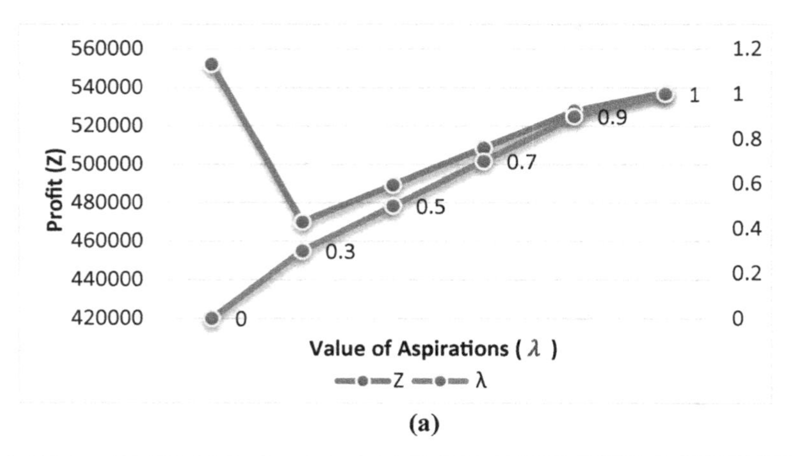

(a)

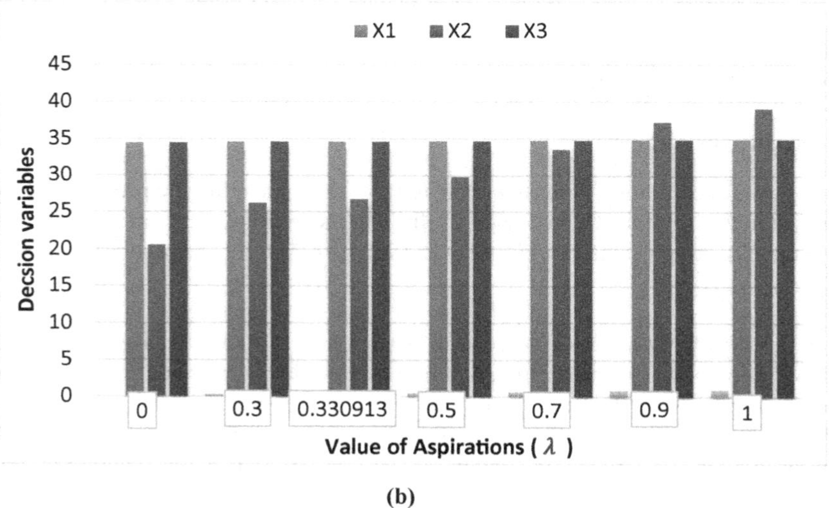

(b)

Fig. 4. (a) Trends of Profit against the value of aspirations. (b) Decision variables vs. aspiration values.

6.2 Decision Making Through AHP

1. After determining the local priorities and checking the consistency at 10%. The global final priority is evaluated and is depicted in Fig. 5.
2. It can be clearly seen that the S1 is the most profitable solution among all the three optimum solutions, due to its overall weight (0.599). It's the best combination for the product mix of the case company with the production of 89.60 tons/ day in which 34.5 tons/ day is for lower GSM, 20.60 tons/ day for medium GSM and 34.5 for higher GSM.

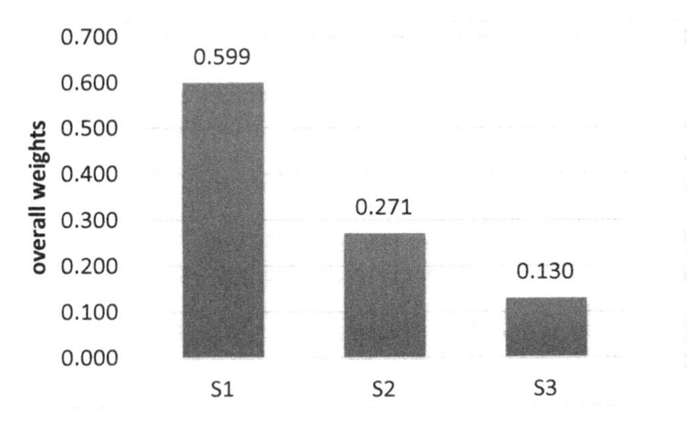

Fig. 5. AHP overall weights for the three optimal solutions.

7 Conclusion

The product mix optimization for maximizing the profit has attracted a lot of attention in recent years. This paper presented an integrated FLP-AHP approach for determining the optimal product mix for an IPPI. The research tends to be a valuable method for improving production planning for three high grades of writing/printing paper with minimal resources. Some conclusions that can be drawn from the research:

- Implementation of fuzzy theory helped in enhancing the profit by 21.35%, which is a significant improvement for a real life situation.
- Integrating AHP into FLP will help managers make better decisions by assisting in identifying the most profitable solution among the available solutions.
- In this study, the authors have considered an example of an IPPI. However, due to the generic nature of FLP-AHP, it can be applied to other manufacturing sectors as well to determine product mix.

Further study can be undertaken by including more variables and constraints that take place during production planning activities, constructing a multi-objective model or considering other fuzzy membership functions like trapezoidal, logistics, S-curve etc. Sustainability factors may also be included to make the model more realistic.

Acknowledgement. This work is in support of the project "Metaheuristics Framework for Multi-objective Combinatorial Optimization Problems (META MO-COP)" reg. no. DST/INT/Czech/P-12/2019.

References

1. Challenges and Opportunities for the Pulp and Paper Industry | SciTech Connect. http://scitechconnect.elsevier.com/challenges-and-opportunities-for-the-pulp-and-paper/. Accessed 27 Nov 2020

2. Jain, R.K.: Compendium of Census survey of Indian Paper Industry, 1st edn. Central Pulp and Paper Institute, Saharanpur (2015)
3. Spitter, J.M., Hurkens, C.A., De Kok, A.G., Lenstra, J.K., Negenman, E.G.: Linear programming models with planned lead times for supply chain operations planning. Eur. J. Oper. Res. **163**(3), 706–720 (2005)
4. Li, Z., Li, Z.: Linear programming-based scenario reduction using transportation distance. Comput. Chem. Eng. **88**, 50–58 (2016)
5. Zadeh, L.A.: Fuzzy sets. Inf. Control **8**, 338–353 (1965)
6. Tanaka, H., Asai, K.: Fuzzy linear programming problems with fuzzy numbers. Fuzzy Sets Syst. **13**(1), 1–10 (1984)
7. Rommelfanger, H.: Fuzzy linear programming and applications. Eur. J. Oper. Res. **92**(3), 512–527 (1996)
8. Bector, C.R., Chandra, S.: Fuzzy Mathematical Programming and Fuzzy Matrix Games, vol. 169. Springer, Berlin (2005). https://doi.org/10.1007/3-540-32371-6
9. Buckley, J.J.: Possibilistic linear programming with triangular fuzzy numbers. Fuzzy Sets Syst. **26**(1), 135–138 (1988)
10. Zimmermann, H.J.: Fuzzy programming and linear programming with several objective functions. Fuzzy Sets Syst. **1**(1), 45–55 (1978)
11. Pendharkar, P.C.: A fuzzy linear programming model for production planning in coal mines. Comput. Oper. Res. **24**(12), 1141–1149 (1997)
12. Wu, Y.K.: On the manpower allocation within matrix organization: a fuzzy linear programming approach. Eur. J. Oper. Res. **183**(1), 384–393 (2007)
13. Abdullah, L., Abidin, N.H.: A fuzzy linear programming in optimizing meat production. Int. J. Eng. Technol. **6**(1), 436–444 (2014)
14. Fang, C.C., Lai, M.H., Huang, Y.S.: Production planning of new and remanufacturing products in hybrid production systems. Comput. Ind. Eng. **108**, 88–99 (2017)
15. Saniuk, S., Saniuk, A.: Decision support system for rapid production order planning in production network. In: Burduk, A., Mazurkiewicz, D. (eds.) ISPEM 2017. AISC, vol. 637, pp. 217–226. Springer, Cham (2018). https://doi.org/10.1007/978-3-319-64465-3_22
16. González Rodríguez, G., Gonzalez-Cava, J.M., Méndez Pérez, J.A.: An intelligent decision support system for production planning based on machine learning. J. Intell. Manuf. **31**(5), 1257–1273 (2019). https://doi.org/10.1007/s10845-019-01510-y
17. Saaty, T.L.: The Analytic Hierarchy Process: Planning, Priority Setting, Resource Allocation. McGraw-Hill Inc., New York (1980)

An Application of BnB-NSGAII: Initializing NSGAII to Solve 3 Stage Reducer Problem

Ahmed Jaber[1,2](✉), Pascal Lafon[1], and Rafic Younes[2]

[1] University of Technology of Troyes UTT, Troyes, France
ahmed.jaber@utt.fr
[2] Lebanese University, Beirut, Lebanon
https://www.utt.fr/, https://ul.edu.lb/

Abstract. The 3 stage reducer problem is a point of interest for many researchers. In this paper, this problem is reformulated to a bi-objective problem with additional constraints to meet the ISO mechanical standards. Those additional constraints increase the complexity of the problem, such that, NSGAII performance is not sufficient. To overcome this, we propose to use BnB-NSGAII [10] method - a hybrid multi-criteria branch and bound with NSGAII - to initialize NSGAII before solving the problem, seeking for a better initial population. A new feature is also proposed to enhance BnB-NSGAII method, called the legacy feature. The legacy feature permits the inheritance of the elite individuals between - branch and bound - parent and children nodes. NSGAII and BnB-NSGAII with and without the legacy feature are tested on the 3 stage reducer problem. Results demonstrate the competitive performance of BnB-NSGAII with the legacy feature.

Keywords: NSGAII · Multi-objective · MINLP · Branch-and-bound · 3-stage reducer

1 Introduction

In [3], the design of the 3 Stage Reducer (3SR) optimization problem has been introduced to illustrate the optimal design framework of the power transmission mechanism. This problem has been a point of interest for many researchers in different domains. Engineering researchers enhance the problem for mechanical engineering applications. In [4], the problem is extended to a mixed variables optimization problem. And recently a similar problem is stated in [5] to illustrate the optimization of the volume and layout design of 3SR. Due to the problem complexity, optimization researchers are interested to test optimization methods on it. In [14], the authors use the 3SR problem to examine the performance of the constraint propagation method.

Supported by organization ERDF, Grand Est and Lebanese University.

B. Dorronsoro et al. (Eds.): OLA 2021, CCIS 1443, pp. 337–349, 2021.
https://doi.org/10.1007/978-3-030-85672-4_25

In this paper, the 3SR problem is reformulated to a bi-objective problem with additional constraints to meet the ISO mechanical standards. Those additional constraints increase the complexity of the problem, such that, the well-known Non-Dominated Sorting Genetic Algorithm 2 (NSGAII) [1] performance is not sufficient.

In [10], the authors enhance the performance of NSGAII by hybridizing it with the multi-criteria branch and bound method [12], the proposed method is called BnB-NSGAII. In this paper, we propose to use the BnB-NSGAII method to initialize NSGAII before solving the 3SR problem, seeking a better initial population. The initial population seeding phase is the first phase of any genetic algorithm application. It generates a set of solutions randomly or by heuristic initialization as input for the algorithm. Although the initial population seeding phase is executed only once, it has an important role to improve the genetic algorithm performance [2].

Furthermore, we propose a new feature to enhance the BnB-NSGAII method, called the legacy feature. The legacy feature permits the inheritance of elite genes between branch-and-bound nodes.

The rest is organized as follows. Section 2 presents the 3SR problem and its complexity. The proposed BnB-NSGAII legacy feature is explained in Sect. 3. The computational results are reported in Sect. 4. Finally, an overall conclusion is drawn in Sect. 5.

2 3 Stage Reducer Problem

The design problem consists in finding dimensions of main components (pinions, wheels and shafts) of the 3 stage reducer (Fig. 1) to minimize the following bi-objective problem:

1. The volume of all the components of the reducer:

$$f_1(\boldsymbol{x}) = \pi \left(\sum_{s=0}^{s=3} l_{a_s} {r_{a,s}}^2 + \sum_{s=1}^{s=3} \left[b_s \frac{m_{\mathrm{ns}}^2}{2} (Z_{s,1}^2 + Z_{s,2}^2) \right] \right) \tag{1}$$

2. The gap between the required reduction ratio $\bar{u}$ and the ratio of the reducer (tolerance):

$$f_2(\boldsymbol{x}) = \frac{1}{\bar{u}} \left| \bar{u} - \prod_{s=1}^{s=3} \frac{Z_{s,2}}{Z_{s,1}} \right|, \; \bar{u} > 1 \tag{2}$$

The problem is designed assuming the following are known:

- The power to be transmitted, P_t and the speed rotation of input shaft N_e.
- The total speed rotation reduction ratio $\bar{u}$, the position of the output shaft from the input shaft position (Fig. 2).
- The dimension of the casing box.

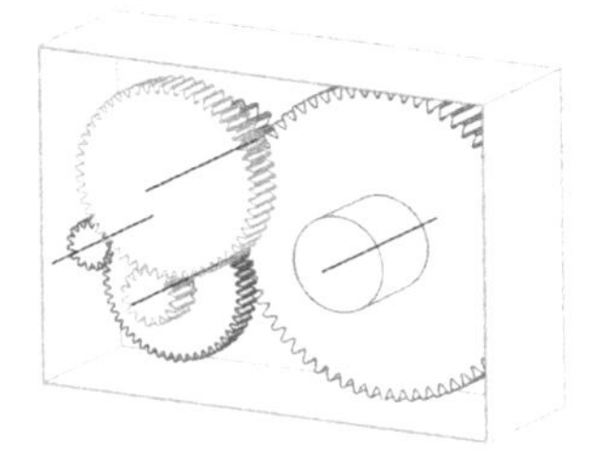

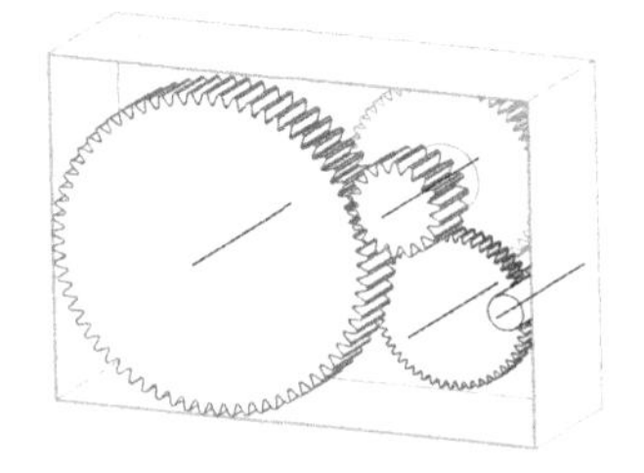

Fig. 1. Front and back view of a 3 stage reducer with closure.

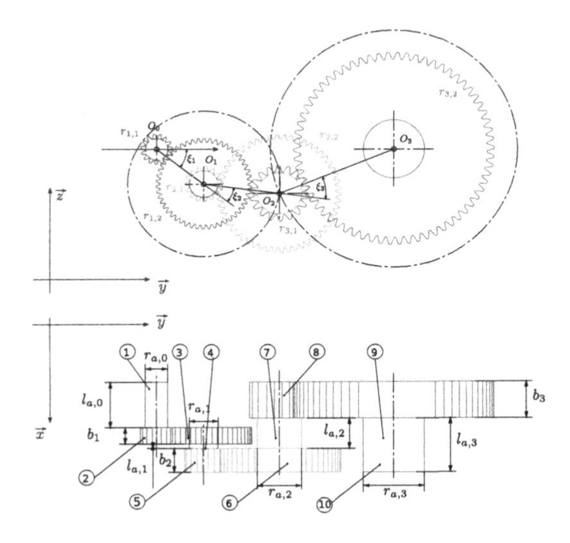

Fig. 2. Detailed view of the 3 stage reducer.

The 3SR problem is formulated with 2 objective functions, 41 constraints (presented in Appendix A), 3 categorical variables (gears modules), 6 integer variables (number of teeth), and 11 continuous variables. Gears modules have 41 possibilities, pinion number of teeth ranges from 14 to 30 and wheel number of teeth ranges from 14 to 150. Hence, the combinatorial space of the 3SR problem consists in $41^3 + (30 - 14)^3 + (150 - 14)^3 \simeq 8.7 \times 10^{14}$. Thus, the problem is considered a mid-sized problem concerning the number of variables and constraints, but, huge combinatorial space.

The additional constraints increase the complexity of the problem. This is noticed by solving the problem using NSGAII with different initial conditions as follows. In first hand, NSGAII is initialized with 1 feasible individual. On the other hand, NSGAII is randomly initialized. Each was run 10 times with the same parameters shown in Table 1. Figure 3(a) shows how many run each method converged to a feasible solution out of 10. Figures 3(a) and 3(b) show that if the initial population contains at least 1 feasible individual, NSGAII converges to a good approximated Pareto front every time. Whilst, if NSGAII is initialized with a random population, NSGAII either fails to converge to a feasible solution, or it converges to a low-quality Pareto front.

Table 1. Parameters used for NSGAII algorithm

Parameters	Value
Cross over probability	0.8
Mutation probability	0.9
Population size	200
Allowable generations	500
Constraint handling	Legacy method [1]
Crossover operator	Simulated binary crossover (SBX) [11]
ETAC	100
Mutation operator	Partially-mapped crossover (PMX) [11]
ETAM	10

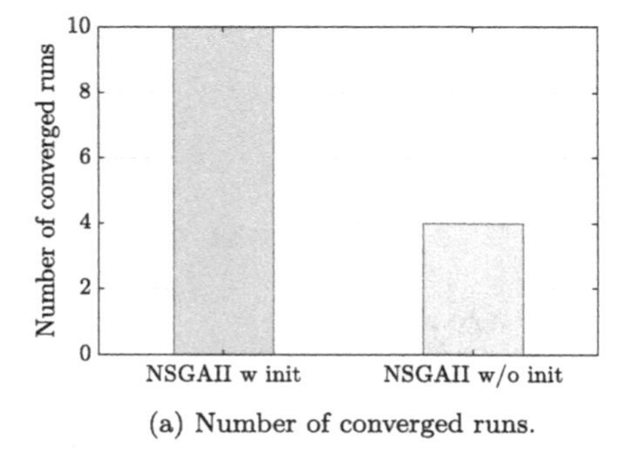

(a) Number of converged runs.

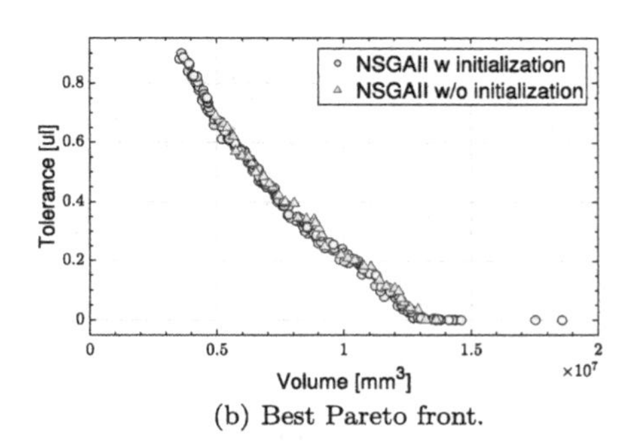

(b) Best Pareto front.

Fig. 3. Results of 3SR problem solved by NSGAII with (blue) and without (red) initial feasible seed. (Color figure online)

Figure 4 shows part of the domain of the 3SR problem explored by NSGAII with feasible initial population. The explored domain shows the complexity of the problem, where both feasible and infeasible solutions share the same domain on the projected objective domain. Moreover, all the feasible solutions are too near to the infeasible ones.

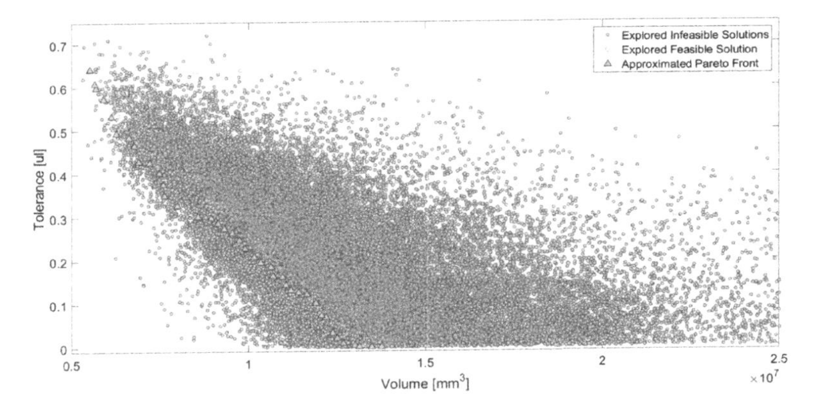

Fig. 4. Explored portion of the domain, showing the 3SR problem complexity.

To enhance the quality of the solution of this problem - and accordingly any similar problem - where feasible solutions are not known, our proposal is first to use BnB-NSGAII proposed in [10] to search for feasible individuals. These individuals are then injected in the random initial population of NSGAII.

3 BnB-NSGAII

In [10] the authors proposed the BnB-NSGAII approach. In this approach, Multi-Criteria Branch and Bound (MCBB) [13] is used to enhance the exploration force of NSGAII by investigating the mixed-integer domain space through branching it to subdomains, then NSGAII bounds each one. In this way, MCBB guides the search using the lower bounds obtained by NSGAII. Our proposal is to furthermore enhance the exploration potential of BnB-NSGAII by adding the legacy feature.

3.1 General Concept of BnB-NSGAII

The general multi-Objective MINLP problem ($\mathcal{P}_{\text{MO-MINLP}}$) is written as

$$
\begin{aligned}
\underset{\boldsymbol{x},\boldsymbol{y}}{\text{minimize}} \quad & \boldsymbol{f}(\boldsymbol{x},\boldsymbol{y}) = f_1(\boldsymbol{x},\boldsymbol{y}),\dots,f_p(\boldsymbol{x},\boldsymbol{y}) \\
\text{subject to} \quad & \\
& c_j(\boldsymbol{x},\boldsymbol{y}) \leq 0,\ j=1,\dots,m \\
& \boldsymbol{x} \in \boldsymbol{X}, \qquad \boldsymbol{X} \in \mathbb{R}^{n_c} \\
& \boldsymbol{y} \in \boldsymbol{Y}, \qquad \boldsymbol{Y} \in \mathbb{N}^{n_i},
\end{aligned} \tag{3}
$$

where p and m are the number of objectives and constraints respectively. $\boldsymbol{X}$ and $\boldsymbol{Y}$ denote the set of feasible solutions of the problem for n_c continuous and n_i integer variables respectively.

$\mathcal{P}_{\text{MO-MINLP}}$ is complex and expensive to solve. The general idea is thus to solve several simpler problems instead. BnB-NSGAII divides $\mathcal{P}_{\text{MO-MINLP}}$ by constructing a combinatorial tree that aim to partition the root node problem - $\mathcal{P}_{\text{MO-MINLP}}$ - into a finite number of subproblems $Pr_1, \dots, Pr_i, \dots, Pr_n$. Where i and n are the current node and the total number of nodes respectively. Each Pr_i is considered a node. Each node is then solved by NSGAII. Solving a node is to determine its lower and upper bounds. The upper bound of a node P_i^N is the Pareto front captured by NSGAII, which is then stored in an incumbent list P^N. Whilst the lower bound is the ideal point P_i^I of the current node.

$$P_i^I = \min f_k(\boldsymbol{x}_i, \boldsymbol{y}_i); \quad k = 1, \dots, p. \tag{4}$$

By solving Pr_i, one of the following is revealed:

- Pr_i is infeasible, means that NSGAII didn't find any solution that satisfies all constraints. Hence, Pr_i is pruned (*fathomed*) by *infeasibility.*
- Pr_i is feasible, but, the current lower bound P_i^I is dominated by a previously found upper bound P^N. Therefore, Pr_i is fathomed by *optimality.*
- Pr_i is feasible, and, P_i^I is not dominated by P^N, $P_i^I \leq P^N$. P^N is then updated by adding P_i^N to it.

In the 3rd case, the combinatorial tree is furtherly branched by dividing Pr_i into farther subproblems, called children nodes. If a node cannot be divided anymore, it is called a leaf node. Leaves are the simplest nodes, since all integer variables are fixed such that $\boldsymbol{y} = \bar{\boldsymbol{y}}$. NSGAII then solve leaves as Multi-Objective continuous Non-Linear problem ($\mathcal{P}_{\text{MO-NLP}}$):

$$\begin{aligned} \underset{\boldsymbol{x}, \bar{\boldsymbol{y}}}{\text{minimize}} \quad & \boldsymbol{f}(\boldsymbol{x}, \bar{\boldsymbol{y}}) = f_1(\boldsymbol{x}, \bar{\boldsymbol{y}}), \dots, f_p(\boldsymbol{x}, \bar{\boldsymbol{y}}) \\ \text{subject to} \quad & \\ & c_j(\boldsymbol{x}, \bar{\boldsymbol{y}}) \leq 0, j = 1, ..., m \\ & \boldsymbol{x} \in \boldsymbol{X}_i, \end{aligned} \tag{5}$$

where $\boldsymbol{X}_i$ denotes the set of feasible solutions of the current node. P_i^N of each leaf is then added to P^N. The overall Pareto front is obtained by removing the dominated elements from P^N.

3.2 BnB Legacy Feature

In NSGAII, the best population is that found in the last generation, since it contains the elite individuals among all the previous generations. In BnB-NSGAII, each node is solved independently. The output of each node is the captured Pareto front only. The last population in the node is thus discarded, although it might be valuable to other nodes.

We propose to permit the legacy between nodes. Where each child node inherits the last population from its parent node. The child node then initializes NSGAII by this population.

The children nodes are subproblems of their parent node. Thus, the boundary of parent node is different than that for the children nodes, $\boldsymbol{Y}_{parent} \neq \boldsymbol{Y}_{child}$. Hence, the population is rebounded before initializing NSGAII. Rebounding the population may lead to the loss of the elite individuals, though some of the elite genes are still conserved.

3.3 An Application of BnB-NSGAII

BnB-NSGAII is characterized by high exploration potential. Thus, in this paper, BnB-NSGAII is used to search for at least one feasible solution for the 3SR problem. For this aim, BnB-NSGAII is properly modified to 1) continue enumeration of the combinatorial tree even if the root node is infeasible. 2) stop whenever a feasible solution(s) is found. Then, NSGAII is called to solve the 3SR problem by initializing it with the feasible solution(s) found as shown in Fig. 5.

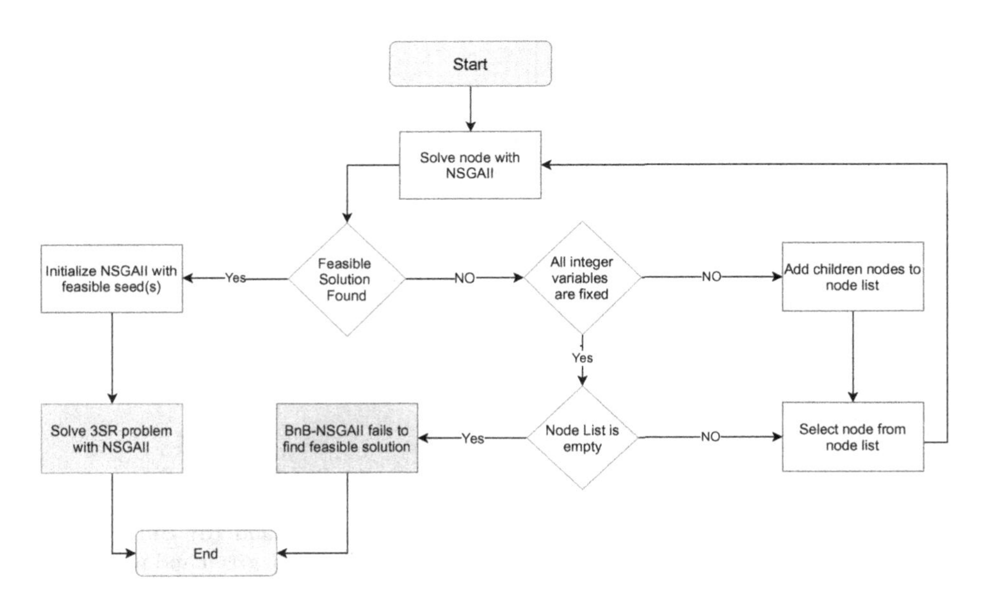

Fig. 5. Flowchart of BnB-NSGAII application.

4 Numerical Experiment

NSGAII and BnB-NSGAII with and without the legacy feature were tested on the 3SR problem. Each method was run 10 times. The test was done using the same parameters for the 3 solvers. Table 1 shows the parameters used in this experiment.

4.1 Results and Discussion

In this experiment, the evaluation of the performance of each method is limited to how many times the method finds at least 1 feasible solution over the 10 runs. Figure 6(a) shows the number of times each method succeeded the test. It can be obviously concluded that BnB-NSGAII legacy method overcomes the performances of NSGAII and BnB-NSGAII. It should be noted that the computational effort is not regarded since all the runs converge within 30 min. Which is considered an acceptable time for such a problem.

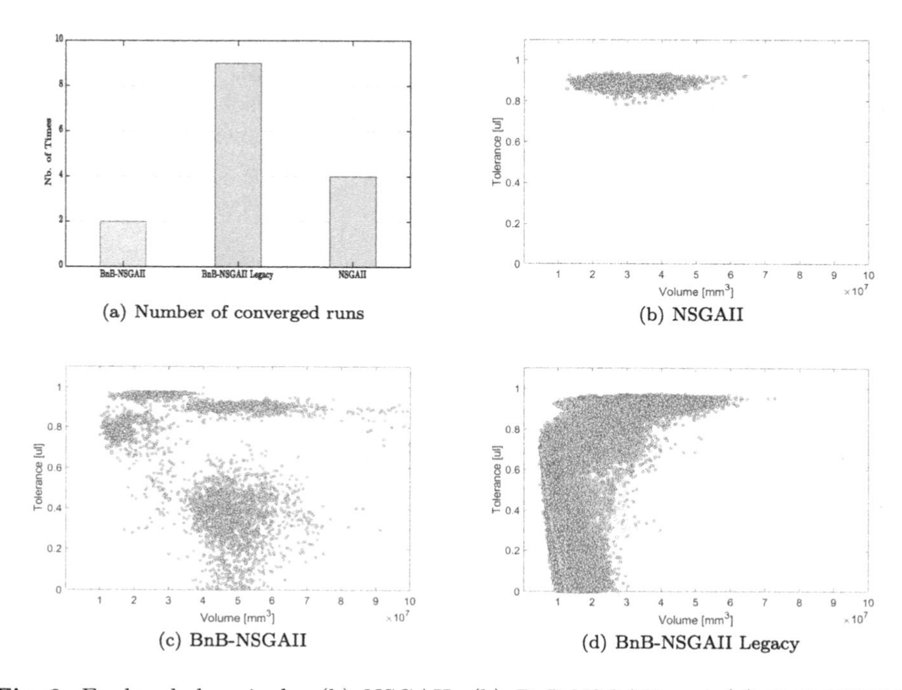

(a) Number of converged runs
(b) NSGAII
(c) BnB-NSGAII
(d) BnB-NSGAII Legacy

Fig. 6. Explored domain by (b) NSGAII, (b) BnB-NSGAII and (d) BnB-NSGAII legacy methods. Feasible and infeasible individuals are plotted in green and red respectively. (Color figure online)

Figure 6(b) shows that NSGAII explored local space of the domain depending on the initial population. While Fig. 6(c) shows that BnB-NSGAII explored random spaces of the domain. Figure 6(d) shows that the legacy feature guides the exploration force of BnB-NSGAII towards the feasible solutions.

5 Conclusion

The 3 stage reducer problem is a point of interest of many researchers, either to use/enhance it for engineering applications, or to examine the performance of optimization methods. The 3SR problem is desirable for such experiments for its complexity.

The 3SR problem was reformulated to a bi-objective problem in this paper to demonstrate a proposed application of BnB-NSGAII. The proposed application is to use BnB-NSGAII as an initiator of NSGAII, where BnB-NSGAII initially seeks feasible individuals before injecting them into the initial population of NSGAII.

BnB-NSGAII was enhanced by adding the legacy feature. The legacy feature is a generic feature that can be implemented in any branch and bound algorithm. Any parameter that is tuned during the node solving process could be the legacy. In this paper, the legacy was the last population in the father node in BnB-NSGAII. The latter was then used to initialize the child node.

The performances of NSGAII and BnB-NSGAII with and without the legacy feature were tested on the bi-objective version of the 3SR problem. Results show that the legacy feature guides the exploration force of BnB-NSGAII leading it to a better solution than that obtained by NSGAII and BnB-NSGAII.

A 3SR Problem Constraints

A.1 Closure Condition

Interference and fitting constraints are adopted from [5]. In [4], the closure condition was expressed with the distance between the terminal point O_3 shown in Fig. 2 and required position of the center of the output shaft. The coordinate of O_3 can be easily compute with the center distance of each stage and the angle ξ_1, ξ_2 and ξ_3. But, if we consider that center distance of each stage allow this closure condition, we can compute the value of ξ_2 and ξ_3. By this way can reduce he number of variables in the optimization problem.

For a given value of ξ_1 and $r_{1,1}$, $r_{1,2}$, center distance of each stage allow a closure if we have:

$$\|\vec{O_1O_3}\| \leq \|\vec{O_1O_2}\| + \|\vec{O_2O_3}\|$$

Assuming the previous condition is true, we can compute the two intersection of circle of center O_1 of $\|\vec{O_1O_2}\|$ radius and circle of center O_3 of $\|\vec{O_2O_3}\|$ radius.

With $a_2 = \|\vec{O_1O_2}\|$ and $a_3 = \|\vec{O_2O_3}\|$ we have:

$$\begin{cases} a_2 \sin\alpha_1 - a_3 \sin\alpha_3 = 0 \\ a_2 \cos\alpha_1 + a_3 \cos\alpha_3 = \|\vec{O_1O_3}\| \end{cases}$$

which give:

$$\cos\alpha_1 = \frac{\overline{O_1H}}{\overline{O_1O_2}} = \frac{a_2^2 - a_3^2 + (O_1O_3)^2}{2a_2\overline{O_1O_3}}$$

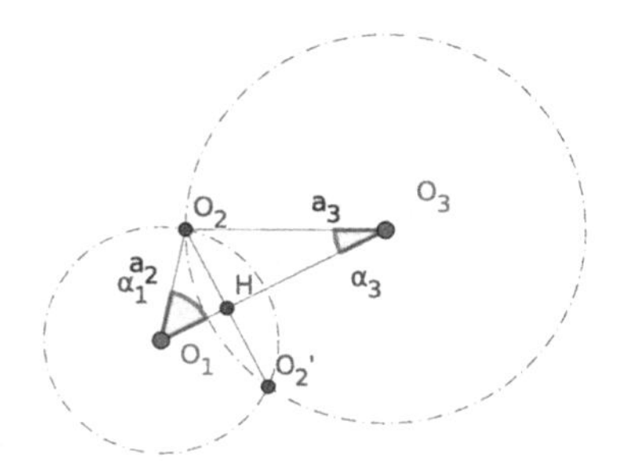

Fig. 7. Gear mesh for each stage.

Knowing α_1, computation of coordinate of O_2 and $O_2^{'}$ is straightforward. If the two position O_2 and $O_2^{'}$ allow the wheel of the 2nd stage to fit in the casing box, then $O_2^{'}$ is preferred for lubrication reason (Fig. 7).

A.2 Mechanical Constraint for One Stage of the Mechanism

Constraints Related to the Gear Pair. Following the recommendation from International Standard ISO 6336, [6–8] we can calculate, knowing the geometry of gear pair, the material and the working conditions the contact and σ_{H} the bending stress σ_{F} in the gear pair. These stresses must be less of equal to the respective permissible value σ_{HP} and σ_{FP}, depending on the material and the working conditions.

From [8] the bending stress σ_{F} is given by (1 for the pinion and 2 for the wheel):

$$\sigma_{\mathrm{F}(1,2)} = \sigma_{\mathrm{F0}} \left(K_{\mathrm{A}} K_{\mathrm{V}} K_{\mathrm{F}\alpha} K_{\mathrm{F}\beta}\right)$$

with $\sigma_{\mathrm{F0}(1,2)}$, the nominal tooth stress:

$$\sigma_{\mathrm{F0}(1,2)} = \frac{F_t}{b m_{\mathrm{n}}} \left(Y_{\mathrm{F}} Y_{\mathrm{S}} Y_{\beta} Y_{\mathrm{B}} Y_{\mathrm{DT}}\right)$$

where:

- F_t: is the tangential load from [6].
- b: is the facewidth.
- m_{n}: is the normal module.

Factors K_{A}, K_{V}, $K_{\mathrm{F}\alpha}$, $K_{\mathrm{F}\beta}$ are related to dynamic ad loading conditions in the gear. Factors Y_{F}, Y_{S}, Y_{β}, Y_{B}, Y_{DT} are related to the geometry effect on stress.

From [8], the permissible bending stress σ_{FP} is given by:

$$\sigma_{\mathrm{FP}} = \frac{\sigma_{\mathrm{FLim}}}{S_{\mathrm{Fmin}}} \left(Y_{\mathrm{ST}} Y_{\mathrm{NT}} Y_{\delta\mathrm{relT}} Y_{R\mathrm{relT}} Y_{\mathrm{X}}\right)$$

with σ_{FLim} is the nominal stress number (bending) from reference test gears [9] and S_{Fmin} the minimal required safety factor. Factors Y_{ST}, Y_{NT}, $Y_{\delta\mathrm{relT}}$, $Y_{R\mathrm{relT}}$,

Y_X are related to the reference test gears and the geometry and material conditions of the gear pair.

From [7] the contact stress is given by (1 for the pinion and 2 for the wheel):

$$\sigma_{H(1,2)} = Z_{(B,D)}\sigma_{H0}\sqrt{K_A K_V K_{H\alpha} K_{H\beta}}$$

with σ_{H0} is the nominal contact stress:

$$\sigma_{H0} = (Z_H Z_E Z_\varepsilon Z_\beta)\sqrt{\frac{F_t}{bd_1}\frac{u+1}{u}}$$

Factors Z_H, Z_E, Z_ε, Z_β are related to the Hertzian theory of contact, and take into account geometry and material in the gear pair.

From [7] the permissible contact stress σ_{HP} is:

$$\sigma_{HP} = \frac{\sigma_{HLim}}{S_{Hmin}}(Z_{NT} Z_L Z_V Z_R Z_W Z_X)$$

with σ_{HLim} is the allowable contact stress number and S_{Hmin} is the minimum required safety factor for surface durability. Factors Z_{NT}, Z_L, Z_V, Z_R, Z_W, Z_X are related to lubrication conditions, surface roughness and hardened conditions and size of the tooth.

So to respect the requirement specification of a given power to be transmitted, the gear pair must respect:

$$\sigma_{F(1,2)} \leq \sigma_{FP}$$
$$\sigma_{H(1,2)} \leq \sigma_{HP}$$

Considering that σ_F is proportional to F_t and σ_H is proportional to $\sqrt{F_t}$ for a given gear pair, we can rewrite these 2 conditions with P_t the power to be transmitted:

$$\frac{\sigma_{FP}}{\sigma_{F(1,2)}} P_t \geq P_t$$
$$\left(\frac{\sigma_{HP}}{\sigma_{H(1,2)}}\right)^2 P_t \geq P_t$$

Usually, some factors are slightly for the pinion and the wheel so transmitted power is different for the pinion (1) and the wheel (2). We will keep the minimal value.

So finally, for the stage number s on the reducer, the following conditions must be fulfilled:

$$\min\left(\frac{\sigma_{FPs}}{\sigma_{F(1,2)s}}\right) P_t \geq P_t \tag{6}$$

$$\min\left(\frac{\sigma_{HPs}}{\sigma_{H(1,2)s}}\right)^2 P_t \geq P_t \tag{7}$$

Following condition must be respected:

- For the transverse contact ratio: $\varepsilon_\alpha \geq 1.3$.
- For the minimal face width: $b \geq 0.1 d_2$
- For the maximal face width: $b \leq d_1$

In order to use pinion with at least $Z_{\min} = 14$ teeth, the value of the profile shift coefficient must be adjusted to avoid gear meshing with the relation:

$$Z_{\min} \geq \frac{2(1-x_1)}{\sin \alpha_{\mathrm{n}}^2} \Rightarrow x_1 \geq 1 - Z_{\min} \frac{\sin \alpha_{\mathrm{n}}^2}{2} \Rightarrow x_1 \geq 0.1812$$

Constraint Related to Shaft's Reducer. In each of the 4 shafts of the mechanism, the transmitted torque produce shear stress. This stress must not exceed the allowable shear of the material of shafts $\tau_{\max}$. We assume here that all the shaft are using the same steel and that all shaft can be consider as beam. So, with $r_{a,0}$, the radius of input shaft, and $r_{a,s}$, $s = 1 \dots 3$ the radius of output shaft of the three stages, we have:

$$\tau_s = \frac{2C_s}{\pi {r_{a,s}}^3} \leq \tau_{\max} \text{ for } s = 1 \dots 3 \tag{8}$$

C_s is the output torque of each stage and C_e the torque on the input shaft, where $Z_{i,1}$ and $Z_{i,2}$ are the number of teeth for pinion ($_1$) and wheel ($_2$) of stage number i:

$$C_s = C_e \prod_{i=1}^{i=s} \frac{Z_{i,2}}{Z_{i,1}}$$

For the input shaft we have:

$$\tau_0 = \frac{2C_e}{\pi {r_{a,0}}^3} \leq \tau_{\max} \tag{9}$$

The total rotation angle between the initial section of the input shaft and the final section of the output shaft is:

$$\theta = \frac{2C_e l_{a,0}}{G\pi {r_{a,0}}^3} + \sum_{s=1}^{s=3} \frac{2C_s l_{a,s}}{G\pi {r_{a,s}}^3}$$

For some reasons (dynamic behaviour of the reducer, ...) this total rotation angle should be limited by a maximal value $\theta_{\max}$.

$$\theta \leq \theta_{\max} \tag{10}$$

References

1. Deb, K., Pratap, A., Agarwal, S., Meyarivan, T.: A fast and elitist multiobjective genetic algorithm: NSGA-II. IEEE Trans. Evol. Comput. **6**(2), 182–197 (2002). https://doi.org/10.1109/4235.996017. http://ieeexplore.ieee.org/document/996017/

2. Deng, Y., Liu, Y., Zhou, D.: An improved genetic algorithm with initial population strategy for symmetric TSP. Math. Probl. Eng. **2015**, e212794 (2015). https://doi.org/10.1155/2015/212794. https://www.hindawi.com/journals/mpe/2015/212794/. ISSN 1024-123X. Hindawi
3. Fauroux, J.C.: Conception optimale de structures cinématiques tridimensionnelles. Application aux mécanismes de transmission en rotation. Ph.D. thesis, INSA de Toulouse, January 1999. https://tel.archives-ouvertes.fr/tel-00006691
4. Fauroux, J.C., Lafon, P.: Optimization of a multi-stage transmission mechanism. IFToMM **2**, 813–817 (2004)
5. Han, L., Liu, G., Yang, X., Han, B.: Dimensional and layout optimization design of multistage gear drives using genetic algorithms. Math. Probl. Eng. **2020**, 3197395 (2020). https://doi.org/10.1155/2020/3197395. Hindawi
6. International Organization for Standardization (Ginebra): ISO 6336-1: Calculation of Load Capacity of Spur and Helical Gears. Part 1. ISO (2019)
7. International Organization for Standardization (Ginebra): ISO 6336-2: Calculation of Load Capacity of Spur and Helical Gears. Part 2. ISO (2019)
8. International Organization for Standardization (Ginebra): ISO 6336-3: Calculation of Load Capacity of Spur and Helical Gears. Part 3. ISO (2019)
9. International Organization for Standardization (Ginebra): ISO 6336-5: Calculation of Load Capacity of Spur and Helical Gears. Part 5. ISO (2019)
10. Jaber, A., Lafon, P., Younes, R.: A branch-and-bound algorithm based on NSGAII for multi-objective mixed integer nonlinear optimization problems. Eng. Optim. 1–19 (2021). https://doi.org/10.1080/0305215X.2021.1904918. Taylor & Francis
11. Maruyama, S., Tatsukawa, T.: A parametric study of crossover operators in pareto-based multiobjective evolutionary algorithm. In: Tan, Y., Takagi, H., Shi, Y., Niu, B. (eds.) ICSI 2017. LNCS, vol. 10386, pp. 3–14. Springer, Cham (2017). https://doi.org/10.1007/978-3-319-61833-3_1
12. Mavrotas, G., Diakoulaki, D.: A branch and bound algorithm for mixed zero-one multiple objective linear programming. Eur. J. Oper. Res. **107**(3), 530–541 (1998). https://doi.org/10.1016/S0377-2217(97)00077-5
13. Mavrotas, G., Diakoulaki, D.: Multi-criteria branch and bound: a vector maximization algorithm for mixed 0–1 multiple objective linear programming. Appl. Math. Comput. **171**(1), 53–71 (2005). https://doi.org/10.1016/j.amc.2005.01.038
14. Yvars, P.A., Zimmer, L.: System sizing with a model-based approach: application to the optimization of a power transmission system. Math. Probl. Eng. **2018**, 1–14 (2018). https://doi.org/10.1155/2018/6861429

The Horizontal Linear Complementarity Problem and Robustness of the Related Matrix Classes

Milan Hladík[1(✉)] and Miroslav Rada[2]

[1] Faculty of Mathematics and Physics, Department of Applied Mathematics, Charles University, Malostranské nám. 25, 11800 Prague, Czech Republic
hladik@kam.mff.cuni.cz

[2] Faculty of Informatics and Statistics, Department of Econometrics, University of Economics, W. Churchill's Sq. 4, 130 67 Prague, Czech Republic
miroslav.rada@vse.cz
https://kam.mff.cuni.cz/~hladik

Abstract. We consider the horizontal linear complementarity problem and we assume that the input data have the form of intervals, representing the range of possible values. For the classical linear complementarity problem, there are known various matrix classes that identify interesting properties of the problem (such as solvability, uniqueness, convexity, finite number of solutions or boundedness). Our aim is to characterize the robust version of these properties, that is, to check them for all possible realizations of interval data. We address successively the following matrix classes: nonnegative matrices, Z-matrices, semimonotone matrices, column sufficient matrices, principally nondegenerate matrices, R_0-matrices and R-matrices. The reduction of the horizontal linear complementarity problem to the classical one, however, brings complicated dependencies between interval parameters, resulting in some cases to higher computational complexity.

Keywords: Linear complementarity · Interval analysis · Special matrices · NP-hardness

1 Introduction

The Linear Complementarity Problem (LCP). The classical LCP problem is a feasibility problem

$$y = Mz + q, \quad y, z \geq 0, \tag{1a}$$

$$y^T z = 0, \tag{1b}$$

Supported by the Czech Science Foundation Grants P403-18-04735S (M. Hladík) and P403-20-17529S (M. Rada).

B. Dorronsoro et al. (Eds.): OLA 2021, CCIS 1443, pp. 350–360, 2021.
https://doi.org/10.1007/978-3-030-85672-4_26

where $M \in \mathbb{R}^{n\times n}$ and $q \in \mathbb{R}^n$ are given and $y, z \in \mathbb{R}^n$ are variables. Condition (1a) is linear, but the (nonlinear) complementarity condition (1b) makes the problem NP-hard [2]. The LCP is called *feasible* if (1a) is feasible, and it is called *solvable* if (1a)–(1b) is feasible. The LCP appears in many optimization and operations research models such as quadratic programming, bimatrix games, or equilibria in specific economies. For more properties and algorithms for LCP see, e.g., the books [4,21].

The Horizontal Linear Complementarity Problem. The horizontal LCP [3] is a slight generalization of LCP, first formulated by Samelson et al. [25]. It reads

$$Ay = Bz + q, \quad y, z \geq 0, \tag{2a}$$

$$y^T z = 0, \tag{2b}$$

where $A, B \in \mathbb{R}^{n\times n}$ and $q \in \mathbb{R}^n$ are given and $y, z \in \mathbb{R}^n$ are variables. Clearly, provided A is nonsingular, we easily reduce the problem to LCP by multiplying A^{-1}

$$y = A^{-1}Bz + A^{-1}q, \quad y, z \geq 0, \tag{3a}$$

$$y^T z = 0, \tag{3b}$$

Otherwise, a reduction is possible only under certain conditions [7,28]. In our context, it is better to consider the horizontal form separately since the form $A^{-1}B$ brings complicated correlations into the matrix. Notice that for the horizontal and other extended forms of LCP, certain LCP-related matrix classes were generalized [26]. For the horizontal LCP, also special algorithms were developed [18,19,29].

Interval Uncertainty. Properties of the solution sets of LCP and horizontal LCP relate with properties of the constraint matrices. In this paper, we study the situation when the matrix entries are not precisely known, but we have interval type uncertainty. A justifications of using intervals for modelling uncertainty is provided in [14,15] and many books, e.g., [20,22].

Formally, an interval matrix is a set

$$\boldsymbol{A} := \{A \in \mathbb{R}^{m\times n};\ \underline{A} \leq A \leq \overline{A}\},$$

where $\underline{A}, \overline{A} \in \mathbb{R}^{m\times n}$, $\underline{A} \leq \overline{A}$, are given matrices and the inequality is understood entrywise. The corresponding midpoint and radius matrices are defined as

$$A_c := \frac{1}{2}(\underline{A} + \overline{A}), \quad A_\Delta := \frac{1}{2}(\overline{A} - \underline{A}).$$

The set of all interval $m\times n$ matrices is denoted by $\mathbb{IR}^{m\times n}$. Following the notation by Fiedler et al. [5], we introduce special matrices in $\boldsymbol{A}$. Given sign vectors $s \in \{\pm 1\}^m$ and $t \in \{\pm 1\}^n$, denote

$$A_{s,t} = A_c - D_s A_\Delta D_t \in \boldsymbol{A},$$

where D_s stands for the diagonal matrix with entries $s_1, \ldots, s_m$ and similarly for D_t. For more results and properties of interval computation, including interval arithmetic, we refer the readers, e.g., to books [20,22].

The LCP with interval uncertainties was investigated in [1,17], among others. They addressed the problem of computing an outer approximation of the solution set of all possible realizations of interval entries. Our goal is different – in this paper, we focus on the interval matrix properties that are related to the LCP.

Problem Statement. Let $\boldsymbol{A}, \boldsymbol{B} \in \mathbb{IR}^{n\times n}$ be given. We consider a class of the horizontal LCP problems with $A \in \boldsymbol{A}$ and $B \in \boldsymbol{B}$. Let $\mathcal{P}$ be a matrix property related to the (horizontal) LCP. We say that $\mathcal{P}$ holds *strongly* if it holds for each $A \in \boldsymbol{A}$ and $B \in \boldsymbol{B}$.

Our aim is to characterize strong versions of several fundamental matrix classes appearing in the context of the (horizontal) LCP. If property $\mathcal{P}$ holds strongly for an interval matrix $\boldsymbol{A}$, then we are sure that $\mathcal{P}$ is provably valid whatever are the true values of the uncertain entries. Therefore, the property holds in a robust sense for the (horizontal) LCP problem.

Notation. We use the shortage $[n] = \{1, \ldots, n\}$. Given a matrix $M \in \mathbb{R}^{n\times n}$ and index sets $I, J \subseteq [n]$, $M_{I,J}$ denotes the restriction of M to the rows indexed by I and the columns indexed by J; it is the empty matrix if I or J is empty. Similarly x_I denotes the restriction of a vector x to the entries indexed by I.

The identity matrix of size n is denoted by I_n, and the spectral radius of a matrix M by $\rho(M)$. The symbol D_s stands for the diagonal matrix with entries $s_1, \ldots, s_n$ and $e = (1, \ldots, 1)^T$ for the vector of ones. The relation $x \gneqq y$ between vectors x, y is defined as $x \geq y$ and $x \neq y$. Inequalities and the absolute value of matrices and vectors are understood entrywise.

2 Particular Matrix Classes

In the following sections, we consider important classes of matrices appearing in the context of the (horizontal) LCP. We characterize their strong counterparts when entries are interval valued. Other matrix properties were discussed, e.g., in [6,11–13,16] and in the context of LCP in Hladík [10].

Basically, due to reduction (3), we will tackle the matrix $A^{-1}B$ for $A \in \boldsymbol{A}$ and $B \in \boldsymbol{B}$. The expression $A^{-1}B$ imposes complicated dependencies between interval parameters, so that is why we have to deal with them carefully. A simple evaluation by interval arithmetic (and estimation of the inverse) leads to (possibly high) overestimation. For the sake of simplicity of exposition, we denote by $\boldsymbol{A}^{-1}\boldsymbol{B}$ the set

$$\{A^{-1}B;\ A \in \boldsymbol{A},\ B \in \boldsymbol{B}\}.$$

Throughout the paper we assume that $\boldsymbol{A}$ is strongly nonsingular. For characterization of nonsingularity and sufficient conditions see [23,24], for instance.

2.1 S-matrices

A matrix $M \in \mathbb{R}^{n\times n}$ is an *S-matrix* if there is $x > 0$ such that $Mx > 0$. The importance of this class is that the LCP is feasible for each $q \in \mathbb{R}^n$ if and only if M is an S-matrix.

Proposition 1. *We have that $\boldsymbol{A}^{-1}\boldsymbol{B}$ is strongly S-matrix if and only if the system*

$$A_{s,e}z = B_{-s,e}x,\ x > 0,\ z > 0. \tag{4}$$

is feasible for each $s \in \{\pm 1\}^n$.

Proof. We want to characterize feasibility of

$$(A^{-1}B)x > 0,\ x > 0$$

for each $A \in \boldsymbol{A}$ and $B \in \boldsymbol{B}$. Substitute $y \equiv Bx$ and $z \equiv A^{-1}y$. Then $Az = y = Bx$ and we reduced the problem to strong solvability of the interval system

$$\boldsymbol{A}z = \boldsymbol{B}x,\ x > 0,\ z > 0. \tag{5}$$

By [5,8], we obtain (4). □

The problem of checking strong S-matrix property is computationally intractable, which justifies the exponential formula 4.

Proposition 2. *Checking strong S-matrix property of $\boldsymbol{A}^{-1}\boldsymbol{B}$ is co-NP-hard even in the case when A is real.*

Proof. By [5], checking solvability of the system

$$|Mx| \le e,\ e^T|x| \ge 1 \tag{6}$$

is NP-hard even on a class of problems with M nonnegative positive definite. This is equivalent to weak solvability (i.e., solvability for at least one realization) of the interval system

$$|Mx| \le e,\ [-e,e]^T x \ge 1,$$

or to weak solvability of

$$Mx \le ey,\ -Mx \le ey,\ [-e,e]^T x \ge y,\ y > 0.$$

By Farkas' lemma, it is equivalent to the situation that the interval system

$$\begin{pmatrix} -M & M & [-e,e] \\ e^T & e^T & -1 \end{pmatrix} \begin{pmatrix} u \\ v \\ w \end{pmatrix} = \begin{pmatrix} 0 \\ -1 \end{pmatrix},\quad u,v,w \ge 0 \tag{7}$$

is not strongly solvable. We claim that strong solvability of this system is equivalent to strong solvability of

$$\begin{pmatrix} 0 & -M & M & [-e,e] \\ 1 & e^T & e^T & -1 \end{pmatrix} \begin{pmatrix} z \\ u \\ v \\ w \end{pmatrix} = \begin{pmatrix} 0 \\ 0 \end{pmatrix}, \quad z,u,v,w > 0. \tag{8}$$

If (8) has a solution (z, u, v, w), then $\frac{1}{z}(u, v, w)$ solves (7). Conversely, let (u, v, w) be a solution of (7). Situation $w = 0$ cannot happen since otherwise the second equation is violated. Thus we have $w > 0$. If $u, v > 0$, then we put $z := 1$ and (z, u, v, w) solves (8). Otherwise we put $u := u + e$, $v := v + e$ and $z := 2n + 1$ and (z, u, v, w) solves (8).

Eventually, we obtained the interval system in the form of (5), where

$$A = \begin{pmatrix} 0 & -M \\ 1 & e^T \end{pmatrix}, \quad \boldsymbol{B} = \begin{pmatrix} -M & [-e,e] \\ -e^T & 1 \end{pmatrix}.$$

Obviously, A is nonsingular. □

2.2 Nonnegative Matrices

Nonnegative matrices are important in the LCP since they represent an efficiently recognizable subclass of copositive matrices.

Proposition 3. *We have that $\boldsymbol{A}^{-1}\boldsymbol{B}$ is strongly nonnegative if and only if the system*

$$A_{s,e}X = B_{-s,e}, \quad X \geq 0 \tag{9}$$

is feasible for each $s \in \{\pm 1\}^n$.

Proof. We need to characterize feasibility of

$$AX = B, \quad X \geq 0$$

for each $A \in \boldsymbol{A}$ and $B \in \boldsymbol{B}$. Thus we arrived at strong solvability of the interval matrix system. Fortunately, for strong solvability, the fact that the system is a matrix equation system makes no harm and we can simply call the characterization from [8], producing (9). □

It is an open problem if checking strong nonnegativity is intractable; we suspect it is.

2.3 *Z*-matrices

A matrix $M \in \mathbb{R}^{n \times n}$ is called a *Z-matrix* if $m_{ij} \leq 0$ for each $i \neq j$. Z-matrices emerge in the context of Lemke's complementary pivot algorithm, because it processes any LCP with a Z-matrix.

Recall that D_y denotes the diagonal matrix with entries $y_1, \ldots, y_n$, which in the following proposition play the role of variables.

Proposition 4. *We have that $\boldsymbol{A}^{-1}\boldsymbol{B}$ is strongly a Z-matrix if and only if the system*

$$A_{s,e}X - A_{-s,e}D_y = B_{s,e}, \quad X \leq 0, \ y \leq 0 \tag{10}$$

is feasible for each $s \in \{\pm 1\}^n$.

Proof. We need to characterize feasibility of

$$AX = B, \quad X_{ij} \leq 0 \ i \neq j$$

for each $A \in \boldsymbol{A}$ and $B \in \boldsymbol{B}$. We express the diagonal of X as a difference of two nonpositive variables, which is an equivalent operation in view of [9]. Thus in matrix form we have $X \mapsto X - D_y$, where $X \leq 0$ and $y \leq 0$. By [8] strong feasibility of this system is equivalent to feasibility of (10) for each $s \in \{\pm 1\}^n$. □

2.4 Semimonotone Matrices

A matrix $M \in \mathbb{R}^{n \times n}$ is *semimonotone* if for each $x \gneqq 0$ there is k such that $x_k > 0$ and $(Mx)_k \geq 0$. By [4], we can state two equivalent conditions of semimonotonicity. First, the LCP has a unique solution for each $q > 0$. Second, for each index set $\emptyset \neq I \subseteq [n]$ the system

$$M_{I,I}x < 0, \quad x \geq 0 \tag{11}$$

is infeasible. From the computational complexity perspective, checking whether M is semimonotone is a co-NP-hard problem [27].

Proposition 5. *We have that $\boldsymbol{A}^{-1}\boldsymbol{B}$ is strongly semimonotone if and only if for each index set $\emptyset \neq I \subseteq [n]$ the system*

$$\overline{A}_{[n],I}z_I + (A_{[n],J})_{-e,s}z_J \leq \overline{B}_{[n],I}x, \tag{12a}$$

$$\underline{A}_{[n],I}z_I + (A_{[n],J})_{e,s}z_J \geq \underline{B}_{[n],I}x, \tag{12b}$$

$$D_s z_J \leq 0, \ z_I < 0, \ x \geq 0 \tag{12c}$$

is infeasible for each $s \in \{\pm 1\}^{|J|}$, where $J = [n] \setminus I$.

Proof. Let $\emptyset \neq I \subseteq [n]$. We need to characterize infeasibility of

$$(A^{-1}B)_{I,I}x < 0, \quad x \geq 0. \tag{13}$$

Substitute $y \equiv B_{[n],I}x$ and $z \equiv A^{-1}y$. Then $Az = y = B_{[n],I}x$. Since $(A^{-1}B)_{I,I} = A^{-1}_{I,[n]}B_{[n],I}$, we can equivalently write (13) as follows

$$Az = B_{[n],I}x, \ z_I < 0, \ x \geq 0.$$

Since this system should be infeasible for each $A \in \boldsymbol{A}$ and $B \in \boldsymbol{B}$, we obtain by [5,8] the characterization (12). □

2.5 Principally Nondegenerate Matrices

A matrix $M \in \mathbb{R}^{n\times n}$ is *principally nondegenerate* if all its principal minors are nonzero. A principally nondegenerate matrix implies that the problem has finitely many solutions (including zero) for every $q \in \mathbb{R}^n$.

Proposition 6. *We have that $\boldsymbol{A}^{-1}\boldsymbol{B}$ is strongly principally nondegenerate if and only if for each index set $\emptyset \neq I \subseteq [n]$ and $s \in \{\pm 1\}^n$ the system*

$$(A_{[n],I})_{s,e} + (A_{[n],J})_{s,e}Z^1 - (A_{[n],J})_{-s,e}Z^2 = (B_{[n],I})_{s,-e}X^1 - (B_{[n],I})_{s,-e}X^2, \tag{14a}$$

$$Z^1, Z^2, X^1, X^2 \geq 0. \tag{14b}$$

is feasible, where $J = [n] \setminus I$.

Proof. Let $\emptyset \neq I \subseteq [n]$ and $k = |I|$. We need to characterize regularity of $(\boldsymbol{A}^{-1}\boldsymbol{B})_{I,I}$. For any particular instance, the system

$$A^{-1}_{I,[n]}B_{[n],I}X = I_k$$

should be feasible. Substitute $Y \equiv B_{[n],I}X$ and $Z \equiv A^{-1}_{J,[n]}Y$. Then $A_{[n],I} + A_{[n],J}Z = Y = B_{[n],I}X$. Thus we arrive at strong solvability of the interval matrix system

$$\boldsymbol{A}_{[n],I} + \boldsymbol{A}_{[n],J}Z = \boldsymbol{B}_{[n],I}X.$$

By [5,8], we obtain the characterization (14). □

2.6 Column Sufficient Matrices

A matrix $M \in \mathbb{R}^{n\times n}$ is *column sufficient* if for every $x \in \mathbb{R}^n$

$$[x_i(Mx)_i \leq 0 \ \forall i] \quad \Rightarrow \quad [x_i(Mx)_i = 0 \ \forall i].$$

Equivalently, by [4], for each pair of disjoint index sets $I, J \subseteq [n]$, $I \cup J \neq \emptyset$, the system

$$\begin{pmatrix} M_{I,I} & -M_{I,J} \\ -M_{J,I} & M_{J,J} \end{pmatrix} x \lneqq 0, \quad x > 0 \tag{15}$$

is infeasible. Notice that the above constraint matrix reduces to $M_{J,J}$ when $I = \emptyset$, and similarly it reduces to $A_{I,I}$ when $J = \emptyset$.

It is known [27] that checking column sufficiency is a co-NP-hard problem, which justifies necessity of inspecting all index sets I, J in (15). In the context of LCP, column sufficiency guarantees that for any $q \in \mathbb{R}^n$ the solution set of the LCP is a convex set (including possibly the empty set).

Proposition 7. *We have that $\boldsymbol{A}^{-1}\boldsymbol{B}$ is strongly column sufficient if and only if the system*

$$\overline{A}_{[n],I}z_I + \underline{A}_{[n],J}z_J \leq \overline{B}_{[n],I}x_I - \underline{B}_{[n],J}x_J, \tag{16a}$$
$$\underline{A}_{[n],I}z_I + \overline{A}_{[n],J}z_J \geq \underline{B}_{[n],I}x_I - \overline{B}_{[n],J}x_J, \tag{16b}$$
$$z_I \leq 0, z_J \geq 0, \ z \neq 0, \ x > 0. \tag{16c}$$

is infeasible for each admissible I, J.

Proof. Let admissible I, J be given. We want to characterize infeasibility of system

$$\begin{pmatrix} (A^{-1}B)_{I,I} & -(A^{-1}B)_{I,J} \\ -(A^{-1}B)_{J,I} & (A^{-1}B)_{J,J} \end{pmatrix} \begin{pmatrix} x_I \\ x_J \end{pmatrix} \lneqq 0, \quad x > 0.$$

Substitute

$$y \equiv B_{[n],I}x_I - B_{[n],J}x_J, \quad z \equiv A^{-1}y.$$

Then $Az = y = B_{[n],I}x_I - B_{[n],J}x_J$, so one can write the system as follows

$$Az = B_{[n],I}x_I - B_{[n],J}x_J, \quad z_I \leq 0, z_J \geq 0, \ z \neq 0, \ x > 0.$$

By means of [5,8], infeasibility of this system for each $A \in \boldsymbol{A}$ and $B \in \boldsymbol{B}$ is characterized by (16). □

Notice that system (14) can easily be expressed as a system of linear inequalities. So checking its feasibility is a tractable problem by means of linear programming (for fixed I, J).

2.7 R_0-matrices

A matrix $M \in \mathbb{R}^{n\times n}$ is an R_0*-matrix* if the LCP with $q = 0$ has only the trivial solution $y = z = 0$. Equivalently, for each index set $\emptyset \neq I \subseteq [n]$, the system

$$A_{I,I}x = 0, \quad A_{J,I}x \geq 0, \quad x > 0 \tag{17}$$

is infeasible, where $J = [n] \setminus I$. The decision problem of a given matrix to be a R_0-matrix is a co-NP-hard [27] problem. If M is an R_0-matrix, then for any $q \in \mathbb{R}^n$ the LCP has a bounded solution set.

Proposition 8. *We have that $\boldsymbol{A}^{-1}\boldsymbol{B}$ is strongly R_0-matrix if and only if system*

$$\underline{A}_{[n],J}z_J \leq \overline{B}_{[n],I}x_I, \tag{18a}$$
$$\overline{A}_{[n],J}z_J \geq \underline{B}_{[n],I}x_I, \tag{18b}$$
$$z_J \geq 0, \ x > 0 \tag{18c}$$

is infeasible for each admissible I, J.

Proof. Let admissible I, J be given. We want to characterize infeasibility of system

$$(A^{-1}B)_{I,I}x = 0, \quad (A^{-1}B)_{J,I}x \geq 0, \quad x > 0.$$

Substitute $y \equiv B_{[n],I}x_I$ and $z \equiv A^{-1}y$. Then $Az = y = B_{[n],I}x_I$ and the system reads

$$Az = B_{[n],I}x_I, \quad z_I = 0, \ z_J \geq 0, \ x > 0$$

or

$$A_{[n],J}z_J = B_{[n],I}x_I, \quad z_J \geq 0, \ x > 0.$$

This system is infeasible for each $A \in \boldsymbol{A}$ and $B \in \boldsymbol{B}$ if and only if system (18) is infeasible; see [5,8]. □

2.8 *R*-matrices

A matrix $M \in \mathbb{R}^{n\times n}$ is an *R-matrix* if for each index set $\emptyset \neq I \subseteq [n]$, the system

$$M_{I,I}x + et = 0, \quad M_{J,I}x + et \geq 0, \quad x > 0, \ t \geq 0 \tag{19}$$

is infeasible w.r.t. variables $x \in R^{|I|}$ and $t \in \mathbb{R}$, where $J = [n] \setminus I$. In the context of the LCP, when M is an R-matrix, then for any $q \in \mathbb{R}^n$ the LCP has a solution.

Despite the (visual) similarity with R_0-matrix, the R-matrix property is much harder in the interval setting and for particular index sets I, J.

Proposition 9. *We have that $\boldsymbol{A}^{-1}\boldsymbol{B}$ is strongly R_0-matrix if and only if system*

$$-\overline{A}_{[n],I}et + (\boldsymbol{A}_{[n],J})_{e,s}z_J \leq \overline{B}_{[n],I}x_I, \tag{20a}$$

$$-\underline{A}_{[n],I}et + (\boldsymbol{A}_{[n],J})_{e,-s}z_J + et \geq \underline{B}_{[n],I}x_I, \tag{20b}$$

$$D_s z_J \geq 0, \ z_J + et \geq 0, \ x > 0, \ t \geq 0 \tag{20c}$$

is infeasible for each admissible I, J and $s \in \{\pm 1\}^{|J|}$.

Proof. Let admissible I, J be given. We want to characterize infeasibility of system

$$(A^{-1}B)_{I,I}x + et = 0, \quad (A^{-1}B)_{J,I}x + et \geq 0, \quad x > 0, \ t \geq 0.$$

Substitute $y \equiv B_{[n],I}x_I$ and $z \equiv A^{-1}y$. Then $Az = y = B_{[n],I}x_I$ and the system reads

$$Az = B_{[n],I}x_I, \quad z_I + et = 0, \ z_J + et \geq 0, \ x > 0$$

or

$$-A_{[n],I}et + A_{[n],J}z_J = B_{[n],I}x_I, \quad z_J + et \geq 0, \ x > 0, \ t \geq 0.$$

This system is infeasible for each $A \in \boldsymbol{A}$ and $B \in \boldsymbol{B}$ if and only if system(20) is infeasible for each $s \in \{\pm 1\}^{|J|}$; see [5,8]. □

3 Conclusion

We considered various matrix classes that appear in the context of the LCP and ensure that the problem has favourable properties (in view of its solvability and properties of the solution set). We fully characterized stability of these matrices on an interval domain and in the case the matrices originate from the horizontal LCP. Practically it brings characterization of robustness of these matrices because whatever are the realizations of the interval data, we are sure that the corresponding property is satisfied.

Several open problems emerged, too. For copositivity and P-matrix property, we presented no closed form characterization and we leave it for future research. Next, notice that many matrix properties are computationally hard to verify even in the real case, so the interval case cannot be easier. Therefore it would be interesting to investigate some polynomially recognizable cases or to come up with suitable sufficient conditions.

References

1. Alefeld, G., Schäfer, U.: Iterative methods for linear complementarity problems with interval data. Computing **70**(3), 235–259 (2003). https://doi.org/10.1007/s00607-003-0014-6
2. Chung, S.J.: NP-completeness of the linear complementarity problem. J. Optim. Theor. Appl. **60**(3), 393–399 (1989)
3. Cottle, R.W.: Linear complementarity since 1978. In: Giannessi, F., Maugeri, A. (eds.) Variational Analysis and Applications, NOIA, vol. 79, pp. 239–257. Springer, Boston (2005). https://doi.org/10.1007/0-387-24276-7_18
4. Cottle, R.W., Pang, J.S., Stone, R.E.: The Linear Complementarity Problem. revised ed of the 1992 original edn. SIAM, Philadelphia, PA (2009)
5. Fiedler, M., Nedoma, J., Ramík, J., Rohn, J., Zimmermann, K.: Linear Optimization Problems with Inexact Data. Springer, New York (2006)
6. Garloff, J., Adm, M., Titi, J.: A survey of classes of matrices possessing the interval property and related properties. Reliab. Comput. **22**, 1–10 (2016)
7. Gowda, M.: Reducing a monotone horizontal LCP to an LCP. Appl. Math. Lett. **8**(1), 97–100 (1995)
8. Hladík, M.: Weak and strong solvability of interval linear systems of equations and inequalities. Linear Algebra Appl. **438**(11), 4156–4165 (2013)
9. Hladík, M.: Transformations of interval linear systems of equations and inequalities. Linear Multilinear Algebra **65**(2), 211–223 (2017)
10. Hladík, M.: Stability of the linear complementarity problem properties under interval uncertainty. Cent. Eur. J. Oper. Res. **29**, 875–889 (2021)
11. Hladík, M.: Tolerances, robustness and parametrization of matrix properties related to optimization problems. Optimization **68**(2–3), 667–690 (2019)
12. Hladík, M.: An overview of polynomially computable characteristics of special interval matrices. In: Kosheleva, O., et al. (eds.) Beyond Traditional Probabilistic Data Processing Techniques: Interval, Fuzzy etc. Methods and Their Applications, Studies in Computational Intelligence, vol. 835, pp. 295–310. Springer, Cham (2020)

13. Horáček, J., Hladík, M., Černý, M.: Interval linear algebra and computational complexity. In: Bebiano, N. (ed.) MAT-TRIAD 2015. SPMS, vol. 192, pp. 37–66. Springer, Cham (2017). https://doi.org/10.1007/978-3-319-49984-0_3
14. Kreinovich, V.: Why intervals? a simple limit theorem that is similar to limit theorems from statistics. Reliab. Comput. **1**(1), 33–40 (1995)
15. Kreinovich, V.: Why intervals? why fuzzy numbers? towards a new justification. In: Mendel, J.M., Omori, T., Ya, X. (eds.) 2007 IEEE Symposium on Foundations of Computational Intelligen, pp. 113–119 (2007)
16. Kreinovich, V., Lakeyev, A., Rohn, J., Kahl, P.: Computational Complexity and Feasibility of Data Processing and Interval Computations. Kluwer, Dordrecht (1998)
17. Ma, H., Xu, J., Huang, N.: An iterative method for a system of linear complementarity problems with perturbations and interval data. Appl. Math. Comput. **215**(1), 175–184 (2009)
18. Mezzadri, F., Galligani, E.: Splitting methods for a class of horizontal linear complementarity problems. J. Optim. Theor. Appl. **180**(2), 500–517 (2019)
19. Mezzadri, F., Galligani, E.: A modulus-based nonsmooth Newton's method for solving horizontal linear complementarity problems. Optimization Letters **15**(5), 1785–1798 (2019). https://doi.org/10.1007/s11590-019-01515-9
20. Moore, R.E., Kearfott, R.B., Cloud, M.J.: Introduction to Interval Analysis. SIAM, Philadelphia, PA (2009)
21. Murty, K.G., Yu, F.T.: Linear Complementarity, Internet Linear and Nonlinear Programming (1997)
22. Neumaier, A.: Interval Methods for Systems of Equations. Cambridge University Press, Cambridge (1990)
23. Rex, G., Rohn, J.: Sufficient conditions for regularity and singularity of interval matrices. SIAM J. Matrix Anal. Appl. **20**(2), 437–445 (1998)
24. Rohn, J.: Forty necessary and sufficient conditions for regularity of interval matrices: A survey. Electron. J. Linear Algebra **18**, 500–512 (2009)
25. Samelson, H., Thrall, R.M., Wesler, O.: A partition theorem for euclidean n-spaces. Proc. Am. Math. Soc. **9**, 805–807 (1958)
26. Sznajder, R., Gowda, M.: Generalizations of P_0- and P-properties; extended vertical and horizontal linear complementarity problems. Linear Algebra Appl. **223–224**, 695–715 (1995)
27. Tseng, P.: Co-NP-completeness of some matrix classification problems. Math. Program. **88**(1), 183–192 (2000)
28. Tütüncü, R.H., Todd, M.J.: Reducing horizontal linear complementarity problems. Linear Algebra Appl. **223–224**, 717–729 (1995)
29. Zheng, H., Vong, S.: On convergence of the modulus-based matrix splitting iteration method for horizontal linear complementarity problems of H_+-matrices. Appl. Math. Comput. **369**(124890), 1–6 (2020)

Incorporating User Preferences in Multi-objective Feature Selection in Software Product Lines Using Multi-Criteria Decision Analysis

Takfarinas Saber[1,2(✉)], Malika Bendechache[1,3], and Anthony Ventresque[1,2]

[1] Lero – the Irish Software Research Centre, Limerick, Ireland
[2] School of Computer Science, University College Dublin, Dublin, Ireland
{takfarinas.saber,anthony.ventresque}@ucd.ie
[3] School of Computing, Dublin City University, Dublin, Ireland
malika.bendechache@dcu.ie

Abstract. *Software Product Lines Engineering* has created various tools that assist with the standardisation in the design and implementation of clusters of equivalent software systems with an explicit representation of variability choices in the form of *Feature Models*, making the selection of the most ideal software product a *Feature Selection* problem. With the increase in the number of properties, the problem needs to be defined as a multi-objective optimisation where objectives are considered independently one from another with the goal of finding and providing decision-makers a large and diverse set of non-dominated solutions/products. Following the optimisation, decision-makers define their own (often complex) preferences on how does the ideal software product look like. Then, they select the unique solution that matches their preferences the most and discard the rest of the solutions—sometimes with the help of some Multi-Criteria Decision Analysis technique. In this work, we study the usability and the performance of incorporating preferences of decision-makers by carrying-out Multi-Criteria Decision Analysis directly within the multi-objective optimisation to increase the chances of finding more solutions that match preferences of the decision-makers the most and avoid wasting execution time searching for non-dominated solutions that are poor with respect to decision-makers' preferences.

Keywords: Feature selection · Software product line · Multi-objective evolution algorithm · Multi-Criteria Decision Analysis

1 Introduction

Software Engineering is divided into multiple domains [1]. One of these domains is Software Product Lines (SPL) which considers groups of related software systems as a whole, rather than dealing with every single one of them separately

B. Dorronsoro et al. (Eds.): OLA 2021, CCIS 1443, pp. 361–373, 2021.
https://doi.org/10.1007/978-3-030-85672-4_27

[2]. Feature Models (FMs) is the most recurrent representation of SPLs. Furthermore, the FM holds a listing of all the possible feature configurations/combinations which could be viewed as constraints. Therefore, making the FM a representation of all valid software products that could be made out the features in the SPL. Building a software product out of a particular SPL requires the selection of features that respect the desired software configuration. With the multiple characteristics/objectives that are interesting to decision-makers in practice (e.g., cost, technical feasibility, or reliability), the problem of finding the 'best' feature configuration is seen as an instance of a *multi-objective optimisation problem* [3,4].

Evolutionary algorithms have long been used to efficiently optimise problems in various domains from Computer Networks (e.g., [5–7]) to Intelligent Transport Systems (e.g., [8]), to Software Engineering, based on analytical/mathematical (e.g., [5,6]) or simulated (e.g., [8,9]) models. Evolutionary algorithms are particularly effective when dealing with multi-objective optimisation problems in software engineering (e.g., [10–13]). This is also the case for multi-objective feature selection in SPL for which the state-of-the-art SATIBEA [3] is an Indicator-Based Evolutionary Algorithm (IBEA) that uses a SAT solver as a mutation operator to correct infeasible solutions.

Multi-objective optimisation techniques result in a set of non-dominated products/solutions from which decision-makers select the product that fits their preferences the most. Given that the number of solutions in the set of non-dominated solutions is often large and that preferences of decision-makers are often complex, decision-makers are usually assisted by Multi-Criteria Decision Analysis (MCDA) tools to accomplish this task [14]. There exist multiple MCDA techniques that take decision-makers' preferences (each of them with its degree of preference expressibility) and return the product that match them the most. We show in this paper that: (i) some MCDA techniques are simplistic and can only handle a limited number of preference types (e.g., only take weights into accounts such as ELECTRE-IV), but they are fast, whereas (ii) other more elaborate MCDA techniques handle larger preference variations (e.g., they enable the use of different utility functions such as PROMETHEE-II), but they are slower and more time-consuming.

In this paper, we aim to include preferences of the decision-makers directly in the multi-objective search process to avoid spending a precious execution time searching for solutions that are (despite being non-dominated) far from decision-makers' preferences. In this paper, we study the effects of using MCDA techniques in the selection process of SATIBEA instead of the Indicator-Based technique (i.e., based on the contribution in Hypervolume of each solution). Particularly, we would like to evaluate the impact in terms of both: (i) the execution time overhead that it would induce, and (ii) quantity of non-dominated solutions matching preferences of decision-makers missed by SATIBEA.

This paper makes the following contributions:

- We propose SAT_MCDA_EA, a hybrid algorithm that includes decision-makers preferences in an MCDA form directly in the evolutionary search process.

- We show that using MCDA techniques as a selection operator has an insignificant impact in terms of execution time overhead in comparison to the execution time taken by one generation of SATIBEA.
- We also show that using MCDA techniques (particularly PROMETHEE-II) enables finding a large number of solutions which better match preferences of decision-makers and that are missed by SATIBEA (despite not outperforming SATIBEA on most of the multi-objective performance metrics).

Combining MCDA techniques with multi-objective evolutionary algorithms has already been attempted in a few recent works (e.g., [15–17]). However, to the best of our knowledge, this is the first time it is attempted in the Software Engineering domain in general and on the multi-objective feature selection in FM in particular.

The remainder of this paper is organised as follows: Sect. 2 presents the background of our study. Section 3 describes some common MCDA techniques and details our SAT_MCDA_EA approach. Section 4 provides our overall set-up and benchmark for multi-objective feature selection in SPL. Section 5 reports the results of our evaluation in terms of execution time overhead and performance of SAT_MCDA_EA against SATIBEA. Finally, Sect. 6 concludes the paper.

2 Background

In this section, we detail two aspects that make up the background of our work.

2.1 Software Product Line Engineering

Software Product Line Engineering is the paradigm that attempts to manage software variations more systematically and provide tools that cover the domain engineering and the application engineering processes with their multiple phases/activities [18]. In SPL, all software artefacts (i.e., variations of the same feature) could be picked and put together to form a particular product as long as they are compatible.

Feature Models is a way to represent an SPL. FMs represent the set of all available features with their variations and incompatibilities (i.e., constraints). Figure 1 shows a toy FM example with ten inter-connected features. It shows, for example, that the final product requires a 'Screen'. It also shows that there exist three 'Screen' types (i.e., 'Basic', 'Colour' or 'High Resolution') and only one of them could be selected for the final product. To build a software product from the SPL, we need to select a subset of features $\mathcal{S} \subseteq \mathcal{F}$ such that constraints of the FM $\mathcal{F}$ are satisfied. Constraints of the FM can be modelled as a satisfiability (SAT) problem for instantiating Boolean variables to true or false (in our case, every variable represents a feature) in a way that satisfies all the constraints. A variable $f_i \in \{\text{true, false}\}$ is set to true if the feature $F_i \in \mathcal{F}$ is picked to be part of $\mathcal{S}$, and false otherwise.

An FM can be represented in a conjunctive normal form (CNF). Therefore, searching for a valid software product in the SPL is equivalent to searching for a feasible solution to the SAT problem. For instance, the FM in Fig. 1 describes the screen alternatives in its SAT model with these clauses: $(Basic \vee Colour \vee High\ resolution) \wedge (\neg Basic \vee \neg Colour) \wedge (\neg Basic \vee \neg High\ resolution) \wedge (\neg Colour \vee \neg High\ resolution)$.

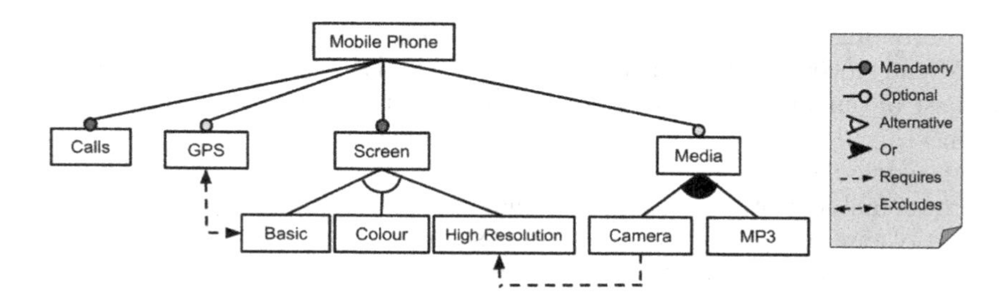

Fig. 1. Example of a feature model

2.2 Multi-Objective Optimisation

Multi-Objective Optimisation (MOO) considers the optimisation of more than two objective functions at the same time. Software products can be seen from various perspectives (e.g., development cost, reliability, performance). Therefore, by considering each of the perspectives as independent objectives, feature selection in SPL is a suitable candidate for MOO [14].

As a meaningful sample case, we use a set of commonly used optimisation objectives in the literature [19–21]:

- *Correctness* – reduce the number of violated constraints.
- *Richness of features* – increase the number of picked features (have products with more functionality, minimisation of its negative value is considered).
- *Features used before* – reduce the number of picked features that were not used before.
- *Known defects* – reduce the number of known defects in picked features.
- *Cost* – reduce the cost of the picked features.

3 State-of-the-Art and Proposed Approach

In this section, we describe the state-of-the-art algorithm SATIBEA and our proposed approach.

3.1 SATIBEA

SATIBEA [3] is an extension to the Indicator-Based Evolutionary Algorithm (IBEA) which guides the optimisation through a quality indicator selection process (in this case, the Hypervolume); a SAT solver has been introduced as a mutation operator to assist IBEA.

Note that there are multiple algorithms designed to address the multi-objective feature selection in SPL problem. Most of these algorithms perform in a similar fashion as SATIBEA (evolutionary algorithm + exact algorithm such as SMT [20] or MILP [21,22]). In this work, we do not compare to them as we do not aim to design an algorithm that is better in terms of multi-objective metrics (even if we report the performance with respect to those metrics below). Instead, our goal is to showcase the fact that including preferences of the decision-makers in the evolutionary search process is worth considering when decision-makers have complex preferences as: (i) it only adds a marginal execution time overhead, and (ii) it finds solutions that are interesting with respect to decision-makers' preferences, but missed by particular IBEA algorithms (in our case SATIBEA).

3.2 Multi-Criteria Decision Analysis

Providing a set of non-dominated solutions, decision-makers explore them to find their preferred one. Given the large size of the non-dominated sets that are obtained after performing the multi-objective optimisation, decision-makers take advantage of MCDA techniques to select the ideal solution with respect to their preferences.

MCDA deals with decision-making constrained by multiple and often conflicting criteria (or objectives or goals). MCDA has been broadly divided into two categories [14]: (i) Outranking Methods: builds a preference relation, and (ii) Multiple Attribute Utility and Value Theory: the 'utility' of every action is scored based on its utility.

In this work, we select three commonly used MCDA techniques: two outranking methods (ELECTRE-IV [23] and PROMETHEE-II [24]) and one Multiple Attribute Utility and Value Theory method (MAUT [25]).

We propose in this paper to substitute the Indicator-Based selection operator in the original SATIBEA algorithm by one of the aforementioned MCDA techniques (i.e., ELECTRE-IV, PROMETHEE-II or MAUT) to create what we call SAT_MCDA_EA. Therefore, we are creating three distinct algorithms under the same umbrella of SAT_MCDA_EA: (i) SAT_ELECTRE-IV_EA, where we use ELECTRE-IV as the selection operator, (ii) SAT_PROMETHEE-II_EA, where we use PROMETHEE-II as the selection operator, and (iii) SAT_MAUT_EA, where we use MAUT as the selection operator.

4 System Set-Up

This section presents the different elements that we have used in our experiments: the dataset, the multi-objective performance metrics, the parameters of the genetic algorithms (i.e., SATIBEA and SAT_MCDA_EA), the parameters we use for the MCDA techniques, and the hardware configuration.

4.1 System and Algorithms Set-Up

We use the implementation of SATIBEA that is made available to us by its creators (implemented in Java) and implement our approach on top of it. We conduct our experiments on a machine with a 4 core CPU (our algorithms use a core at a time though) and 16 GB of RAM. We ran all our algorithms and determined the average results over 30 runs for each instance.

We use the same parameters for SATIBEA as those defined by its authors (e.g., population size: 300, crossover rate: 0.8, mutation rate of each feature selection: 0.001, and solver mutation rate: 0.02). We also use the same parameters as SATIBEA for our SAT_MCDA_EA approach. Furthermore, we define addition parameters for the MCDA techniques to simulate preferences of decision-makers. Note that the chosen preferences are only selected to showcase different capabilities of each MCDA method. Therefore, it will be worth performing a more robust analysis with different kinds of preferences and a full parameters sweeping for each of these MCDA methods in a future work.

- ELECTRE-IV: requires a parameter triplet (optimisation threshold, preference threshold, and indifference threshold) for every objective. We set these triplets to (5, 6, 5), (3, 4, 3), (0.1, 0.3, 0.1), (1, 2, 1) and (3, 4, 3) for Correctness, Richness of features, Feature used before, Known defects, and Cost.
- PROMETHEE-II: requires a parameter pair (weight and preference function) for each objective. We set equal weights for all objectives and set their preference functions to Level, Linear, Linear, Level, and Gaussian for Correctness, Richness of features, Feature used before, Known defects, and Cost.
- MAUT: only requires one parameter per objective (weight) that we set equally for all the objectives.

Based on the parameters that each of the MCDA techniques requires, we see that PROMETHEE-II is the most expressive between them as it enables decision-makers to design their own custom utility function for each objective and feed it to the MCDA.

4.2 Dataset

For our experiments, we use the five of the largest open source FMs we could find [20]. Table 1 shows the version and the size of each of the FMs that we consider in our experiments. The table also reports the number of features and the size of the SAT problem necessary to represent the FM in a conjunctive normal form (in terms of number of variables and number of clauses). Similarly to the SATIBEA paper [3], we set the execution time on the Linux Kernel to 1,200s. For the other datasets, we use smaller execution times based on the convergence time of SATIBEA [19,26].

Table 1. Versions and characteristics of the feature models used in our experiments.

Dataset	Version	#Features	#Variables	#Clauses	Time (s)
Linux kernel	2.6.28.6	5,701	6,888	343,944	1,200
eCos	20100825	1,244	1,244	3,146	50
Fiasco	2011081207	300	1,638	5,228	200
FreeBSD	8.0.0	1,396	1,396	62,183	200
μClinux	3.0	616	1,850	2,468	100

4.3 Multi-objective Performance Metrics

To assess the performance of our algorithms we use 5 multi-objective performance metrics: 4 quality metrics (Hypervolume, Epsilon, Generation Distance, and Inverted Generation Distance) and 1 diversity metric (Spread).

- Hypervolume (HV): computes the volume (measured in k dimensions of the problem's search space) that is dominated by the Pareto front (to maximise).
- Epsilon (ϵ): evaluates the smallest distance that is needed for every solution in Pareto front to dominate the Reference front (to minimise).
- Generation Distance (GD): evaluates the smallest distance needed for every solution in Pareto front to dominate the Reference front (to minimise).
- Inverted Generation Distance (IGD): evaluates average distance between every solution in Reference front and its closest solution in Pareto front (to minimise).
- Spread (S): computes the solutions' distribution to evaluate their extent spread in Pareto front (to maximise).

5 Evaluation

5.1 Execution Time Overhead

One of the major issues that kept designers of evolutionary algorithms away from using MCDA techniques within the search process is the excessive execution time that these techniques require. More researchers and practitioners favour less time-consuming indicator-based methods. This is even more true with problems that are only given a few seconds as a total optimisation time budget. In this section, we evaluate the overhead execution time that is introduced by the use of MCDA techniques. We compare the execution time of MCDA techniques to the execution time needed to evolve a full generation and also to the execution time of the default indicator-based method (in our case, the Hypervolume).

Table 2 shows the average execution time in millisecond over 30 iterations of the second generation of SATIBEA (the generation following the evolution of the randomly generated initial population) using the default indicator-based

Table 2. Average execution time (ms) of the second generation of SATIBEA, indicator-based selection, and MCDA selection methods.

Dataset	Generation	Indicator-Based	ELECTRE-IV	MAUT	PROMETHEE-II
Linux Kernel	53,788	30.50	1.71	62.75	101.23
eCos	1,235	30.22	1.93	60.33	114.82
Fiasco	12,477	44.49	1.42	59.68	149.04
FreeBSD	12,742	29.57	1.56	71.28	127.30
uClinux	1,197	31.6	1.55	58.09	96.44

(Hypervolume). The table also shows the average execution time of each particular selection technique from Indicator-Based, to the three considered MCDA techniques (i.e., ELECTRE-IV, MAUT, and PROMETHEE-II).

We clearly see that the execution time of a full SATIBEA generation is very large in comparison to the execution time of the different selection operators (148 times larger on average than the largest selection time per instance). A single generation takes on average 531, 11, 84, 100, and 12 times larger execution times than the most time-consuming selection process (in this case, using PROMETHEE-II) on the instances Linux Kernel, eCos, Fiasco, FreeBSD and uClinux respectively. This is a clear indication that using any of the studied MCDA techniques is less likely to add a significant execution time overhead. The execution time of the section process is particularly insignificant when dealing with the large instances (Linux Kernel, Fiasco and FreeBSD).

We see that with the exception of ELECTRE-IV, MCDA techniques (i.e., MAUT and PROMETHEE-II) necessitate a larger execution time than the default Indicator-Based selection. This is one of the main reasons why the simplistic weighted-sum is the de-facto go to in absence of a pure multi-objective optimisation (keeping objectives separate with no aggregation). However, we notice in our usecase that the order by which the execution time of these MCDA techniques exceed the Indicator-Based selection is rather small ($\sim$0.9 and $\sim$2.5 more execution time on average for MAUT and PROMETHEE-II respectively).

Therefore, we could claim that from an execution time perspective and in the context of multi-objective feature selection in large software product lines such as the ones studied in our paper, decision-makers should no longer be reluctant to provide their preferences in advance to be embedded in the multi-objective optimisation process.

5.2 Multi-objective Performance Metrics

Knowing that using MCDA techniques in the multi-objective optimisation process does not add a significant execution time overhead is good, but obtaining improved results is better –despite not being the most important in our case as our goal is to find more solutions that match decision-makers' preferences. Therefore, we would like to evaluate the impact of our approach in terms of performance and quantify it using the different multi-objective metrics seen in Sect. 4.

Table 3 shows the average performances achieved by SATIBEA and SAT_MCDA_EA techniques (i.e., SAT_ELECTRE-IV_EA, SAT_MAUT_EA, SAT_PROMETHEE-II_EA) with respect to the quality metrics HV, IGD, GD, Epsilon and Spread. We put in bold the best achieved performances per instance and per metric. We also put (*) when results are not statistically significant between SATIBEA and the best performing SAT_MCDA_EA technique (p-value < 0.05 when evaluated using the non-parametric two-tailed Mann-Whitney U test).

Table 3 clearly shows that SATIBEA achieves the best performances on the metrics HV and IGD on all instances. SATIBEA also achieves the best performances on Epsilon in 4 out of 5 instances on average. This is a clear indication that SATIBEA maintains its supremacy with regards to very important multi-objective performance metrics. This is quite understandable as SATIBEA's aim by design is to cover most of the search space, which yields better multi-objective quality metrics performances. However, SAT_MCDA_EA algorithms target solutions that better match the predefined preferences of the decision-makers and leave large parts of the search space unprobed, which yields low multi-objective quality metrics performances.

Table 3 also shows that SATIBEA does not always achieve the best results with respect to the Spread metric. SAT_ELECTRE-IV_EA achieves the best performance on Spread on 3 out of 5 instances on average. Although, Spread is a secondary metrics and should not be interpreted alone without the other quality metrics. Looking at SAT_ELECTRE-IV_EA's performance in terms of HV, we see that it is poor, which reduces the importance of its Spread performance.

Table 3 also shows that SATIBEA is not achieving the best GD on any instance (achieved by SAT_PROMETHEE-II_EA). This is an indication that most of the solutions that are found by SAT_PROMETHEE-II_EA are non-dominated by the solutions found by the other algorithms. However, given that the performance of SAT_PROMETHEE-II_EA in terms of HV is poor, we can deduce that its solutions are not diverse enough. While this might seem negative, we believe that this is a good characteristic. Decision-makers would rather be provided with several non-dominated solutions that are similar and better match their preferences, rather than a set of non-dominated solutions covering a larger space, but match their preferences less. Furthermore, SAT_MAUT_EA also achieves a better performance than SATIBEA in terms of GD on 3 out of 5 instances on average.

5.3 SAT_MCDA_EA's Strictly Non-Dominated Solutions

With SAT_PROMETHEE-II_EA and SAT_MAUT_EA achieving good GD performances, we would like to measure the ratio of non-dominated solutions found by SAT_MCDA_EA algorithms, but missed by SATIBEA. We gather all non-dominated solutions found over all iterations by each algorithm and perform a pairwise non-dominance comparison. Table 4 shows the ratio (in percentage) of

Table 3. Comparison of the average performances achieved by SATIBEA and the various SAT_MCDA_EA algorithms.

Dataset	Metric	SATIBEA	SAT_ELECTRE-IV_EA	SAT_MAUT_EA	SAT_PROMETHEE-II_EA
Linux Kernel	HV	**0.136**	0.124	0.123	0.134
	IGD	**0.010**	0.016	0.016	0.012
	GD	0.030	0.130	0.012	**0.007**
	ϵ	**1982**	2047	2051	1991
	S	1.16	**1.24**	1.21	1.19
eCos	HV	**0.252**	0.206	0.188	0.085
	IGD	**0.0071**	0.0072	0.008	0.016
	GD	0.0722	3.8714	0.0935	**0.0031**
	ϵ	**147**	260	217	149
	S	**1.51***	1.30	1.33	1.55
Fiasco	HV	**0.195**	0.133	0.132	0.124
	IGD	**0.009**	0.022	0.024	0.018
	GD	0.065	0.237	0.076	**0.008**
	ϵ	277	917	950	**171**
	S	**1.58**	1.14	1.16	1.27
FreeBSD	HV	**0.24**	0.18	0.18	0.08
	IGD	**0.006**	0.011	0.012	0.018
	GD	0.091	0.156	0.066	**0.004**
	ϵ	**133**	303	308	498
	S	1.21	**1.23***	1.20	1.21
uClinux	HV	**0.893**	0.89	0.891	0.805
	IGD	**0.054**	0.055	0.056	0.060
	GD	0.043	0.016	0.015	**0.012**
	ϵ	**598***	611	604	1199
	S	1.067	**1.229**	1.198	1.003

Table 4. Ratio (in per cent) of strictly non-dominated solutions found over the 30 iterations by SATIBEA using one of the MCDA methods in comparison with the solutions found by SATIBEA when using the default Indicator-Based method.

Dataset	SAT_ELECTRE-IV_EA vs SATIBEA	SAT_MAUT_EA vs SATIBEA	SAT_PROMETHEE-II_EA vs SATIBEA
Linux Kernel	40	41	66
eCos	33	42	90
Fiasco	27	59	94
FreeBSD	26	48	92
uClinux	5	34	73

solutions found by each SAT_MCDA_EA that are strictly non-dominated (neither equal nor dominated) by any solution found by SATIBEA.

Table 4 confirms our assumption that many solutions found by SAT_MAUT_-EA and SAT_PROMETHEE-II_EA are strictly non-dominated by those found by SATIBEA. We see that SAT_PROMETHEE-II_EA finds the largest number of solutions non-dominated by those found by SATIBEA (~83% non-dominated solutions on average, and 94% on Fiasco). Therefore, if decision-makers have a prior knowledge of what makes a good software, they are better off using PROMETHEE-II as a selection operator. While this will not yield optimal multi-objective metrics, it will yield more solutions matching their preferences.

6 Conclusion and Future Work

In this paper, we proposed using MCDA techniques directly within the multi-objective search process by employing them as the selection operator. We have evaluated their impact both in terms of induced execution time overhead and in terms of quality of the obtained solutions. We have seen that using the MCDA techniques introduces a non-significant overhead execution time with respect to the execution time of the other operators that make up the evolution. However, we have also seen that using the MCDA techniques within the search process impacts negatively the performance of the algorithm with respect to various multi-objective performance metrics with the exception of GD. We have confirmed that the SAT_MCDA_EA algorithms perform particularly well with respect to GD as they find a large number of solutions that match their preferences but that are not dominated by the solutions found by SATIBEA. The insight obtained from this study encourages us to deepen the investigation of combining MCDA techniques with the multi-objective feature selection in SPL.

Acknowledgement. This work was supported, in part, by Science Foundation Ireland grants No. 13/RC/2094_P2 (Lero) and 13/RC/2106_P2 (ADAPT).

References

1. Ramirez, A., Romero, J.R., Ventura, S.: A survey of many-objective optimisation in search-based software engineering. J. Syst. Softw. **149**, 382–395 (2019)
2. Metzger, A., Pohl, K.: Software product line engineering and variability management: achievements and challenges. In: FSE, pp. 70–84. ACM (2014)
3. Henard, C., Papadakis, M., Harman, M., Le Traon, Y.: Combining multi-objective search and constraint solving for configuring large software product lines. In: ICSE, pp. 517–528 (2015)
4. Yadav, H., Chhikara, R., Kumari, A.C.: A novel hybrid approach for feature selection in software product lines. Multimed. Tools Appl. **80**(4), 4919–4942 (2021). https://doi.org/10.1007/s11042-020-09956-6
5. Lynch, D., Saber, T., Kucera, S., Claussen, H., O'Neill, M.: Evolutionary learning of link allocation algorithms for 5G heterogeneous wireless communications networks. In: GECCO, pp. 1258–1265 (2019)

6. Saber, T., Fagan, D., Lynch, D., Kucera, S., Claussen, H., O'Neill, M.: A hierarchical approach to grammar-guided genetic programming: the case of scheduling in heterogeneous networks. In: Fagan, D., Martín-Vide, C., O'Neill, M., Vega-Rodríguez, M.A. (eds.) TPNC 2018. LNCS, vol. 11324, pp. 225–237. Springer, Cham (2018). https://doi.org/10.1007/978-3-030-04070-3_18
7. Saber, T., Fagan, D., Lynch, D., Kucera, S., Claussen, H., O'Neill, M.: A multi-level grammar approach to grammar-guided genetic programming: the case of scheduling in heterogeneous networks. Genet. Program Evol. Mach. **20**(2), 245–283 (2019). https://doi.org/10.1007/s10710-019-09346-4
8. Saber, T., Wang, S.: Evolving better rerouting surrogate travel costs with grammar-guided genetic programming. In: IEEE CEC, pp. 1–8 (2020)
9. Bendechache, M., et al.: Modelling and simulation of elastic search using cloudsim. In: DS-RT, pp. 1–8 (2019)
10. Saber, T., Gandibleux, X., O'Neill, M., Murphy, L., Ventresque, A.: A comparative study of multi-objective machine reassignment algorithms for data centres. J. Heuristics **26**(1), 119–150 (2020). https://doi.org/10.1007/s10732-019-09427-8
11. Saber, T., Delavernhe, F., Papadakis, M., O'Neill, M., Ventresque, A.: A hybrid algorithm for multi-objective test case selection. In: CEC, pp. 1–8 (2018)
12. Saber, T., Ventresque, A., Brandic, I., Thorburn, J., Murphy, L.: Towards a multi-objective VM reassignment for large decentralised data centres. In: UCC, pp. 65–74. IEEE (2015)
13. Saber, T., Ventresque, A., Gandibleux, X., Murphy, L.: GeNePi: a multi-objective machine reassignment algorithm for data centres. In: Blesa, M.J., Blum, C., Voß, S. (eds.) HM 2014. LNCS, vol. 8457, pp. 115–129. Springer, Cham (2014). https://doi.org/10.1007/978-3-319-07644-7_9
14. Mjeda, A., Wasala, A., Botterweck, G.: Decision spaces in product lines, decision analysis, and design exploration: an interdisciplinary exploratory study. In: VaMoS, pp. 68–75. ACM (2017)
15. Mohammed, A., Harris, I., Soroka, A., Nujoom, R.: A hybrid MCDM-fuzzy multi-objective programming approach for a G-resilient supply chain network design. Comput. Ind. Eng. **127**, 297–312 (2019)
16. Kapsoulis, D., Tsiakas, K., Trompoukis, X., Asouti, V., Giannakoglou, K.: Evolutionary multi-objective optimization assisted by metamodels, kernel PCA and multi-criteria decision making techniques with applications in aerodynamics. Appl. Soft Comput. **64**, 1–13 (2018)
17. Jafarian-Namin, S., Kaviani, M.A., Ghasemi, E.: An integrated MOEA and MCDM for multi-objective optimization (case study: control chart design). In: IEOM (2016)
18. Horcas, J.M., Pinto, M., Fuentes, L.: Software product line engineering: a practical experience. In: SPLC, pp. 164–176 (2019)
19. Saber, T., Brevet, D., Botterweck, G., Ventresque, A.: Is seeding a good strategy in multi-objective feature selection when feature models evolve? Inf. Softw. Technol. **95**, 266–280 (2018)
20. Guo, J., et al.: SMTIBEA: a hybrid multi-objective optimization algorithm for configuring large constrained software product lines. Softw. Syst. Model. **18**(2), 1447–1466 (2019). https://doi.org/10.1007/s10270-017-0610-0
21. Saber, T., Brevet, D., Botterweck, G., Ventresque, A.: Reparation in evolutionary algorithms for multi-objective feature selection in large software product lines. SN Comput. Sci. **2**(3), 1–14 (2021). https://doi.org/10.1007/s42979-021-00541-8

22. Saber, T., Brevet, D., Botterweck, G., Ventresque, A.: MILPIBEA: algorithm for multi-objective features selection in (evolving) software product lines. In: EvoCOP, pp. 164–179 (2020)
23. Govindan, K., Jepsen, M.B.: ELECTRE: a comprehensive literature review on methodologies and applications. Eur. J. Oper. Res. **250**(1), 1–29 (2016)
24. Brans, J.-P., De Smet, Y.: PROMETHEE methods. In: Greco, S., Ehrgott, M., Figueira, J.R. (eds.) Multiple Criteria Decision Analysis. ISORMS, vol. 233, pp. 187–219. Springer, New York (2016). https://doi.org/10.1007/978-1-4939-3094-4_6
25. Allah Bukhsh, Z., Stipanovic, I., Klanker, G., O'Connor, A., Doree, A.G.: Network level bridges maintenance planning using multi-attribute utility theory. Struct. Infrastruct. Eng. **15**(7), 872–885 (2019)
26. Brevet, D., Saber, T., Botterweck, G., Ventresque, A.: Preliminary study of multi-objective features selection for evolving software product lines. In: Sarro, F., Deb, K. (eds.) SSBSE 2016. LNCS, vol. 9962, pp. 274–280. Springer, Cham (2016). https://doi.org/10.1007/978-3-319-47106-8_23

Author Index

Printed in the United States
by Baker & Taylor Publisher Services